AF606440

SPACE TELESCOPE SCIENCE INSTITUTE

SYMPOSIUM SERIES: 11

Series Editor S. Michael Fall, Space Telescope Science Institute

THE HUBBLE DEEP FIELD

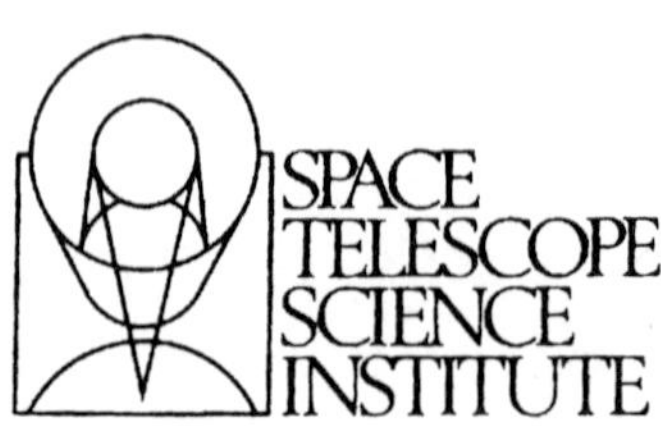

Other titles in the Space Telescope Science Institute Symposium Series.

1 Stellar Populations
Edited by C. A. Norman, A. Renzini and M. Tosi 1987 0 521 33380 6
2 Quaser Absorption Lines
Edited by C. Blades, C. A. Norman and D. Turnshek 1988 0 521 34561 8
3 The Formation and Evolution of Planetary Systems
Edited by H. A. Weaver and L. Danly 1989 0 521 36633 X
4 Clusters of Galaxies
Edited by W. R. Oegerle, M. J. Fitchet and L. Danly 1990 0 521 38462 1
5 Massive Stars in Starbursts
Edited by C. Leitherer, N. R. Walborn, T. M. Heckman and C. A. Norman
1991 0 521 40465 7
6 Astrophysical Jets
Edited by D. Burgarella, M. Livio and C. P. O'Dea
1993 0 521 44221 4
7 Extragalactic Background Radiation
Edited by D. Calzetti, M. Livio and P. Madau 1995 0 521 49558 X
8 The Analysis of Emission Lines
Edited by R. E. Williams and M. Livio 1995 0 521 48081 7
9 The Collision of Comet Shoemaker–Levy 9 and Jupiter
Edited by K. S. Noll, H. A. Weaver and P. D. Feldman
1996 0 521 56192 2
10 The Extragalactic Distance Scale
Edited by M. Livio, M. Donahue and N. Panagia
1997 0 521 59164 3

The Hubble Deep Field

Proceedings of the Space Telescope Science Institute Symposium,
held in Baltimore, Maryland
May 6–9, 1997

Edited by

MARIO LIVIO
Space Telescope Science Institute, Baltimore

S. MICHAEL FALL
Space Telescope Science Institute, Baltimore

PIERO MADAU
Space Telescope Science Institute, Baltimore

Published for the
Space Telescope Science Institute

PUBLISHED BY THE PRESS SYNDICATE OF THE UNIVERSITY OF CAMBRIDGE
The Pitt Building, Trumpington Street, Cambridge CB2 1RP, United Kingdom

CAMBRIDGE UNIVERSITY PRESS
The Edinburgh Building, Cambridge CB2 2RU, United Kingdom
40 West 20th Street, New York, NY 10011–4211, USA
10 Stamford Road, Oakleigh, Melbourne 3166, Australia

First Published 1998

Printed in the United States of America

Typeset by the author.

A catalog record for this book is available from the British Library

Library of Congress Cataloging in Publication data is available

ISBN 0 521 63097 5 hardback

Contents

Participants

Abraham, Roberto	Royal Greenwich Observatory
Allen, Ron	Space Telescope Science Institute
Aloisi, Alessandra	University of Bologna
Bahcall, John	Institute for Advanced Study
Balogh, Mike	University of Victoria
Barger, Amy	University of Hawaii
Berlind, Andreas	Ohio State University
Bershady, Matthew	Penn State University
Blandford, Roger	California Institute of Technology
Boldt, Elihu	NASA/Goddard Space Flight Center
Bond, Howard	Space Telescope Science Institute
Borne, Kirk	NASA/Goddard Space Flight Center
Brown, Michael	University of Arizona
Bruzual, Gustavo	C.I.D.A.
Burg, Claudia-Angelica	Arizona State University
Burgarella, Denis	Laboratoire d'Astronomie Spatiale
Burns, Dixie	Ohio State University
Calzetti, Daniela	Space Telescope Science Institute
Campos, Ana	Observatorio Astronomico Nacional
Cavaliere, Alfonso	Universita di Roma
Cohen, Judith	California Institute of Technology
Connolly, Andrew	The Johns Hopkins University
Conti, Alberto	Ohio State University
Cowie, Lennox	Institute for Astronomy
Davies, Jonathan	University of Wales Cardiff
de la Varga, Ana	Hamburger Sternwarte
Dennefeld, Michel	Institute d'Astrophysique de Paris
Devost, Daniel	University Laval
Dickinson, Mark	Space Telescope Science Institute
Disney, Mike	University College, Cardiff
Donahue, Megan	Space Telescope Science Institute
Dressel, Linda	RJH Scientific, Inc.
Duerbeck, Hilmar	Astronomisches Institute/ST ScI
Efstathiou, George	University of Oxford
Egami, Eiichi	Max Planck Institute für Extraterrestrische Physik
Eisenhardt, Peter	JPL/Caltech
Ellis, Richard	University of Cambridge
Elson, Richard	University of Florida
Faber, Sandra	University of California
Fall, Mike	Space Telescope Science Institute
Fasano, Giovanni	Osservatorio Astronomico di Padova
Ferguson, Harry	Space Telescope Science Institute
Fernandez Soto, Alberto	SUNY/Stony Brook
Fluke, Christopher	University of Melbourne
Frank, Juhan	Louisiana State University/ST ScI
Frei, Zslot	University of Pennsylvania
Freudling, Wolfram	European Southern Observatory
Fritze-von Alvensleben, Uta	NASA/Goddard Space Flight Center

Fruchter, Andy	Space Telescope Science Institute
Fullton, Laura	Space Telescope Science Institute
Gardner, Jonathan	NASA/Goddard Space Flight Center
Gaudi, Scott	Ohio State University
Giavalisco, Mauro	Carnegie Observatories
Gibbons, Rachel	Space Telescope Science Institute
Gonzalez, Rosa	Space Telescope Science Institute
Gracia, Javier	Universidad Autonoma de Sinaloa
Green, William	Hopping Green Sams & Smith
Griffiths, Richard	Carnegie Mellon University
Gronwall, Caryl	Wesleyan University
Groth, Edward	Princeton University
Gull, Theodore	NASA/Goddard Space Flight Center
Gursky, Herbert	Naval Research Laboratory
Gwyn, Stephen	University of Victoria
Hauser, Mike	Space Telescope Science Institute
Heap, Sara	NASA/Goddard Space Flight Center
Hook, Richard	SR-ECF/European Southern Observatory
Hudson, Mike	University of Victoria
Illingworth, Garth	University of California
Im, Myungshin	Space Telescope Science Institute
Iskudarian, Sophey	Byurakan Observatory
Kaiser, Mary Beth	The Johns Hopkins University
Kao, Lancelot	The University of Chicago
Kauffman, Guinevere	Max Planck Institute für Astrophysik
Kawaler, Steven	Iowa State University
Kellerman, Kenneth	National Radio Astronomy Observatory
Kinney, Anne	Space Telescope Science Institute
Koemiesberger, Gloria	Instituto de Astronomia
Lanzetta, Kenneth	SUNY/Stony Brook
Lilly, Simon	University of Toronto
Lin, Huan	University of Toronto
Livio, Mario	Space Telescope Science Institute
Lombardi, Marco	Scuola Normale Superore
Lucas, Ray	Space Telescope Science Institute
Madau, Piero	Space Telescope Science Institute
Magris, Gladis	C.I.D.A.
Malkan, Matt	University of California
Marleau, Francine	University of California
Martin, Crystal	Space Telescope Science Institute
Martini, Paul	Ohio State University
Meurer, Gerhardt	Space Telescope Science Institute
Minniti, Dante	Lawrence Livermore National Laboratory
Miralles, Joan-Marc	Observatoire Midi-Pyrénées
Moeller, Claudia	Universitäts-Sternwarte Göttingen
Muxlow, Thomas	Nuffield Radio Astronomy Laboratory
Nagamine, Kentaro	Princeton University
Noguchi, Masafumi	Tohoku University
Ortiz Gil, Amelia	SUNY/Stony Brook

Osmer, Patrick	Ohio State University
Ostriker, Jeremiah	Princeton University Observatory
Panagia, Nino	Space Telescope Science Institute
Partridge, Bruce	Haverford College
Peebles, James	Princeton University
Phillips, Andrew	University of California
Pogge, Richard	Ohio State University
Pozzetti, Lucia	Universita di Bologna/ST ScI
Pritchet, Chris	University of Victoria
Ramadurai, S.	Tata Institute of Fundamental Research
Ratnatunga, Kavan	Carnegie Mellon University
Rhodes, Jason	Princeton University
Rich, Michael	Columbia University
Richards, Eric	University of Virginia
Romano, Patrizia	Ohio State University
Rowan-Robinson, Michael	Imperial College
Sahu, Kailash	Space Telescope Science Institute
Sandage, Allan	Carnegie Observatories
Sawicki, Marcin	University of Toronto
Schreier, Ethan	Space Telescope Science Institute
Schweizer, Francois	Carnegie/DTM
Seitter, Waltraut	Astronomisches Institute/ST ScI
Shair, Fred	Jet Propulsion Laboratory
Silk, Joseph	University of California
Sirianni, Marco	Space Telescope Science Institute
Smith, Eric	NASA/Goddard Space Flight Center
Solai, Jeyakumar	Tata Institute of Fundamental Research
Stage, Michael	California Institute of Technology
Steinmetz, Matthias	University of Arizona
Stephens, Andrew	Ohio State University
Storrie-Lombardi, Lisa	Carnegie Observatories
Szalay, Alexander	The Johns Hopkins University
Szokoly, Gyula	The Johns Hopkins University
Thronson, Harley	NASA Headquarters
Vaisanen, Petri	Harvard-Smithsonian Center for Astrophysics
van den Bergh, Sidney	Dominion Astrophysical Observatory
Visvanathan, Natarajan	Mount Stromlo & Siding Spring Observatories
Vogeley, Michael	Space Telescope Science Institute
Vogt, Nicole	University of California
Voit, Mark	Space Telescope Science Institute
Wamsteker, Willem	ESA IUE Observatory
Wang, Zhong	Smithsonian Astrophysical Observatory
White, Simon	Max Planck Institute für Astrophysik
Williams, Bob	Space Telescope Science Institute
Wilner, David	Harvard-Smithsonian Center for Astrophysics
Windhorst, Rogier	Arizona State University
Woodgate, Bruce	NASA/Goddard Space Flight Center
Woods, Eric	Harvard University
Yahil, Amos	SUNY/Stony Brook

Yi, Sukyoung	NASA/Goddard Space Flight Center
Zepf, Steve	University of California
Zimmerman, Robert	The Sciences Magazine
Zirbel, Esther	Haverford College

Preface

It has been said about the Hubble Deep Field, that never in the history of astronomy has so much research effort been put into a completely blank piece of the sky! Yet, the papers presented in this volume demonstrate dramatically that this has been a worthwhile endeavour. These papers represent the invited talks that were presented in the Symposium "The Hubble Deep Field," which was held at the Space Telescope Science Institute, May 6–9, 1997. By now, it has become abundantly clear that the Hubble Deep Field and all the related research that it has inspired, represent a huge step forward in our understanding of the universe at high and intermediate redshifts.

We thank Sharon Toolan and Ron Meyers of ST ScI for their help in preparing this volume for publication.

Mario Livio, S. Michael Fall, Piero Madau
Space Telescope Science Institute
Baltimore, Maryland
May, 1997

Beginnings of observational cosmology in Hubble's time: Historical overview

By ALLAN SANDAGE

Observatories of the Carnegie Institution of Washington

1. Prologue

1.1. *A History of a Name*

When the organizers of this symposium asked if I could talk on history, it was not clear if they hoped for a history of HST or a scientific history of why the telescope has been named for Hubble. There is a history to both subjects.

In the early days of planning, when the telescope was a three meter dream, it was initially called the LOT for Large Orbiting Telescope. This brought forth several objections because a cadre of adventurous astronomers had urged a site on the moon. The word "orbiting" was said to block such a plan. Consequently, the name was changed in the late 1960s to LST for Large Space Telescope.

That name was still used as late as 1974 in all the planning and in the several major symposia held in a first lobbying effort, both by industry and by the scientists, to sell the telescope. One such important symposium was held in Washington at the Sheraton-Park Hotel from January 30 to February 1, 1974. The meeting was organized by F. Peter Simmons who had become Project Manager for LST at the McDonald-Douglas Astronautics Company after his earlier role at the Grumman Aerospace Corporation as Director of Astronomy for the highly successful Orbiting Astronomical Observatories (OAO). Simmons was later to play an even larger role in coordinating and organizing a major lobbying response, both in Congressional committees and in industry in the mid 1970s, to gain support for Lyman Spitzer's (1946) early suggestion for a space telescope.

It was at the 1974 Washington symposium where much of the science and the necessary technology for the project was first publicly laid out in awesome detail. Many of the future stars of the enormously complicated project, both astronomers and engineers, spoke. The names of the industry affiliations from which the technical scientists and engineers came included Ball Brothers, Bendix, Boeing, Convair, General Dynamics, Grumman, Itek, Lockheed, Martin Marietta, McDonnell-Douglas, Perkin-Elmer, and TRW, showing the wide industry interest in the project. Representatives from NASA Headquarters and from Goddard and Marshall were also there.

Among the astronomers were Lyman Spitzer (in absentia, see Spitzer 1997), Robert O'Dell (LST project scientist), Nancy Roman (chief, astronomy/relativity, NASA), Laurence Fredrick, Margaret Burbidge, Robert Danielson, Ivan King, John Bahcall, George Herbig, Gerry Neugebauer, and Harlan Smith.

John Naugle, Associate Administrator for Space Science, NASA headquarters, gave the sobering epilogue where he outlined the major hurdles to be conquered as seen in 1974. Among the many important cautions he gave, one of the most central was: "From what you have heard over the past few days, it is quite clear that we are smart enough technically to build the Large Space Telescope now." [However] "scientists must recognize that where they are dependent upon public support for their endeavors, they must communicate the importance of their endeavors to the public—the knowledge they have gained and its importance. This enables the public to participate, in many cases vicariously, in these activities. If scientists devote perhaps one-tenth of the creative energy devoted to understanding the universe to explaining to the public the reasons for and the importance of what they are doing, then I think the problems that we have in obtaining support for basic research will disappear."

One of the grand purposes of the present workshop is to do just that.

For various reasons, mostly to do with the arcane art of political persuasion, the name of the telescope was again changed simply to ST when the aperture was reduced to 2.4 meters in the late 1970s. The rationale given was that the word "large" was too strong, suggesting not only an ultimate instrument but also an ultimate price, thereby possibly jeopardizing a future really big space telescope. However, the change of name was again opposed by those who argued that LST was the appropriate name, standing as it did for the Lyman Spitzer Telescope. The dream might not have become reality without Spitzer's vision, and of course, also not without the near ineffable genius of the engineers and scientists and the remarkable ability of industry. This symposium is for all of you who have made it possible for us.

1.2. *Hubble's legacy*

Why then was the telescope eventually named for Edwin Hubble? That too is appropriate, but much farther back in history. Simply, Hubble had manufactured the foundations upon which a large part of the present work of the telescope on cosmology is centered.

In only 12 years from 1924 to 1936, Hubble brought to an almost modern maturity the four foundations of observational cosmology, even as its principles are practiced today.

(1) He proved that nebulae are galaxies by identifying the content of NGC 6822, M33, and M31 (Hubble 1925, 1926a, 1929a) to be stars similar to those in the Milky Way.

(2) From an early beginning in 1922, he perfected the galaxy classification system (Hubble 1926b, 1936c) that undoubtedly contains clues to galaxy formation and evolution. Hubble's proposal, now universally adopted, was more systematic than that of Lundmark (1926, 1927), but there are obvious similarities, especially as to names. Lundmark introduced three groups as "amorphous ellipticals," "true spiral," and "magellanic cloud types." A flavor of a rivalry between these two giants is seen in Lundmark's (1927) footnote rebutting Hubble's (1926b) perhaps unjustified attack on Lundmark, also in a footnote.

(3) He organized existing data on redshifts and apparent magnitudes (Hubble 1929b) of nearby galaxies into a believable redshift-distance relation, searched for throughout the 1920s as the "de Sitter effect" by many others (Wertz [the European Hubble without a telescope], Truman, Silberstein, Lundmark) but without success, and seen in the early data as adumbrations by Lemaitre (1927, 1931) and Robertson (1928). Hubble, with Humason, then greatly extended the velocity-distance relation into the "remote" expansion field (Hubble & Humason 1931, 1934: Humason 1936; Hubble 1936a, 1937, 1953).

(4) He made a massive observational program of galaxy counts for the N(m) function, from which he attempted to measure the curvature of space (Hubble 1934, 1936b,c, 1937, 1953).

More detail on the history of these developments is the subject of this review. Most emphasis is placed on galaxy counts (item 4) as buttressed by data on redshifts and magnitudes (item 3) as needed for the interpretation. A few comments on the role of the abnormal galaxy morphology at faint magnitudes in the HDF, (item 2), closes the review.

2. The 1934–1936 N(m) count campaign

2.1. *The observational data from the 1934 campaign*

Building on the work of Fath (1914) as analyzed by Seares (1925), work that was based on galaxy counts using 60-inch reflector plates taken for The Mount Wilson Catalogue of Photographic Magnitudes in Selected Areas 1–139 (Seares, Kapteyn, & van Rhijn, 1930, hereafter the Mount Wilson Catalog), Hubble (1926b) used all existing data to show that the "white nebulae" increased in number as log N(m)$\sim$0.6m. This is the requirement for a uniform (homogeneous) distribution in depth in Euclidean space, regardless of any form of the distribution of absolute magnitudes (the luminosity function) as long as the integral of that function over luminosities is finite and if there are no effects on the magnitudes with distance (absorption, redshift, etc.). It is also the expected form in the limit of zero redshift, even using the modern (Mattig) equations that correctly describe the distribution (section 4).

The observational data available in 1926 was spotty and not well calibrated in magnitudes. Beginning in 1927, Hubble undertook a major survey with the Mount Wilson 60 and 100-inch telescopes to carry the survey of galaxies at increasing distances by extending the magnitude coverage beyond $m_{pg} = 16.7$ which was the effective limit of his 1926 study.

Hubble (1934) completed the massive observational program in 1934 in which he counted 44000 galaxies in an area of 650 square degrees on 1283 plates in a systematic sampling in both Galactic hemispheres. The result was a definitive study of the average properties of galaxy distribution, both in depth (for homogeneity) and around a significant fraction of the sky (for isotropy). In this major paper, Hubble (a) confirmed that galaxies continue to increase in numbers to the faintest limits surveyed (the ultimate organizational hierarchy appeared to have been reached), (b) there is a strong latitude effect showing absorption by the Galaxy, (c) the "zone of avoidance" is mapped in greater detail than was possible by Seares (1925), (d) the frequency distribution of the numbers of galaxies per square degree, when the counts on each of the 1283 plates were reduced to standard conditions (for the latitude effect, for different exposure times for the plates, for different seeing, for distance-to-center of each plate due to coma, etc), shows a normal error (Gaussian) distribution in log N, *not in N itself* (his Fig. 7, 1934).

This last discovery was one of the first indications of the tendency of galaxies to cluster, and was so noted by Hubble. In the 1934 paper he wrote:

[The log normal, rather than a straight N normal distribution is] "the feature [that] serves as a description and a measure of the tendency to cluster. It is clear that the groups and clusters are not superposed on a random (statistically uniform) distribution of isolated nebulae, but that the relation is organic."

While it is not clear what he meant by the last phrase of being organic, it is known (his comment once to me) that he knew that the distribution of the growth of bacteria in petri dishes in the laboratory show a log normal distribution of counts, and that after some time, clusters or colonies describe the mature distribution across the face of the dishes. (See Saslaw 1989; Saslaw and Hamilton 1984; Crane and Saslaw 1986; Coleman and Saslaw 1990; Karasev 1982, for modern discussions of the importance of a *log normal* rather than a direct normal distribution for the question of clustering).

Important as the 1934 paper was, no reliable apparent magnitudes could be attached to the counts. Indeed, the data were given as log N(E) (Hubble's Fig. 2), where E are the various exposure times of the photographic plates taken in the program.

As a final step, approximate conversion to magnitudes was then made by considering the reciprocity failure "Schwarzschild p exponent" in I$\sim$$E^p$ for the intensity (I) and exposure time (E). In this way Hubble could assert that the galaxy counts continued to increase approximately as would be required as N(m)$\sim$0.6m if galaxies are distributed homogeneously in Euclidean space in the absence of all effects of redshift.

2.2. *The 1936 observational campaign*

In a most important paper two years later, Hubble (1936b) made an attempt to reduce the count data to a reliable system of magnitudes and to push the counts to a fainter limit.

It is important to point out that none of the photometry was done on individual galaxies as is done today. Rather, the "limiting" magnitude of plates taken with a particular exposure time, "reduced to standard conditions" ("full" photometric development, particular seeing conditions, particular emulsion batch, particular telescope, and standard photometric conditions) was estimated, based on comparisons using standard *stars*. The next step was to estimate the difference in the limiting magnitude between stars and the in-focus galaxy images. This was accomplished by a series of experiments (Hubble 1932, 1936b), among which were out-of-focus images of stars made to resemble galaxy images of particular sizes. In this way, the "limiting magnitude" of galaxies was estimated on the "standard condition plates." These are the magnitudes listed for each of the log N values in Table IV of Hubble (1936b).

There are two major problems with this procedure. (1) The apparent magnitudes of the stars used as standards had large systematic errors starting as bright as $m_{pg} = 16$, conclusively demonstrated only as late as 1950 (cf. Stebbins, Whitford, and Johnson 1950, see later). (2) Stars in the Mount Wilson Catalog used as standards reached magnitudes only as faint as $m_{pg} \sim 18.5$ in most of the Selected Areas, considerably brighter than what was needed by Hubble for his deepest counts. Hubble (1936b) writes "the estimation of the limiting magnitude for 2-hour exposures necessarily involved considerable extrapolation." The limit for his faintest counts was eventually listed as $m_{pg} = 21.03$.

Furthermore, even as Hubble's survey work was proceeding in 1934, Baade, whose main Mount Wilson duties were to determine scale errors in the Mount Wilson Catalog, was discovering substantial errors in Selected Area 68 which was the principal Area used by Hubble in his 1929 study of M31. The deviations from a Pogson scale began as bright as $m_{pg} \sim 17$. Baade's methods were still photographic, but now, using platinum neutral half filter methods (e.g., Weaver 1946; Stock and Williams 1962), his results were a substantial advance over the multiple exposure plus graded diaphragm methods used by Seares for the Mount Wilson Catalog† between 1910 and 1925.

Hubble's (1936b) final table of log N(m) values at faint magnitudes shows five data points for the counts at m_{pg} magnitudes of 18.47, 19.0, 19.4, 20.4, and 21.03, plotted as Fig. 1 here from Fig. 1 of Hubble. The analysis for the curvature of space and Hubble's answer whether the redshift is a true Friedmann-Lemaitre expansion depended on these five points.

Plotted are the integral counts as the log of the number of galaxies per square degree that are brighter than apparent magnitude m. The line labeled "Uniform Distribution" has a slope of dlog N(m)/dm = 0.6. The five points show a shallower slope. The

† Baade never published his new photometry in any complete detail, although he did summarize his corrections to Hubble's M31 magnitude scale in the paper announcing the resolution of the disk of M31 into stars (Baade 1944). Baade needed the faint magnitudes, transferred from his new scale in SA 68, to estimate that the resolved stars in the M31 disk had absolute magnitudes of $M_{pg} \sim -1.5$ and therefore that they are similar to globular cluster stars at the top of the giant branch. This connection played a central role in Baade's development of the population concept (Sandage 1986).

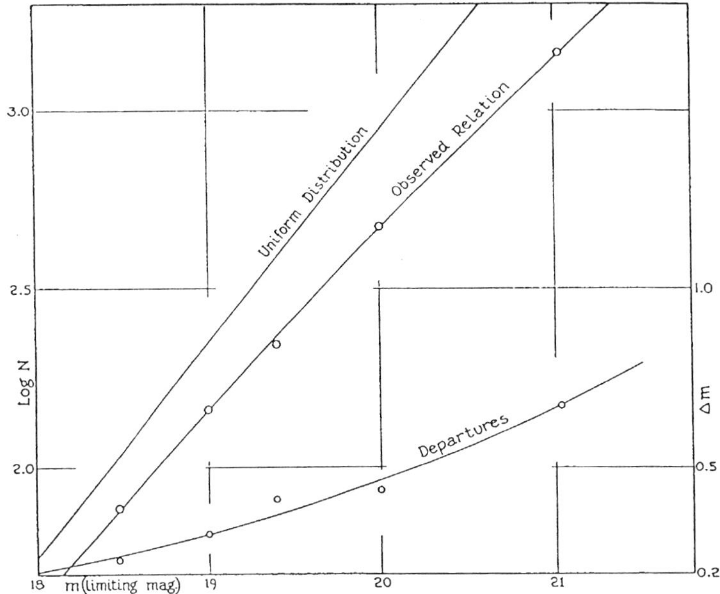

FIGURE 1. Hubble's final formulation of the log N(m) integral count-magnitude relation upon which his subsequent analysis of spatial curvature and the question of the reality of the expansion was based. N(m) is the number of galaxies per square degree brighter than apparent magnitude m. Diagram from Fig. 1 of Hubble (1936b).

departures of the five points from the "Uniform Distribution" line is shown as the lower curve. It is this "departure" curve that comprise the entire data set discussed by Hubble concerning the curvature of space and the reality of the expansion.

2.3. *Hubble's interpretation*

We now come to one of the most remarkable episodes in all of science. Hubble's (1936b) detailed analysis of Fig. 1 is a most fascinating study of how an interpretation, without caution concerning possible systematic errors, led to a conclusion that the systematic redshift effect is probably not due to a true Friedmann-Lemaitre expansion, but rather to an unknown, then as now, unidentified principle of nature. Indeed, even in the abstract to this 1936 paper on the "Effects of Redshift on the Distribution of Nebulae" Hubble concluded: "The high density suggests that the expanding models are a forced interpretation of the data." His belief that the expansion probably is not real persisted even into his final 1953 paper which was the Darwin lecture of the RAS, given in May of the year he died in September. What were the steps leading to this conclusion that, in today's climate, seems so remarkable?

Redshifts, ubiquitous for all galaxies everywhere, decrease the received flux. The larger the redshift, the larger is the decrease. This effect is one of the two principal reasons for the observed departure of the data in Fig. 1 from the "uniform distribution" supposition. The second is the departure of the intrinsic geometry from Euclidean, measured by "curvature" in the sense introduced by Gauss (1827, 1873). Hubble considered both of these effects.

There is no redshift information in Fig. 1. Hence, a relation between redshift and apparent magnitude is required to change the N(m) relation to the more fundamental N(z) function that is needed to make the calculations. And, because the counts in

Figure 1 are for field galaxies, the (m.z) relation for clusters as determined by Hubble and Humason (1931) and Humason (1936) had to be replaced by data for field galaxies. The required (m,z) relation was set out in Fig. 2 of Hubble (1936b), based on data of Hubble & Humason (1934).

Using the field galaxy (m,z) ridge-line relation of log cz = 0.2m + 0.77, the N(m) data of Fig. 1 could be changed to an N(z) relation, but only after correcting the observed magnitudes for the technical effects of redshifts, expressed as a K term, both selective and neutral (more later).

Armed now with a corrected N(z) relation, an assumption must be made concerning the relation between "distance" and redshift so that the N(z) relation could be changed to an N(r) relation, assumed to be proportional to the volume contained within "distance" r, leading to the geometry that is either Euclidean if (vol$\sim r^3$), or non-Euclidean if otherwise.

Note the extremely complicated multiple steps and, further, the questionable approximation of replacing distribution functions by mean values; viz. the counts in m are changed to N(z) and the K(z) term is changed to K(m) via an assumed *mean* (m,z) relation rather by using a luminosity distribution function for absolute magnitudes that takes into account the intrinsic spread in m at a given z.

A few of the intricate details of Hubble's procedure are set out in brief in the next section. Here we only summarize his conclusions, based on his analysis of the corrections to apparent magnitudes due to the effects of redshifts.

As is now well known, if redshifts are due to a true expansion, the required correction to the observed apparent magnitudes are by two factors of (1 + z) for the so called number effect (the number of photons received per second from a receding source) plus the energy effect [each photon is degraded in energy by the redshift, again by (1 + z)]. Hubble concluded (see the next section) that if two factors of (1 + z) are applied to his Fig. 1 data, then the curvature correction needed to make the data conform to the "Uniform Distribution" condition would have to be enormous, giving a very small, high density, large curvature universe, so small and of such high density that Hubble believed that the procedure gave impossible results. He continued to write his conclusion to the end, calling into question the reality of the expansion that required the second factor of (1 + z) correction for the "number effect."

To make understandable the language of Hubble's analysis, the K correction as used by Hubble is the selective effect (plus the bandwidth effect)† of shifting a galaxy spectrum through the blue photographic band pass "plus" either the one or two factors of 2.5 log (1 + z). Hubble expressed the total effect as K = B × z.

Hubble determined B from the observations using the "departure" curve in Fig. 1. His program was then to compare this B(observed) with a theoretical B* calculated using the assumption of either a true expansion or not (either one or two factors of 1 + z), plus the selective term found by shifting an assumed galaxy spectrum through the photometric band pass (the selective part of the K term; see Humason et al. 1956, Appendix B; Oke and Sandage 1968). In what follows, the argument hinges on the comparison of B and B*.

† The spectrum is stretched by multiplying each rest wavelength by a factor of 1 + z. By so doing, compensation must be made for the decreased bandwidth of the stretched spectrum over the fixed sensitivity pattern of the plate, filter, and telescope photometric functions. I mistakenly have written (Sandage 1995, Lecture 2) that Hubble neglected the bandwidth correction. However, a detailed examination of his calculated K corrections for a blackbody with T = 6000° K (Hubble 1936b, his Tables V and VI) shows that he did not neglect the bandwidth term and that my comment in the Saas Fee Lectures is incorrect.

Hubble's analysis (page 533 of the 1936b paper) of his "departure" curve gave B = 2.94. His *calculated* K term (assuming a black body galaxy spectrum of T = 6000° K) was either B* = 3.0 for no expansion, or B* = 4.0 for a real expansion (energy plus number effect). Clearly, only the no-expansion solution fitted Hubble's putative B = 2.94 departure data in Fig. 1.

Many conclusions were made from this result, not only concerning the reality of the expansion but also concerning the consequences of a real expansion for a second-order term in the velocity-distance relation as the measurement of deceleration, the value of the space curvature, and the question of evolution of galaxy absolute magnitudes in the look-back time. Several of the direct quotes concerning these issues are of interest for the work of the present workshop.

On his page 542: "It is evident that the observed result, B = 2.94, is accounted for if redshifts are not velocity shifts. The comparison is based on an effective temperature, T_o of 6000°, but the uncertainties cover the range down to about $T_o = 5750$. The interpretation is consistent with the data [but only if]—the *expansion and spatial curvature are either negligible or zero*." (Emphasis added).

Concerning the redshift-distance relation; page 38 of Hubble (1937): "The inclusion of recession factors would displace all the points [in the Hubble diagram of redshift vs. apparent magnitude of great clusters—his Fig. 1 of the 1937 reference] to the left [higher redshifts at a given magnitude], thus destroying the linearity of the law of redshifts". [N.b., this is not correct when the appropriate Mattig relations for the (m,z) Hubble diagram are used; see later].

For the conclusion on the reality of the expansion (Hubble 1936b, page 553):—"if redshifts are not primarily due to velocity shifts, the observable region loses much of its significance. The velocity-distance relation is linear; the distribution of nebulae is uniform; there is no evidence of expansion, no trace of curvature, no restriction of the time scale." Page 553/4: "The unexpected and truly remarkable features are introduced by the additional assumption that redshifts measure recession. The velocity-distance relation deviates from linearity by the exact amount of the postulated recession. The distribution departs from uniformity by the exact amount of the recession. The departures are compensated by curvature which is the exact equivalent of the recession. Unless the coincidences are evidence of an underlying necessary relation between the various factors, they detract materially from the plausibility of the interpretation.—the small scale of the expanding model, both in space and time is a novelty, and as such will require rather decisive evidence for its acceptance."

From his Darwin lecture (Hubble 1953): "When no recession factors are included, the law will represent approximately a linear relation between red-shifts and distance. When recession factors are included, the distance relation is expected to be—non-linear in the sense of accelerated expansion" [sic, not the correct sign; the word must clearly be *decelerated*, as he in fact wrote twice earlier in 1936 and 1937] . "—[If no recession factor is included] the 'age of the universe' is likely to be between 3000 and 4000 million years, and thus [again with no recession factor] comparable with the age of rock in the crust of the Earth."

Concerning the second-order term in the velocity-distance relation: (1936b), page 546: "Since the second-order term [with recession factors included] is definitely positive, the possible models are restricted to those in which the rate of expansion has been diminishing during the past several hundred million years." And again in (1937, page 43): "The chief significance of the term for cosmological theory lies in the positive sign [of the redshift vs. distance correlation]. The rate of expansion of the universe has been slowing down, at least for the past several hundred million years. The 'age of the universe' is considerably shorter than that permitted by the linear law."

Finally, as to evolution of luminosity in the look-back time (Hubble 1936b, page 543). "As for the constancy of nebular luminosities, the question is whether or not luminosities of spirals change materially (say 10%, or 0.1 mag) during [the look-back time]. —very few students will hesitate to adopt the assumption that systematic variation in so short a [time] interval will be inappreciable." Reasons are then given in the remainder of the paragraph, none of which would pass today's referees armed with the present knowledge of population synthesis and stellar evolution.

The clearest proof that Hubble maintained these views concerning the reality of the expansion until the end is the style of argument in his 1953 Darwin Lecture, seen in particular from Fig. 1 of that lecture. No recession factor was put to the K-correction magnitudes in the abscissa of the m,z (Hubble) diagram. This diagram was the first rendering in the literature of that central cornerstone of observational cosmology using new data from the Palomar 200-inch reflector.

The several arguments set out above were of particular importance for the Palomar program on these matters that followed from 1953 through the 1980s. Reasons why Hubble's conclusions should be changed emerged slowly from this program, not only because of new observations based on photoelectric photometry, but also from a much deeper understanding of the interface between the observations and the theory (section 4).

3. Why Hubble's program failed

What then was wrong with Hubble's 1936 analysis of the count data in Fig. 1 that led him to his remarkable conclusion of no expansion?

There were five problems. (1) Incorrect K term values as a function of redshift because galaxy spectra have a much cooler color temperature than 6000°. (2) The apparent magnitude scale used by Hubble via the Selected Area magnitudes, even as partially corrected by Baade in the late 1930s, was wrong. (3) Hubble's assumption that "distance" is given by cz/H for large redshifts is not correct, but known only after the Mattig revolution (section 4). (4) The assumption that uniform spatial distribution requires log N(m) to increase as ~0.6 m for large redshifts is also wrong according to the theory of Friedmann spaces, again shown by the new Mattig equations. (5) The assumption of constant luminosity and/or density evolution at high redshifts is evidently wrong as shown by the large excess in the counts (a fact that would have been discovered by Hubble from his counts if he had kept the true expansion assumption) shown not only by the modern N(m) counts, but also by the strange galaxy morphology at the faintest HST levels (section 7).

These points are reviewed in order.

3.1. *Enter Greenstein*

The 1936 analysis by Hubble had already begun to unravel by a devastating paper by Greenstein (1938), in which he showed that the color temperature of M31 was only 4200° K rather than 6000°. Shifting a black body spectrum through the m_{pg} pass bands gave selective K corrections plus *either* one or two factors of 2.5 log (1 + z) that were no where near the B = 2.94 determined from the "departure" observations by Hubble. Hence, nothing worked in *any* interpretation of Hubble's 1936b counts. Greenstein's conclusion was: "From λ6500 to 3900 Å, [the spectrum of M31] closely resembles that of a black body of temperature 4200°.—The effect of such low temperatures on the present interpretation of counts of extragalactic nebulae is serious. It seems improbable that the effect of the redshift on the apparent magnitudes of nebulae, found by Hubble, can be interpreted either as a velocity or as a nonvelocity shift."

3.2. *Modern K term*

However, the problem with the K-term was even more serious because it was soon realized, via the non-existent Stebbins-Whitford (1948) effect and its explanation (Oke and Sandage 1968), that galaxy spectral energy distributions (SED) are very poorly approximated as black bodies, primarily because of the important 4000 Å break. Improvement of the SEDs that had been measured by Oke and Sandage was made by Whitford (1971). Oke and Sandage had observed only the central regions of five giant E galaxies using a spectrum scanner at the Cassegrain focui of both the Mount Wilson 60 and 100-inch reflectors. Whitford could observe a much larger fraction of the total E-galaxy light with his spectrum scanner at the Lick Crossley reflector because its shorter focal length and therefore smaller focal plane scale. The radial color gradient of E galaxies, becoming bluer from the center to the outside, explained the 10% difference between the two studies.

The resulting mean SEDs, shown in Fig. 2, permitted, for the first time, the calculation of realistic K(z) corrections for the effects of shifting the mean E-galaxy SED through various photometric pass bands. Since then, a large industry of K term calculations has developed, not only for E galaxies but for all Hubble types. A more modern history with entrance to the extensive literature is given elsewhere (Sandage 1988, 1995).

The result of Whitford's calculations (his Table 3), approximated as the first term of a power series in z, is

K_B = 7.1z if no recession (only the energy effect term), and

K_B = 8.1z if the number effect term for real recession is included.

Clearly, neither of these cases are consistent with Hubble's putative 1936 requirement that B_{mpg} = 2.94.

3.3. *Corrections to the magnitude scale*

As mentioned earlier, in one of the most important papers of the decade, Stebbins, Whitford, and Johnson (1950) showed that corrections to the m_{pg} magnitudes in the Mount Wilson Catalog for SA 57, SA 61, and SA 68 begin as bright as m_{pg} = 16 and increase with increasing faintness. They further showed that the North Polar Sequence, long the principal source of faint standards in the decades before 1950 (Seares 1915, 1922a,b; Seares and Humason 1922) was accurate in scale to better than the 0.1 mag level to the limit of its tabulation.

Later work extended the result to additional selected areas where it was shown that the needed corrections were generally ubiquitous and in some areas reached 1.5 mag at m_{pg}(Seares) = 18.5. Examples for four Selected Areas are in Figure 3, taken from an early summary of a systematic program to determine the corrections to the faintest level of the Catalog (Sandage 1983, 1998) in a dozen Selected Areas. The corrections begin near m_{pg} = 15 and increase to an average of 0.7 mag at m_{pg} = 18.

4. The Mattig revolution 1958–1959

Now we come to the heart of the problem with Hubble's interpretation, but more importantly to the watershed for practical cosmology itself in a fundamental development that changed the field.

Reading most of the many papers on observational cosmology before the early 1960s, nowhere does one see the modern approach of solving the Friedmann equation that

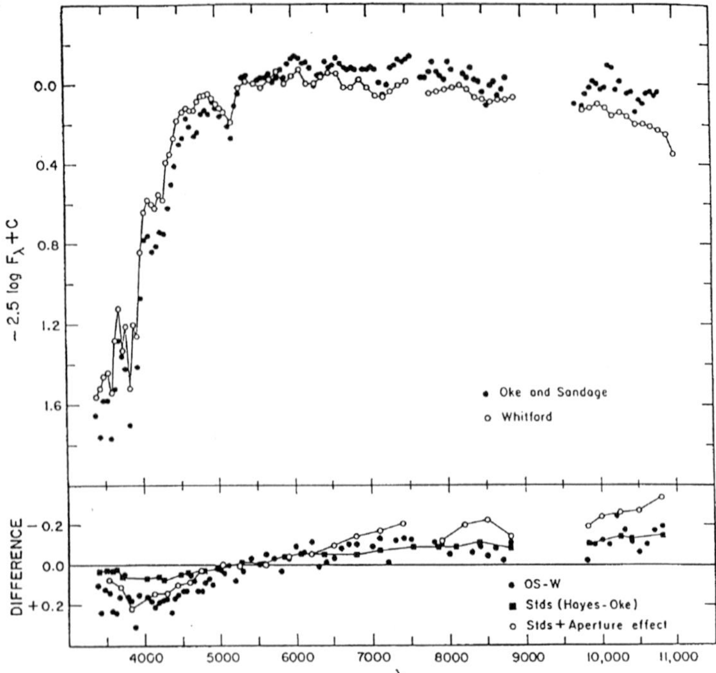

FIGURE 2. Mean spectral energy distribution of giant E galaxies as measured by Oke and Sandage (1968) for the very central regions and by Whitford (1971) for larger scanning apertures. These are the SEDs used for the first reliable calculation of the selective part of the K corrections in B, V, and R pass bands for E galaxies. The calculated K term in B using these SRDs was very much larger than the m_{pg} K correction used by Hubble (1936b). Diagram from Whitford (1971).

describes the development of the scale factor R(t) with time, and how the closed form of R(t) for arbitrarily high z is to be used to obtain the exact equations necessary to interpret the observations, valid for all redshifts. What in fact is the correct equation for the interval "distance," r, as a function of redshift?

Before the correct equations became known in the late 1950s, all relations involving observed magnitudes, angular diameters, redshifts, spatial volumes, and the consequently N(m) counts were given in Taylor series expansions in z, using only $R(t_o)$ and the first several derivatives of R(t) about the present epoch. The only assumption on R(t) was that it is a smooth enough function for the Taylor expansion to mean something. These series expansions, while good for small redshifts, were worthless for redshifts larger than perhaps 0.3, which was in fact near the limit of redshifts known even in the late 1950s.

What is remarkable about this situation is that the Friedmann equation and its solution never entered most of these papers at the interface between theory and observation. Examples are the marvelously complicated series-expansion papers by Davidson (1959a,b, 1960), and even more remarkably, the first edition of the famous text book by McVittie (1956).

The developments that began the modern era were the derivations of all the relevant equations using the Friedmann equation to give the explicit solutions of R(t) for all t (i.e., for arbitrarily high redshifts). Remarkably, the complete development is set out in two short papers by Mattig (1958, 1959).

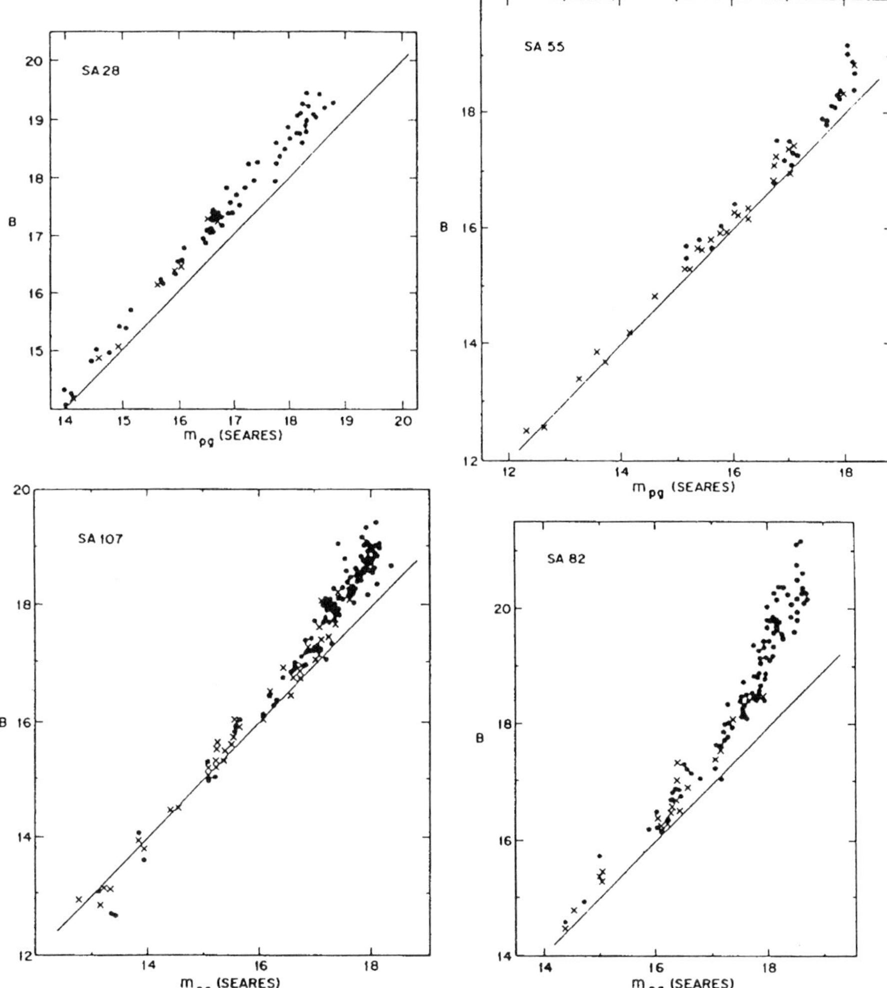

FIGURE 3. Comparison of the apparent magnitudes in the Mount Wilson Catalog of Selected Areas with the listed m_{pg} magnitudes by Seares et al. (1930). The modern B magnitudes have been measured photoelectrically in a program at Mount Wilson in the 1970s. Diagram from Sandage (1983).

In the first he derives the famous r(z,q_o) relation

$$r(z, q_o) = \frac{1c}{R_o H_o q_o^2 (1+z)} \left\{ q_o z + (q_o - 1) \left(\sqrt{1 + 2q_o z} - 1 \right) \right\} \quad . \tag{4.1}$$

The paper is only three pages long. The second, in which he derives the volume elements (both Euclidean and noneuclidean) also as functions of z and q_o is done in only two pages.†

Yet these two papers changed the subject. An example is the second edition of McVittie's (1965) text book where every equation in practical cosmology is based on Mattig,

† In a later paper Mattig (1968) introduced closed equations valid for a finite cosmological constant.

in contrast to the 1956 first edition where everything is in series expansions of R(t), sans even the Friedmann equation as a guide.

The important point for our purposes is the comparison of Mattig's exact theory with Hubble's intuitive assumption for the relation between interval "distance"† and redshift.

Hubble's assumption for an ($R_o r$, z) relation was,

$$R_o r = cz/H_o \quad . \tag{4.2}$$

The correct (Mattig 1958) equation for any q_o and z is equation (4.1) above, for which the two most interesting special cases are for $q_o = 0$ in a zero density (massless) universe, and $q_o = 1/2$ for flat space-time curvature. The relevant equations are

$$R_o r = \frac{cz}{H_o(1+z)}\left(1+\frac{z}{2}\right) \quad , \tag{4.3}$$

for $q_o = 0$, and

$$R_o r = \frac{2c}{H_o}\left(1-\frac{1}{\sqrt{1+z}}\right) \quad , \tag{4.4}$$

for $q_o = 1/2$, both of which differ from Hubble's assumption in equation (4.2) except near the $z = 0$ limit where all equations have the same z dependence. Figure 4 shows the differences between equations (4.2), (4.3), and (4.4) for progressively larger redshifts. These differences translate into different predictions of how the proper volume elements vary with redshift in geometries of arbitrary curvature.

The step beyond equation (4.1) in Mattig's derivation is to use the Robertson (1938) equation for how the received flux of a source of absolute luminosity, L, varies with redshift. This equation relates the redshift and the received bolometric luminosity, l, with L and with the interval "distance" $R_o r$ from equation (4.1) as

$$l = \frac{L}{4\pi(R_o r)^2(1+z)^2} \quad , \tag{4.5}$$

to finally obtain the predicted count-brightness N(m) relation for galaxies distributed uniformly in space.‡ This prediction also differs from Hubble's assumption that log N(m)$\sim$ 0.6m for a space with zero curvature.

Fig. 5 shows the Mattig (1959) predictions for the N(m) count distribution in the ideal case of a single fixed absolute luminosity (a delta function luminosity distribution) for the counted galaxies, bolometric magnitudes (i.e., corrected for K dimming, both selective and non-selective), and no evolution, either luminosity or density, in the look-back time. The differences between the Hubble assumption (the upper dashed line) and the exact Mattig predictions of equations (4.3) and (4.4) above are evident in Fig. 5.

In this and the previous section we have seen how (1) errors in the K term evaluation, (2) errors in the apparent magnitude scale, and (3) errors in the precepts of the proper equations for "distances" and the consequent variation of apparent magnitude with distance for arbitrarily high redshifts affected Hubble's conclusions. The two re-

† See McVittie (1974) for an interesting discussion of why "distance" is an ambiguous concept in cosmology, differing depending on the parameters used to measure it. Because of this ambiguity, it is necessary that no equations that connect observables should contain "distance" in their final form. The equations must all be constituted to contain only the directly observed parameters of flux, angular size, and redshift (Sandage 1988).

‡ The crucial Robertson equation (4.5) was in contention for many years, with arguments of various subtleties occurring between such giants as de Sitter, von Laue, Tolman, Vogt, and others as to whether the factors of (1 + z) in the denominator should be as shown or rather only the first power, or even to the third power. Robertson gave the definitive proof of equation (4.5) in the cited reference.

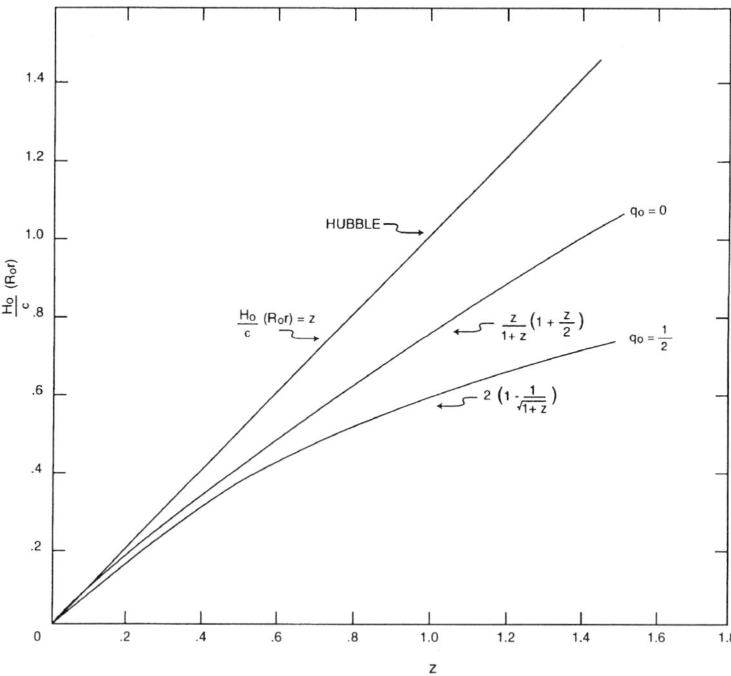

FIGURE 4. Comparison of Hubble's assumption (upper straight line) of how the parameter "distance" that is needed in the correct cosmological equations relating volume to redshift differs from the correct "distance"-redshift relations for two values of the deceleration parameter, as calculated from the Mattig (1958) equation for $r(z,q_o)$.

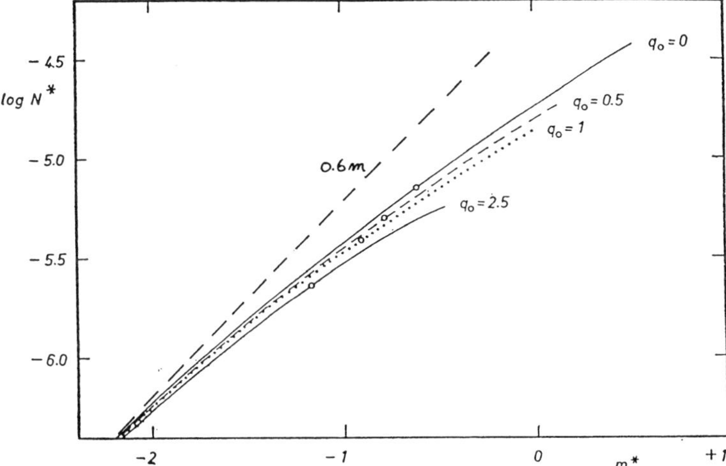

FIGURE 5. Comparison of Hubble's assumption of the expected log N(m)~0.6m relation (upper dashed line) with the correct Mattig (1959) prediction (no evolution) in closed form, based on the solution of the Friedman equation, rather than on previous formulations via series expansions in z. Diagram from Mattig (1959) with Hubble's assumption put in as the dashed line.

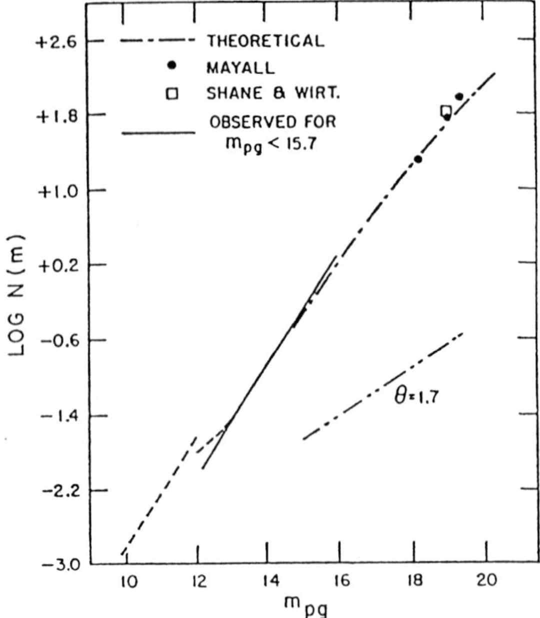

FIGURE 6. Summary of the N(m) count distribution as the data existed in 1972, before the major modern observational campaigns had begun. Diagram for Sandage, Tammann, and Hardy (1972). See this reference for a discussion of (1) the North Galactic Anomaly shown as the discontinuity at bright magnitudes, and (2) the meaning of the $\theta = 1.7$ line as a prediction of a hypothetical hierarchical distribution in depth.

maining considerations concerning these conclusions are (a) a comparison of Hubble's N(m) count data with the modern counts, and (b) replacing Hubble's analysis that used mean values for absolute magnitude and the apparent magnitudes at given redshifts by proper distribution functions of (m,z) and (M, morphological type) luminosity functions. Consider first the modern counts.

5. The modern count campaigns

5.1. *The post Hubble era (1970–1997)*

The last major program begun by Hubble three years before his death was a comprehensive plan to repeat his 1934–1936 N(m) campaign using the newly completed 48-inch Palomar Schmidt telescope, commissioned in September 1948. The wide field and superb imaging quality of this miracle telescope offered substantial advantages over the coma-affected and small area photographic plates used in the 1934–1936 work.

The first plates for Hubble's newly planned galaxy count survey were taken in the spring of 1949, before the systematic Palomar Sky Survey was begun. To assist in the work, I had been sent up as a graduate student by Greenstein to the Mount Wilson office from Cal Tech. I can only now suppose that Greenstein's selection from among other graduate students in the newly created (1948) Cal Tech department of astronomy, was based on my previous experience in survey photography, star counting, magnitude transferral from Selected Areas by photographic means, and obtaining magnitude distribution functions of the counts in areas of similar absorption, gained in a junior-senior undergraduate thesis program at the University of Illinois in the Perseus and Taurus

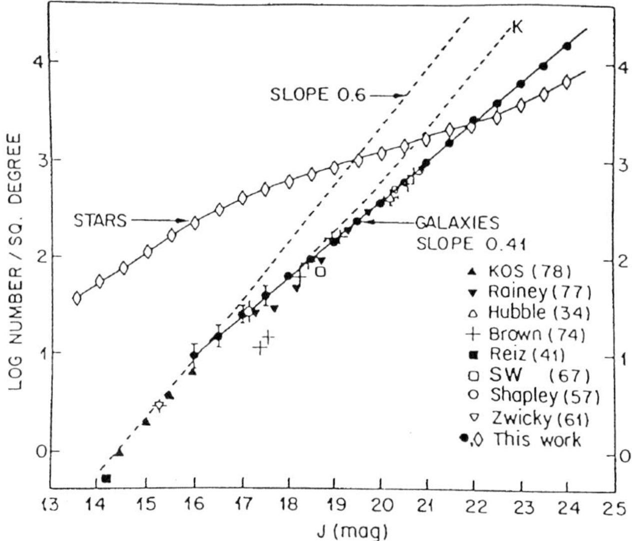

FIGURE 7. An early summary of the major count campaigns extending to the faint J magnitudes of J = 24. The departure of the slope from 0.6 beginning at J = 17 is consistent with the early theoretical curve fitted to the data in Fig. 6. Diagram from Tyson and Jarvis (1979).

dark cloud regions of the Milky Way. (An analysis was later published by another who had no relation to the work).

The technique to be employed for Hubble's new galaxy-count program was again to count to successively fainter magnitudes on a series of plates of graded exposure times, with later calibration of the various limiting magnitudes by determining the relation of the plate limits for stars and for galaxies—identical to the procedures in 1934–1936, but with a modern stellar magnitude scale in Selected Area 57 as the standard sequence.

The project was not completed for two principal reasons: (1) The conviction arose that a definitive new study must be based on the measurement of individual magnitudes for each galaxy rather than by simply counting to a plate limit. The technical means via areal photometry to do this were not in hand in the 1950s. (2) The solution to the problem of aperture corrections via growth curves to obtain either a "total" magnitude or a correction to obtain it had not then been solved (Humason et al. 1956, Appendix A). (Adequate standard profile templates had not been measured in the early 1950s). Therefore, Hubble's new count program was never begun in earnest after his death in 1953.

It was only 30 years later, beginning in the late 1970s that the modern photometric methods of individual galaxy photometry was made possible by the two-dimensional detectors and computer technology. A summary of several of the many galaxy count programs using these modern methods, completing Hubble's 1950 hopes, is the subject of this section.

The state of the count-magnitude relation just at the end of the 1930–1975 exploratory period is shown in Fig. 6 (Sandage, Tammann, & Hardy 1972), combining Hubble's faint data with those of Mayall (1934), the bright counts of Zwicky to B ~15.7, and the vast, wide area, counts of Shane and Wirtanen (see Shane 1975), but again counting to a plate limit.

One of the first papers in the post-exploratory era (individual magnitudes determined and account taken of total vs. isophotal magnitudes) was that of Tyson and Jarvis (1979),

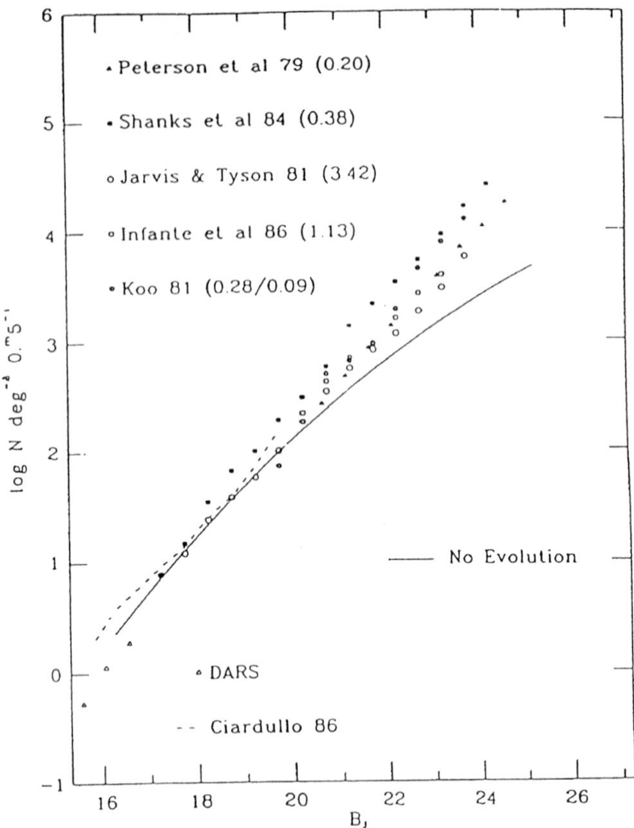

FIGURE 8. Summary by Ellis showing not only the range of differences in the counts from different modern group campaigns, but also the beginning of the now well known large excess of the counts fainter than $B_J = 20$, discovered by Kron (1980), relative to the predicted no evolution "standard" Mattig model. The B_J magnitudes are related to the standard Johnson B by J = B–0.1. Diagram from Ellis (1986).

shown in Fig. 7. They combined their data with many previous surveys to give a comprehensive summary reaching J(Gunn) = 24. The deviation of log N(J) from a slope of 0.6 is clearly seen beginning near J = 17, similar to that shown in Fig. 6 and initially discovered by Hubble (Fig. 1). An interesting point from Fig. 7 is that the surface density of galaxies exceeds that of stars for $J > 22$.

Fig. 8 shows the state of the work nearly a decade later, taken from a summary by Ellis (1986). The two principal features of this diagram are (a) the comparison of individual data points from various surveys, shown to illustrate the general accuracy of the count data at better than 0.2 dex in log N(m), and (b) the comparison of the data with the predicted counts from the Mattig formulation using proper distribution functions for (m,z), for the luminosity function (M,T), for K(z, sed), and for an assumption of the morphological mix of field galaxies (see next section). The excess of the counts relative to the predicted no-evolution Mattig curve is evident. It was discovered by Kron (1970) and is a feature of all subsequent surveys, forming a major subject of this workshop.

However, the degree of the excess is a function of wavelength, also discovered by Kron (1980) from his progressive blueing of the color distribution fainter than $b_j = 22$. A summary of counts in the three band passes of b_j, r, and K is in Fig. 9 from the review

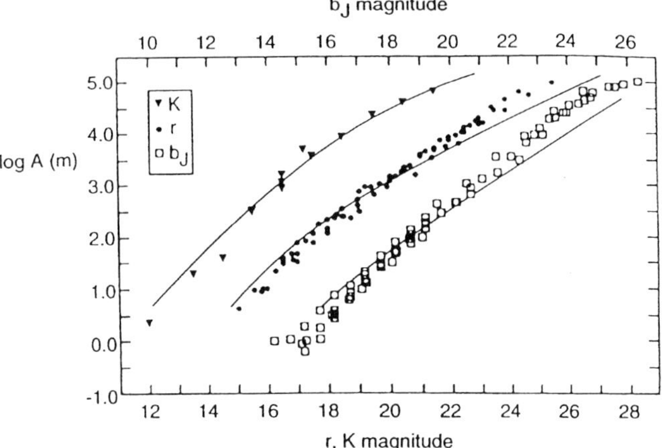

FIGURE 9. Summary of differential counts in b_J, r, and K reviewed by Koo and Kron (1992). A(m) is the number of galaxies per square degree per magnitude interval in the magnitude interval from m 1/2. The predicted Mattig theoretical non-evolution curves are shown for comparison.

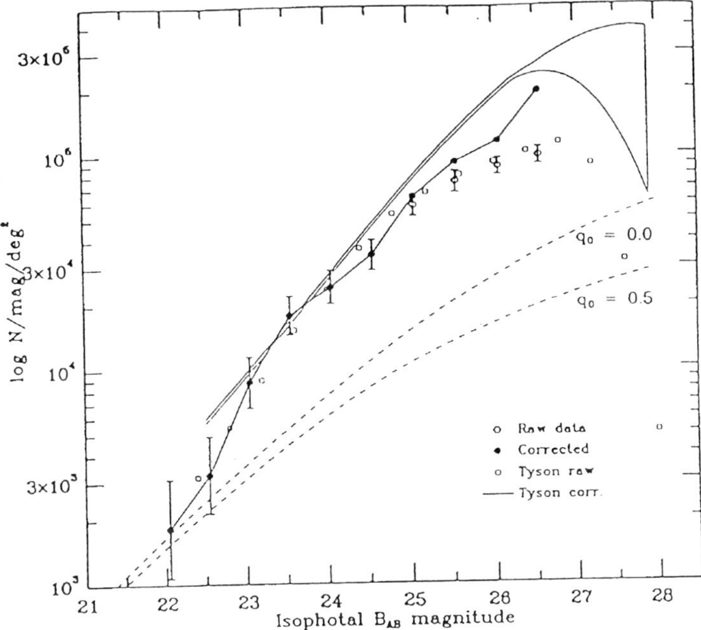

FIGURE 10. Summary of the very deep differential [i.e., A(m)] counts showing again the strong excess of the faint counts over the no-evolution predictions for two Mattig models. Diagram from Lilly, Cowie, & Gardner (1991).

by Koo and Kron (1992). The important point is that the counts in K show no excess relative to the predicted N(m), whereas the counts in b_j have the pronounced excess.

A particularly clear representation of the large excess at the faintest levels is in Fig. 10 from Lilly, Cowie, and Gardner (1991) in the blue pass band. Again, the predicted N(m) no-evolution relations, calculated with all the complications summarized in section 6, are shown as the dotted curves for two values of the curvature parameter.

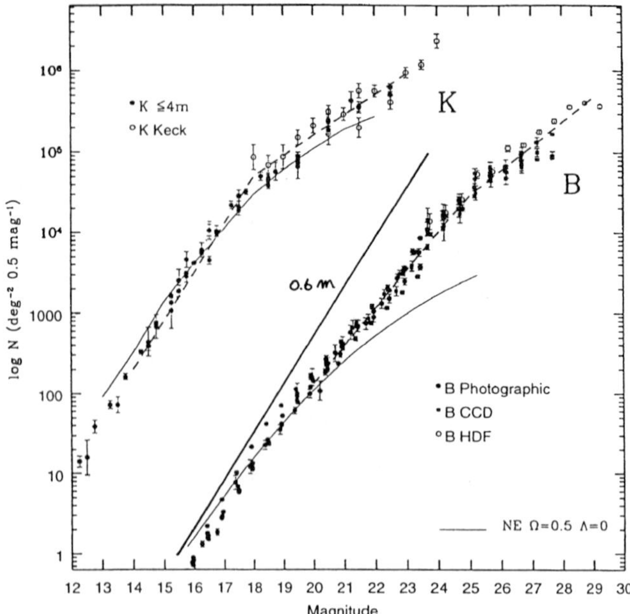

FIGURE 11. Count data from Ellis (this volume) but with the 0.6m line drawn to emphasize, that despite the excess relative to the no-evolution Mattig predicted N(m), the counts are still less numerous than Hubble's pre-Mattig assumption that log N(m)~0.6m.

A similar summary in B and K is given by Ellis in this volume. His diagram is reproduced as Fig. 11 here, but with the 0.6m line, as in Hubble's pre-Mattig assumption, drawn, showing that the counts, even in B, are less numerous than the 0.6 relation, as indeed first discovered by Hubble in 1936. Nevertheless, the counts are larger than the theoretical no-evolution predictions, especially in B.

5.2. *Comparison with Hubble 1936*

Finally, it is of interest to compare the modern counts in B, as read from Figs. 7–11 and averaged, with Hubble's 1936 five-data-point N(m) relation from Fig. 1. The comparison is shown in Fig. 12. The remarkable feature is that Hubble got the slope almost right. The main difference is only in the zero point of the magnitudes. Hubble's magnitudes are ~0.6 mag brighter than the modern values for the same integral count. Remarkably, the effect is in fact explained simply by the known corrections to the magnitude scales in the Mount Wilson Catalog, similar to those shown in Fig. 3.

Note that the counts in Fig. 12, both for the Hubble and the modern data, are as observed (but reduced to "standard conditions"). No K corrections have been applied for the effects of redshift. Hence, whatever the explanation is why Hubble interpreted his counts as requiring the no-expansion model, that explanation does not lie in an error in the 1936 N(m) count data themselves where the slope (not the zero point) is important. In previous sections we have set out the reasons for Hubble's conclusion, based on his inappropriate precepts about the theory (K term and the Mattig exact equations for the z, volume and the m,z relations). Any errors in Hubble's conclusions evidently do not rest on the count data themselves.

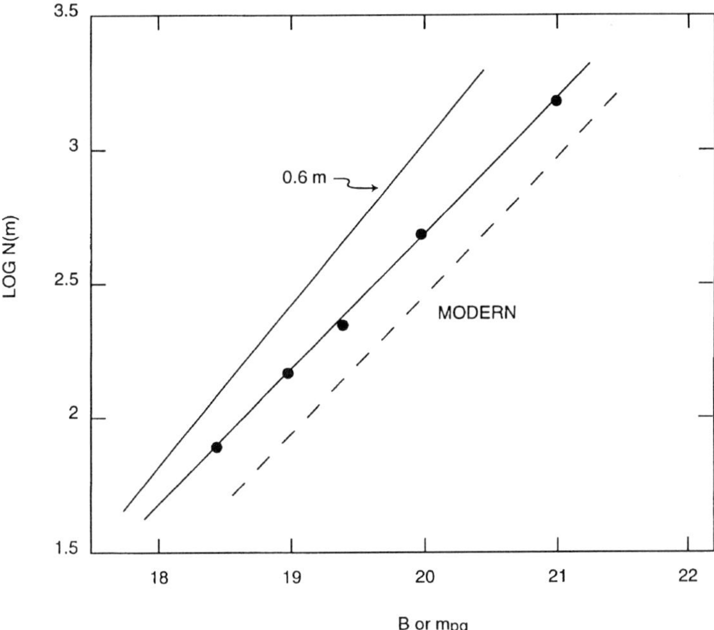

FIGURE 12. Comparison of Hubble's 1936 counts (from Fig. 1) with the mean of the modern counts discussed in the text.

6. The steps needed to make a proper calculation of N(m)

6.1. *The input parameters*

The theoretical predictions of the expected no-evolution N(m) relations as shown in Figs. 8–11 are based on integrations of various distribution functions together with the Mattig N(z, q_o) volume relations for any geometry (e.g., Sandage 1995, lectures 1–3 for a summary).

There are five needed ingredients. (1) The m = f(z) distribution giving the spread of apparent magnitude at given redshifts must be known. Typical distributions at various redshifts are set out in Fig. 13, taken from Koo and Kron (1992). (2) This distribution is the integral over morphological type of the type-dependent luminosity function as log N(M, type). An early example of such a luminosity function is shown in Fig. 14, based on both the Virgo Cluster survey and the local field (Binggeli, Sandage, and Tammann 1988) as defined by the Revised Shapley-Ames Catalog. (3) K corrections for any given band pass and morphological type, K(z, band pass, type), are needed to convert the Mattig equation in m(bol) to observed m(z). The SEDs for each galaxy type are needed to calculate these K corrections. Such calculations are given, among others, by Colman, Wu, and Weedman (1980), and Yoshii and Takahara (1988). (4) The morphological type mix at each z is needed so as to add the separate calculations of N(m, type) to produce the composite N(m) over all types. A number of studies of the type mix as function of z are summarized elsewhere (Sandage 1988, section 7.2). (5) Faith, courage, and daring are needed to end with any conviction that the predicted N(m) describes reality, or, if the prediction disagrees with the observed N(m, band-pass) relations, that evolution (either density or luminosity, or both) as a function of both z and morphological type is required to make agreement. Much faith, courage, daring, and conviction have been displayed in

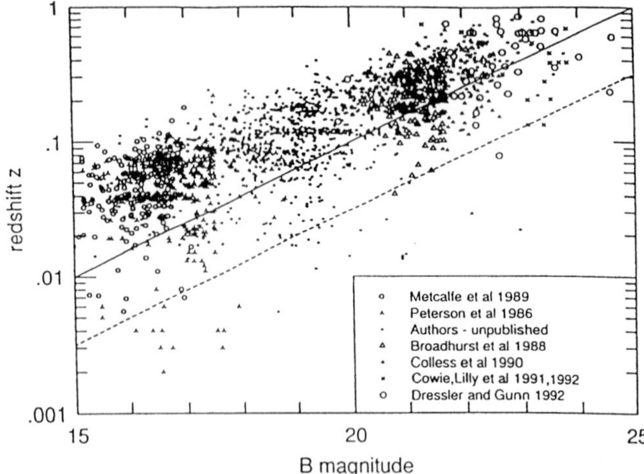

FIGURE 13. The distribution of redshift at given B apparent magnitudes from the various listed surveys. Diagram from Koo and Kron (1992).

the many current Journal papers concerning evolution in the look-back time. We shall see more of all four this week.

Will these be viewed in several future decades in the same way that most of the current community views Hubble's 1936–1953 conviction that the redshift-distance relation is not due to expansion?

6.2. *A conservative view*

As a caution against the view that radical evolutionary changes in the look-back time are needed to explain the excess in the counts, and therefore that we have strong direct evidence for a radically different universe at large redshifts, we quote from the quiet voices of Koo and Kron (1992) where they write:

> "To account for part of these data, different groups have proposed modifications to the conventional picture of mild luminosity evolution for $z < 1$. Examples include adoption of a large cosmological constant; substantial merging at low redshifts; a vast increase in bursting activity at moderate look-back times; or entirely new populations of galaxies that were present at high redshift but absent today. We have instead taken a more conservative stance and asked whether all the data might be found to be consistent with a simpler picture in which the cosmological constant is zero, the number of galaxies is conserved over time, and the shape of the luminosity function for each galaxy class is constant. Notwithstanding the new redshift data, we have argued that the combined uncertainties in the models and in the data so far preclude the necessity of more exotic assumptions."

7. A change of morphology at high z—"Mergers" or ELS with noise?

Perhaps the most persuasive evidence for substantial evolution in the look-back time is currently provided by an aspect of observational cosmology also pioneered by Hubble (1926b, 1936c) through his invention of the galaxy classification system. We shall hear much at this workshop of the seemingly drastic difference in galaxy morphology of the

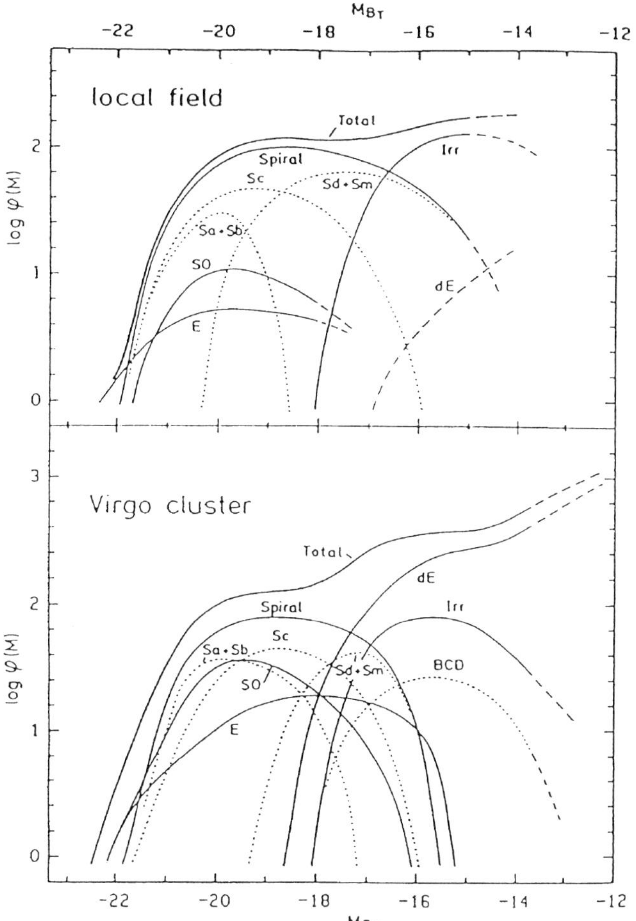

FIGURE 14. Type-dependent luminosity function (H–o = 50) needed to calculate a realistic no-evolution N(m) predicted count-magnitude relation. Diagram from Binggeli et al. (1988).

galaxies in the HDF images fainter than B = 22 compared with the morphology of nearby samples.

Several years ago, at the invitation of Jerome Kristian, PI of a deep survey GTO project, I classified the faint galaxies on a series of stacked very deep HST frames reaching B = 29 for two fields that were part of the more extended Groth fields that flank the HDF position. As is now well known from other such studies, I also found that the percentage of abnormal galaxies between V = 22 and V = 29 is as high as 50% of the total compared with the less than 5% abnormality rate in the nearby sample in, for example, the Revised Shapley-Ames Catalog (Sandage & Tammann 1987). The actual percentages were as follows. Normal spirals, in the sense of the present-day Hubble sequence, are relatively rare at 20%. Normal appearing red ellipticals are present in the significant numbers of 25% of the total. The remaining 55% were divided between mildly abnormal (shreds, interacting multiple components with tides, conglomerants consisting of condensations in a common envelope), and highly irregular forms similar to the local galaxies LMC, SMC, NGC 1313, NGC 1156, NGC 4449, and the pair NGC 4485/4490 for example (see the Carnegie Atlas of Galaxies for illustrations).

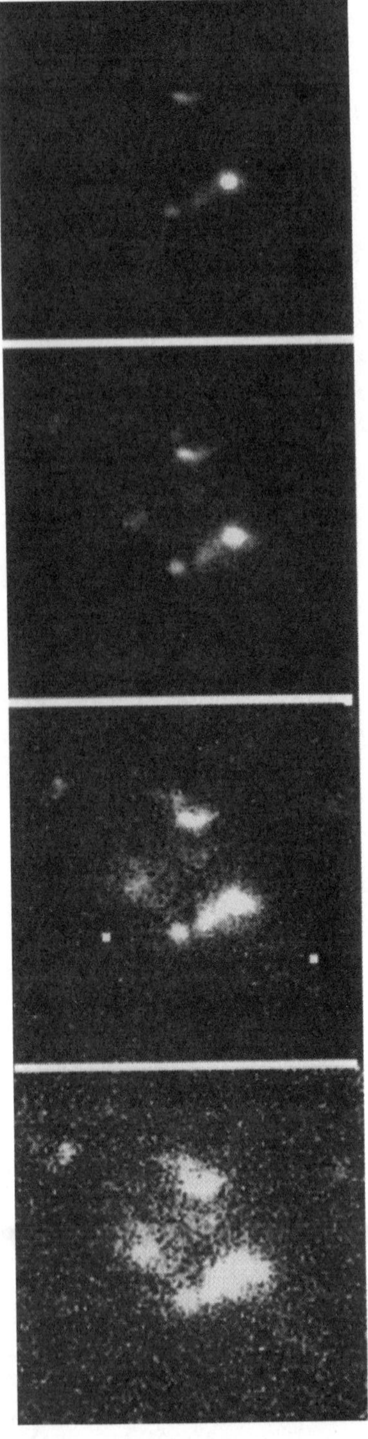

FIGURE 15. Four frames with different stretch of a small part of the Kristian et al. deep field showing a conglomerant structure the appears to be an envelope with density fluctuations that are collapsing onto the central E-like compact core seen in the top frame. The process may be an example of an ELS collapse formation with density fluctuation noise.

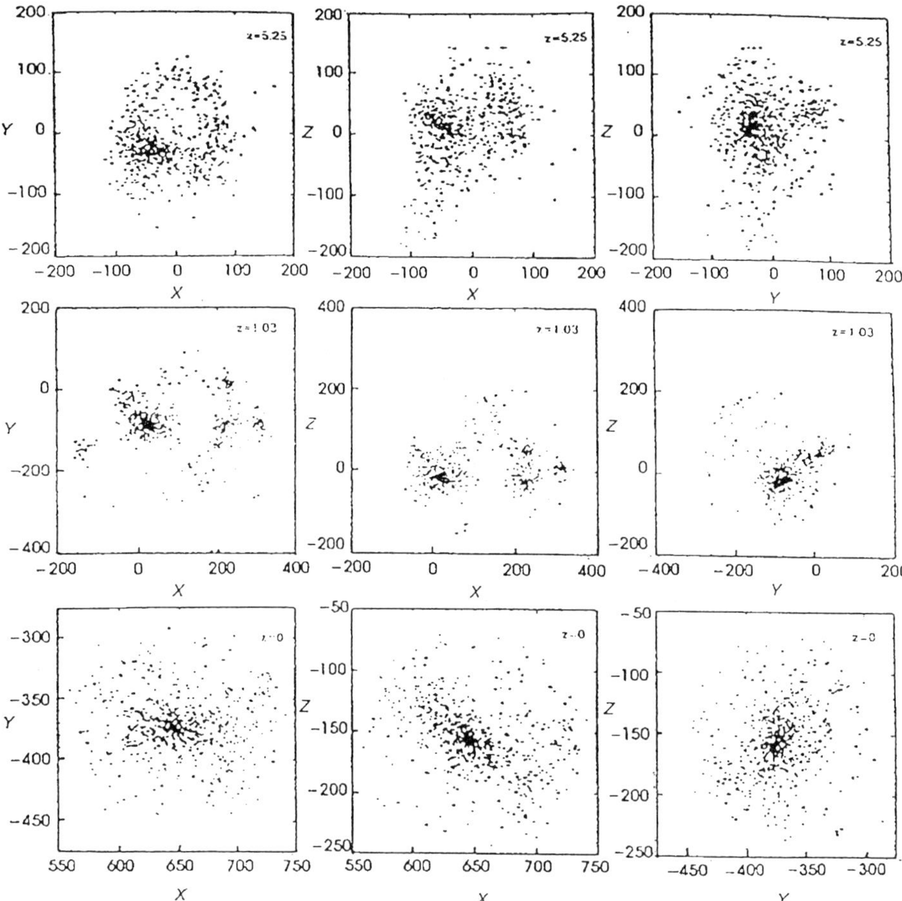

FIGURE 16. N-body simulation by Zurek, Quinn, & Salmon (1988) of a collapse protogalaxy at three different epochs after the initial collapse, as measured by the redshift (look-back time). Projections on the X, Y, Z axes are shown. The description could be ELS with noise.

The significantly larger fraction of peculiar galaxies at the faintest magnitudes is nowadays often attributed to the formation process of galaxies at the moderate redshifts of $z > 2$, meaning look-back times as short as $\sim 2/3\ H_o^{-1}$. If so, we have rather direct pictures of how in fact protogalaxies at these intermediate look-back times have become morphologically normal within even the last 2/3 of the time since their protogalactic creation.

The high percentage of multiple forms that are clearly interacting provide a clue to the galaxy formation process. A particularly good example of a conglomerant is shown in Fig. 15 from one of the Kristian fields. Four frames are shown with increasing stretches into the digital record. Only the compact E-like core is shown in the top frame. Fainter components of the structure are shown in the succeeding three frames, showing what appears to be a number of individual components collapsing within a common envelope onto the main E-like core. This could either be interpreted as the "merger" of separate fragments "falling into equilibrium," which is the substance of the conjecture of Searle and Zinn (1978), or as condensations of initial density contrasts within a large, overarching collapsing envelope that was a single protogalaxy, as in the collapse picture of

Eggen et al. (1962, ELS). In the latter case, the formation process is often said to be ELS with "noises" that are the density fluctuations in a lumpy ELS extended envelope that must have been present in the earliest stages of galaxy formation (Sandage 1990). Such a configuration as a function of time is well shown in an early n-body simulation by Zurek, Quinn, and Solman (1988) shown in Figure 16, or as Fig. 26 of Saslaw (1985) in a larger context.

8. Philosophy vs. practicality

In most of the talks we shall hear in these four days we shall catch echoes from the past, many based on Hubble's initial developments in the span of 14 years from 1922 to 1936. The themes will be morphological classifications, redshifts, counts, evolution, galaxy formation, star formation rates, and world models. Are these important? Return to Naugle's statement quoted at the beginning about informing the public so as to assess the importance to society of the HST mission.

There are, of course, no practical applications directly from the work. All attempts to sell scientific progress of the nature we are dealing with in terms of any practical spin-off applications have always been singularly unconvincing to many. The importance of cosmology in human affairs and culture lies elsewhere. Indeed, if the spin-offs were to be the justification for the program, better to spend the money directly on them rather than to cloth them in the far-out studies of cosmology. No, the justification must rest on the nature of the inquiries themselves.

The problem of selling such esoteric studies as astronomy to the populous at large may be as old as Adam. Socrates said it best in his understatement to Glaucon in Book VII of the Republic, circa 370 BC, with the exchange:

Socrates: Shall we make astronomy the next study? What do you say?

Glaucon: Certainly. A working knowledge of the seasons, months, and years is beneficial to everyone, to commanders as well as farmers and sailors.

Socrates: You make me smile, Glaucon. You are so afraid that the public will accuse you of recommending unprofitable studies.

Of course, for emphasizing the philosophy of the work rather than its practicality, Socrates was forced by the politicians of Athens to drink hemlock. The charge was corrupting the young.

As most of the participants in this workshop are the young, the old folk spouting philosophy in these four days on such "unprofitable" studies as cosmology may themselves be forced to drink hemlock come Friday noon as the ultimate protection against their further corrupting activities.

REFERENCES

BAADE, W. 1944 *ApJ* **100**, 137.

BINGGELI, B., SANDAGE, A., AND TAMMANN, G. A. 1988 *Ann. Rev. A&A* **26**, 509.

COLEMAN, G. D., WU, C. C., & WEEDMAN, D. W. 1980 *ApJS* **43**, 393.

COLEMAN, P. H., & SASLAW, W. C. 1990 *ApJ* **353**, 354.

CRANE, P., & SASLAW, W. C. 1986 *ApJ* **301**, 1.

DAVIDSON, W. 1959a *MNRAS* **119**, 54.

DAVIDSON, W. 1959b *MNRAS* **119**, 665.

DAVIDSON, W. 1960 *MNRAS* **120**, 271.

EGGEN, O. J., LYNDEN-BELL, D., & SANDAGE, A. 1962 *ApJ* **136**, 748.

ELLIS, R. 1986 in *Observational Cosmology, IAU Symp. 124*, (eds. A. Hewitt, G. Burbidge, & L. Z. Fang). p. 367. Reidel.

FATH, E. A. 1914 *AJ* **28**, 75.

GAUSS, C. F. 1828 *Comm. Soc. Sci. Gottingen* **Vol. VI**.

GAUSS, C. F. 1873 *Werke* (Gottingen), **Vol. IV**.

GREENSTEIN, J. L. 1938 *ApJ* **88**, 605.

HUBBLE, E. 1922 *ApJ* **56**, 162.

HUBBLE, E. 1925, *ApJ* **62**, 409 (NGC 6822).

HUBBLE, E. 1926a *ApJ* **63**, 236, (M33).

HUBBLE, E. 1926b *ApJ* **64**, 321 (classification).

HUBBLE, E. 1929a *ApJ* **69**, 103, (M31).

HUBBLE, E. 1929b, *Proc. Nat. Acad. Sci.* **15**, 168 (redshift-distance).

HUBBLE, E. 1932 *ApJ* **76**, 106.

HUBBLE, E. 1934 *ApJ* **79**, 8 (galaxy distribution).

HUBBLE, E. 1936a *ApJ* **84**, 270.

HUBBLE, E. 1936b *ApJ* **84**, 517 (curvature).

HUBBLE, E. 1936c *Realm of the Nebulae*. Yale Univ. Press.

HUBBLE, E. 1937 *Observational Approach to Cosmology*. Clarendon Press.

HUBBLE, E. 1953 *MNRAS* **113**, 658 (Darwin Lecture).

HUBBLE, E., AND HUMASON, M. L. 1931 *ApJ* **74**, 43.

HUBBLE, E., AND HUMASON, M. L. 1934 *Proc. Nat. Acad. Sci.* **20**, 264.

HUMASON, M. L. 1936 *ApJ* **83**, 10.

HUMASON, M. L., MAYALL, N. U., & SANDAGE, A. 1956 *AJ* **61**, 97.

KARASEV, B. V. 1982 *Soviet Astr. Letters* **8**, 284.

KOO, D., & KRON, R. 1992, *Ann Rev. A&A* **30**, 613.

KRON, R. 1980 *ApJS* **43**, 305.

LEMAITRE, G. 1927 *Ann. Soc. Sci. Bruxelles* **47**, 49.

LEMAITRE, G. 1931 *MNRAS* **91**, 483 (a translation of his 1927 paper).

LILLY, S. J., COWIE, L. L., & GARDNER, J. P. 1991 *ApJ*, **369**, 79.

LUNDMARK, K. 1926 *Ark. Math. Astr. Phys., Ser. B.* **Vol. 19**, No. 8.

LUNDMARK, K. 1927 *Medd. Astr. Obs. Uppsala* **No. 30**.

MAYALL, N. U. 1934 *Lick Obs. Bull.* **16**, 177 (No. 458).

MATTIG, W. 1958 *Astron. Nachr.* **284**, 109.

MATTIG, W. 1959 *Astron. Nachr.* **285**, 1.

MATTIG, W. 1968, *Zs. f. Ap.* **69**, 418.

MCVITTIE, G. C. 1956 *General Relativity and Cosmology*. Chapman & Hall.

McVittie, G. C. 1965 *General Relativity and Cosmology, 2nd edition.* Chapman & Hall.

McVittie, G. C. 1974 *Q.J. RAS* **15**, 246.

Oke, J. B. & Sandage, A. 1968 *ApJ*, **154**, 21.

Robertson, H. P. 1928 *Phil. Mag.* **5**, 835.

Robertson, H. P. 1938 *Zs. f. Ap.* **15**, 69.

Sandage, A. 1983 in *Kinematics, Dynamics, & Structure of the Milky Way,* (ed. W. L. H. Shuter). p. 315. Reidel.

Sandage, A. 1986 *Ann. Rev. Astron. Astrophy.* **24**, 421.

Sandage, A. 1988 *Ann. Rev. Astron. Astrophy.* **26**, 561.

Sandage, A. 1990 *Journ. RASC* **84**, 70.

Sandage, A. 1995 in *The Deep Universe: Inventing the Past, Saas Fee Lectures* (ed. B. Binggeli & R. Buser). Springer.

Sandage, A. 1998 *PASP*, in press.

Sandage, A. & Tammann, G. A. 1987 *A Revised Shapley-Ames Catalog of Bright Galaxies.* Carnegie Institution of Washington. Pub. No. 635

Sandage, A., Tammann, G. A., & Hardy, E. 1972, *ApJ*, **172**, 253.

Saslaw, W. C. 1985 *Gravitational Physics of Stellar and Galactic Systems.* p. 187. University of Cambridge Press.

Saslow, W. C. 1989 *ApJ* **341**, 588.

Saslaw, W. C., & Hamilton, A. J. 1984 *ApJ* **276**, 13.

Seares, F. H. 1915 *ApJ*, **41**, 206.

Seares, F. H. 1922a it ApJ, **56**, 97.

Seares, F. H. 1922b, *Trans IAU Rome* **Vol. 1**, 71.

Seares, F. H. 1925 *ApJ* **62**, 168.

Seares, F. H. & Humason, M. L. 1922 *ApJ* **56**, 84.

Seares, F. H., Kapteyn, J. C., & van Rhijn, P. J. 1930 *Mount Wilson Catalogue of Photographic Magnitudes in Selected Areas 1–139.* Carnegie Institution of Washington, Pub. No. 402.

Searle, L., & Zinn, R. 1978 *ApJ* **225**, 357.

Shane, C. D. 1975 in *Galaxies and the Universe* (ed. A. Sandage, M. Sandage, & J. Kristian). Chapt. 16. Univ. Chicago Press.

Spitzer, L. 1946 reprinted in *Astron. Quart.* **7**, 131 (1990).

Spitzer, L. 1997 in *Dreams, Stars, and Electrons* (ed. L. Spitzer & J. P. Ostriker). p. 395. Univ. Princeton Press.

Stebbins, A. E. & Whitford, A. E. 1948 *ApJ* **108**, 413.

Stebbins, J., Whitford, W. E., & Johnson, H. L. 1950 *ApJ* **112**, 469.

Stock, J., & Williams, A. D. 1962 in *Astronomical Techniques* (ed. W. A. Hiltner). p. 374. Univ. Chicago Press, Chapt. 17.

Yoshii, Y., & Takahara, F. 1988 *ApJ* **326**, 1.

Tyson, J. A. & Jarvis, J. F. 1979 *ApJ* **230**, L153.

Weaver, H. F. 1946 *Popular Astronomy* **Vol. 54**, 211, 287, 339, 389, 451, 504.

Whitford, A. E. 1971 *ApJ* **169**, 215.

Zurek, W. H., Quinn, P., & Salmon, J. K. 1988 *ApJ* **330**, 519.

The Hubble Deep Field: Introduction and motivation

By RICHARD S. ELLIS

Institute of Astronomy, Madingley Road, Cambridge CB3 0HA, England,
Email: rse@ast.cam.ac.uk

Although it is just over a year since the data was made public, the HDF exposure has stimulated considerable progress towards our understanding of the faint galaxy population. I present a brief personal account of the history of faint galaxy studies culminating in the HDF, and describe what I consider to be the main highlights thus far from this remarkable image. The HDF has given considerable impetus to studies of galaxy evolution and this has led to the emergence of a convincing empirical framework. Further exploitation of deep HST images in conjunction with ground-based 2-D spectroscopy will assist in the physical understanding of the evolutionary processes involved.

1. Introduction

We're here to celebrate and discuss scientific results from the Hubble Deep Field (Williams et al. 1997). Most would agree that this exposure represents an observational landmark in the long arduous path of exploring and understanding the Universe of faint galaxies. Indeed, it is difficult to remember a single observation in astronomy that has influenced our subject so quickly. Moreover, its full impact may not yet be realised. In this workshop, we will debate the significance of the conclusions so far derived and learn of new developments that follow directly from the HDF. This remarkable image has acted as an inspiration to many astronomers because, exceptionally, we were granted immediate access to the data. Those working with other facilities, such as ISO (Rowan-Robinson, this volume) and the VLA (Kellermann, this volume) have been quick to follow the example by concentrating their deepest exposures on this same field.

As well as explaining what I consider to be the main extragalactic highlights from the HDF at this point, largely to set the scene for the more detailed articles that follow, I will recall some of the earlier work which inserted pieces of the jigsaw that we now recognise more clearly via the HDF. Of course, HDF has also had significant impact in the non-extragalactic area. I'm glad this is well covered in the workshop but won't attempt to review progress in those important areas.

2. How the HDF came to be

How did the HDF come to be and why has it been so successful? Bob Williams convened an advisory committee comprising a number of extragalactic scientists who met here on 31st March 1995. Our brief was to be visionary and to consider the optimum science that would emerge from a significant allocation of Director's Discretionary Time. Those present will remember a rather rambling discussion which focused ultimately on one or two long exposures, both as a scientific mission and a public legacy for HST. My own notes from that meeting indicate much disagreement on details: the number of fields (north and south or just one?), the number of filters (surprisingly only one or two... nobody argued for more so far as I can recall), where to point (interesting area with a distant QSO or cluster, or a blank field?). Some of us questioned whether the community should at least be allowed to demonstrate whether it had a smarter idea than those of the gathered 'experts'. The latter proposition, unsurprisingly, did not achieve much support! What I am trying to say, in a way that does not insult my fellow committee members, is that we hardly prescribed HDF at that meeting. The credit lies with Bob Williams and his team at ST ScI who turned a very sketchy idea into a carefully-planned series of observations.

Contrary to what some might imagine, the reason HDF has been so successful is not solely the depth of the image; images almost as deep have been obtained from the ground (cf. Metcalfe et al. 1995, Smail et al. 1995). The major step forward was the combination of the image quality only HST can deliver and, foremost, the multi-passband nature of the data. The use of 4 strategic filters, whose relative exposure times were carefully balanced, has been particularly successful. The UV and blue exposure times were prohibitively long for most guest observers and surprisingly little multicolour data had been obtained with HST prior to the HDF. Add to that the public interest in the beautiful colour images, free access to the data, the galvanising effect of this large investment on astronomy's premier facility on other telescopes and you have the ingredients of the success of the HDF.

3. The highlights

So what are the most important scientific results from HDF so far? I only have sufficient time to sketch what I consider to be the five most significant results whilst giving due credit to earlier workers who established the foundations of what has emerged. I have to be a bit selective so I present this as a personal account rather than a comprehensive review.

3.1. *The flattening of the count slope N(m)*

The quest to take deeper images of the sky motivated the HDF at its most basic level but this quest has a long and distinguished history. Sandage (1995) discusses the classical work, whereas Koo & Kron (1992) and Ellis (1997) review the more recent observations. The modern era begins with the commissioning of the wide-field prime focii on our national 4-m telescopes in the 1970's. Combining fine-grain emulsions and automatic measuring machines, Kron, Kibblewhite and Tyson laid the foundations of image processing of faint galaxy images (Kron 1978, Peterson et al. 1979, Tyson & Jarvis 1979). In an era when the photographic plate is so often disregarded, it is salutory to note how much of our observational achievement was established from photographic plates. Only now, after 20 years, are giant CCD cameras rivaling the combined depth and field of view. Of particular note for this meeting is Koo's thesis (1981) where, in an early version of the HDF, four-colour photographic photometry was analysed in the context of photometric redshifts to demonstrate enhanced star formation as a function of look-back time.

The importance of pushing deeper was obviously recognised. Tyson (1988) was the first to attempt ambitious long CCD exposures developing, with Jarvis, Valdes, Seitzer and others, the relevant observing and processing technologies. The steep blue count slope first found by Kron (1978) seemed to continue and many of us imagined we might soon hit the confusion limit. The suggestion that the count slope flattened below the Olber's limit, $dlogN/dm = 0.4$, came tentatively from Lilly et al. (1991) and later, with greater confidence, from the very deep exposures taken by Metcalfe et al. (1995) including the *Herschel Deep Field*, a friendly ground-based rival of the HDF.†

† I should add, somewhat topically, that Tom Shanks looked like doing marvelously well out of the recent British election as his Herschel Deep Field blue galaxies were adopted as a slogan by

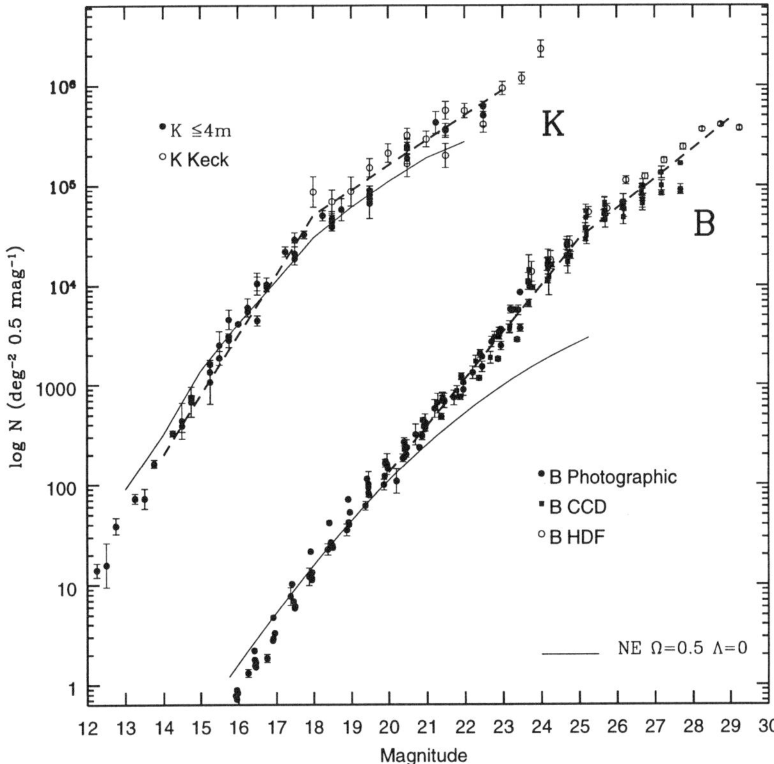

FIGURE 1. Differential number magnitude counts in the B-band (including those derived from the HDF) and K-band (including recent Keck determinations using the Keck telescope). The K counts have been offset by +1 dex for clarity. The two power law slopes (dashed lines) indicate the point beyond which the integrated night sky brightness begins to converge. Solid lines refer to no evolution predictions in the Einstein-de Sitter case (see Ellis 1997 for further details).

The flattening was dramatically confirmed in the HDF counts and overcounting multi-component galaxies as separate units (Colley et al. 1996) would make the faint slopes even flatter. The bulk of the received extragalactic light must therefore come from the point of inflexion—an apparent magnitude ($B \simeq 25$) within spectroscopic reach where the mean redshift is modest ($z \simeq 1$). Importantly, the same effect has been seen in the infrared at $K \simeq 18$ (Gardner et al. 1993, Moustakas et al. 1997). Very few of the faint K-selected galaxies are not visible in the optical suggesting the result is a fundamental feature of galactic history.

3.2. *The small angular sizes of the faintest galaxies*

On first seeing the HDF image (in a national newspaper) I was struck by the amount of blank sky it contained. Together with the flat count slope, an important secondary conclusion arising from this simple observation is the small angular sizes of the faintest sourccs.

the Conservative party whose emblem was projected on them in a national newspaper. However, the colour of the British sky has since switched dramatically from blue to red!

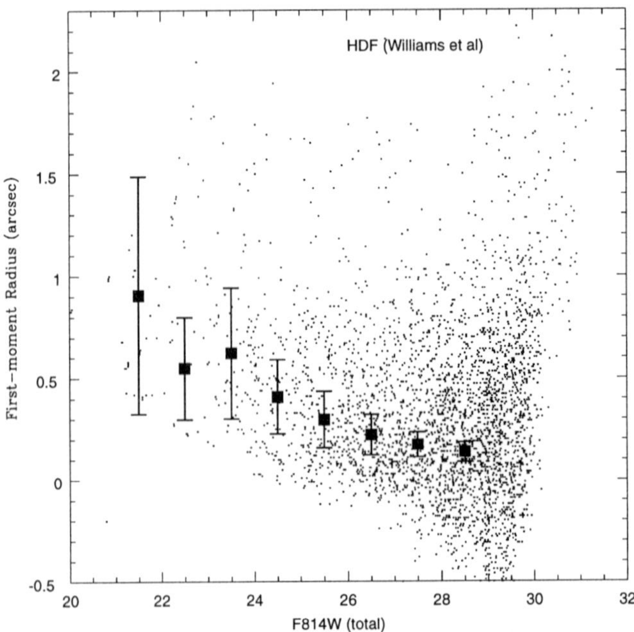

FIGURE 2. Correct intensity-weighted first moment radius versus apparent magnitude with population median and standard deviations for sources in the HDF.

This is a newer result mainly due to HST but one whose interpretation is perhaps less straightforward (see Ferguson, this volume). Most of the early ground-based photographic and CCD data was taken in what is now considered to be mediocre seeing. Astronomers at the NTT and CFHT showed the importance of improving the dome seeing and demonstrated that most of the accessible faint blue galaxies brighter than the count slope 'break point' ($B < 25$) were resolved (Giraud 1992, Colless et al. 1994). The first deep Keck images (Smail et al. 1995) and HST Medium Deep Survey images (Roche et al. 1996) suggested half-light radii of < 0.3 arcsec at fainter limits. HDF has extended this trend to considerably fainter limits (Fig. 2).

The result has a simple interpretation in the context of hierarchical galaxy formation since, beyond $z = 1$, a small angular size corresponds to a physically small source ($\simeq$ 2–3 h^{-1} kpc) regardless of the cosmological model. On the other hand, surface brightness losses and effects due to band shifting may conspire to reduce extended sources to apparently point-like HII regions at high redshift. Coaddition of representative cases suggests this is unlikely to be the case. However, NICMOS images may give a more representative indication of the physical sizes involved.

3.3. *The increasing fraction of irregular and multiple component systems*

The 1980s also saw the first deep redshift surveys which provided quantitative evidence for galaxy evolution brighter than the break points in Fig 1. Progress followed the new technology of multi-object spectrographs, from plug-plate fibre systems (Hill et al. 1980), robot positioners (Parry & Sharples 1988) to multislit spectrographs such as the Cryogenic Camera (Butcher 1982), LDSS-1/2 (Colless et al. 1992, Allington-Smith et al. 1994) and MOSIS (LeFevre et al. 1994). Such surveys revealed a rapidly declining population

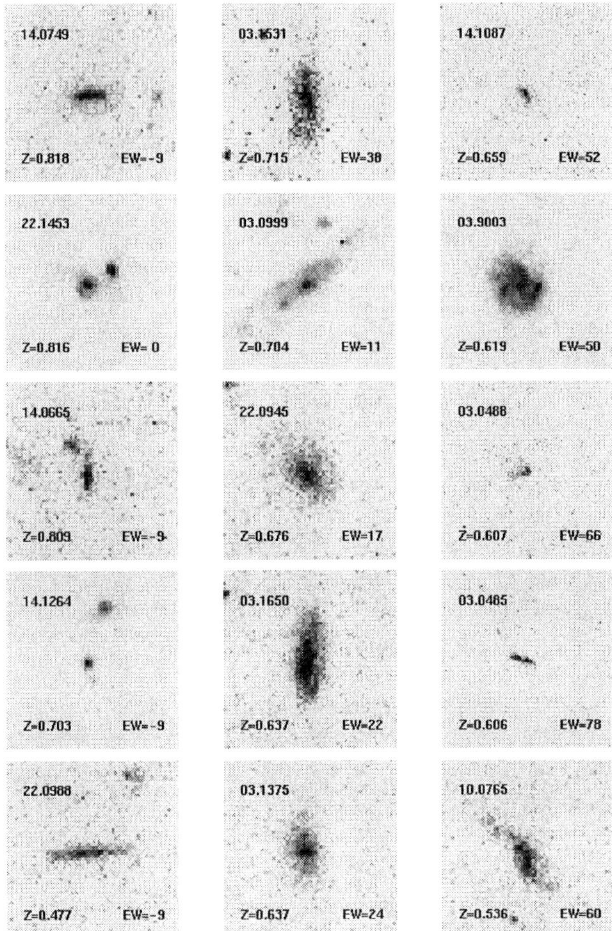

FIGURE 3. A mosaic of HST images for galaxies with irregular morphology selected from the sample of Brinchmann et al. (1997). Labels refer to rest-frame [O II] equivalent widths and redshifts.

of star-forming galaxies over $0 < z < 1$ which seem to be responsible for the bright count excess. What are these rapidly-evolving galaxies?

The Medium Deep Survey (Griffiths et al. 1994, Windhorst et al. 1995) analysed parallel WFPC-2 data for $\simeq 30$ fields and presented the morphological mixture as a function of apparent magnitude. Although normal galaxies are seen in numbers consistent with approximately constant co-moving space densities, these authors were struck by the rising fraction of faint system with irregular morphology; many are suggestive of merging systems. Glazebrook et al. (1995b) and Driver et al. (1995) argued that rapid evolution was almost exclusively occurring in the 'irregular/peculiar/merger' category—admittedly rather a catch-all for non-regular systems whose physical nature remains unclear. Abraham et al. (1996a,b) introduced a more quantitative basis for faint galaxy morphologies and discussed how to allow for redshift-dependent distortions in extended the analysis, using the HDF, to I = 25. They claimed few of the faintest galaxies could be shoe-horned into the classical Hubble sequence.

The linking of redshifts and HST morphologies has been slow to emerge. An international group, based on the original CFRS and LDSS teams, have now amassed over

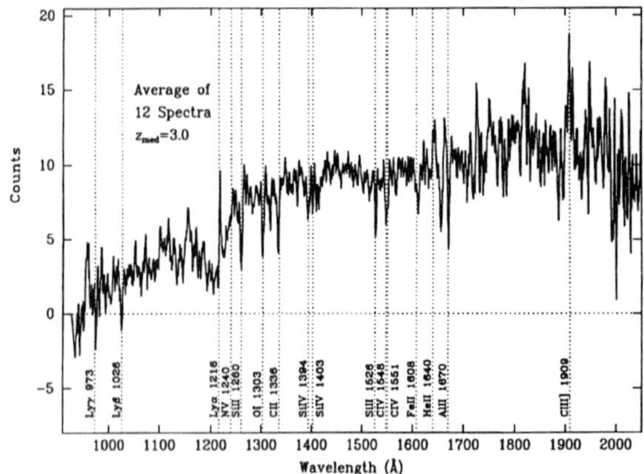

FIGURE 4. Mean spectrum of the U dropouts in the HDF and flanking fields from the analysis of Lowenthal et al. (1997).

300 faint galaxies to $I < 22$ or $B < 24$ for which extended WFPC-2 imaging and spectroscopic redshifts are available. This was not feasible to construct with the Medium Deep Survey because of the mismatch between the small WFPC-2 field and the larger ones of the ground-based multiobject spectrographs. Brinchmann et al. (1997) examined in some detail the possible effects of redshift on the perceived morphology. They found a rapid rise with redshift in the fraction of galaxies with irregular morphologies which they claim cannot be due to k-correction or surface brightness effects. This trend represents a major component of the evolution seen over $0 < z < 1$. However, it is clear the evolution in blue luminosity density need not come entirely from this population (Lilly, this volume) and, moreover, it is not obvious that the declining fraction of galaxies with irregular sources is matched by a compensating growth of disks or spheroidal galaxies over the same period as might be expected in simple hierarchical pictures (Baugh et al. 1997).

3.4. *The location of high redshift galaxies with modest star formation rates*

The most important feature of the HDF was the dedication of a significant fraction of the Discretionary Time to deep UV and blue exposures useful for locating the high redshift galaxies. Using photometry to constrain the redshift distribution from the effect of the Lyman limit was first attempted by Guhathakurta et al. (1990). They showed that the bulk of the $R < 25$ sources most probably had redshifts $z < 3$. Steidel & Hamilton (1992), Steidel et al. (1996a) later demonstrated prior to the HDF via their own imaging and Keck spectroscopy in QSO fields with Lyman limit absorption line systems. The broader significance of this work was amplified and extended to lower redshift using the shorter wavelengths available with the HDF. Although Giavalisco et al. (1996) had the first HST images of the $z > 2.8$ galaxies, the quality of the HDF images selected to be beyond $z = 2.3$ by a similar technique was considerably superior (Ellis 1996, Fig. 3).

An interesting feature of this remarkable population is its relatively low abundance (compared to the huge number of foreground systems) and the modest inferred star formation rate derived from the ultraviolet continuum flux. Other speakers will elaborate on progress in this area and no doubt the precise star formation rates and the possibilities of misinterpretation will be raised. Enough spectra have now been taken (Fig. 4) for us to realise they are mostly metal poor precursors but of what kind of galaxy and with what mass is not yet clear. The crucial point for now is that via these and other observations, the volume density and spectral characteristics of star-forming galaxies at $z > 2.3$ have become available and confirmed what was suspected from the counts, sizes and morphologies: we are probably looking at the first era of star formation in some category of galaxy. Most of the activity which produced the regular Hubble sequence occurred at lower redshift.

3.5. *Constraints on the redshift distribution of the faintest systems*

The multicolour HDF data has led to a resurrection of interest in estimating redshifts from colours. The technique goes back to Baum (1962), Koo (1985) and Loh & Spillar (1986). It is interesting to note that the method of photometric redshifts was heavily criticised by some at that time even though, in the case of Loh & Spillar, more than 4 filters were employed.† However, I think some people missed the point. In my opinion, the motivation is not to predict a precise redshift as a substitute for a spectrum, but rather to use the method to secure an overall statistical distribution $N(z)$. Unfortunately the predominantly blue SEDs make this difficult to achieve using optical data at $1 < z < 2$ because of the paucity of spectral discontinuities. Comparisons amongst the various HDF photometric redshift catalogues (Ellis 1997) shows no convincing evidence that the redshift resolution with 4 optical filters is any better beyond $z \sim 1$ than that associated with the Lyman limit moving through the filters. Of course, that is already important information! Moreover, the addition of near-infrared data should improve the situation considerably (Lanzetta, this volume). The first results from Connolly et al. (1997) are particularly interesting and suggest a mean redshift to $J = 23.5$ that is surprisingly low, in rough agreement with that inferred to $R = 25.5$ from the lensing inversion technique of Ebbels et al. (1997) when applied through rich well-constrained clusters such as Abell 2218 (Kneib et al. 1996) (Fig. 5).

4. The upshot

The above results, which of course by no means come exclusively from HDF (but have benefitted enormously from it), point to a remarkably simple observational synthesis discussed by Fall et al. (1996), Madau et al. (1996) and Madau (1997a,b). This will no doubt be widely discussed during the meeting. The observational data point to a remarkably recent era of major star-formation as delineated by the radiation we can see. Although different techniques are used to delineate the total star formation rate per comoving volume and extrapolation is necessary beyond the magnitude limits probed at the various redshifts, progress is already being made to overcome these limitations. Supporting the argument that we have witnessed the construction of galaxies directly with HST over $1 < z < 3$ are the flattening of the faint K-band counts (which precludes the existing of a significant population of highly reddened sources) and the rapid growth in mean physical size, and in the $0 < z < 1$ era, the declining fraction of irregular and multi-component galaxies in conjunction with the luminosity function changes witnessed in the redshift surveys. Completely independent support comes from trends with redshift in the gas content (Storrie-Lombardi et al. 1996) and metallicity (Pettini et al. 1994) of the QSO absorption line population. There are many uncertainties in each of these measures but the synthesis of so many results from different directions is quite compelling.

† I remember at a conference in Erice there were separate dining arrangements for those who believed in photometric redshifts!

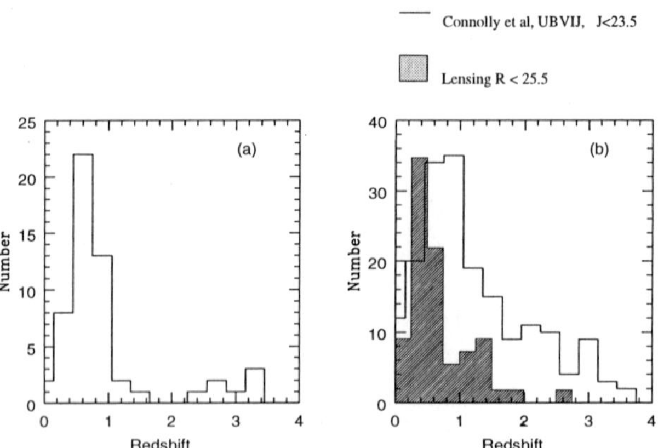

FIGURE 5. Redshift distributions of very faint galaxies: (a) using Keck spectroscopy in the HDF but to no formal magnitude limit, (b) photometric redshifts to $J = 23.5$ utilising HDF 4-colour photometry and ground-based JHK photometry from the analysis of Connolly et al. (1997). The shaded distribution is that inferred to $R = 25.5$ from gravitational lensing viewed through the well-constrained cluster Abell 2218 (Ebbels et al. 1997). Regardless of the technique, a surprisingly large fraction of the population beyond the break point in the counts (Fig. 1) has $z < 1$.

We should remember that this an *empirical* picture; it does not guarantee an unique physical interpretation. Much attention has been given recently to a supposed theoretical triumph in explaining it on the basis of hierarchical models (Baugh et al. 1997). I think we have to put this result in perspective. Perhaps it is not surprising that hierarchical models can be arranged to approximately fit the above observations given the redshift of peak star formation activity is surely sensitive to the precise way in which feedback is implemented in the models.† However, there are some remaining puzzles. Foremost, in the hierarchical models, we expect the growth of large disks to have occurred relatively recently (Efstathiou, this volume) and this should mirror the decline in the abundance of systems with irregular morphology (Baugh et al. 1997). The presence of well-formed massive galaxies to at least $z = 0.8$, with approximately the present comoving density (Lilly, this volume), suggests a more complex interpretation may be required. And, of course, many suspect that the optically-detected trends may be only a lower limit to the star formation energy density that occurs over all wavelengths (Rowan-Robinson, this volume).

5. Where next?

I would conclude that the much-heralded 'synthesis' of theory and data is premature. We have also only just scratched on the surface of the HDF data. Indeed, much of the recent progress is based merely on the *integrated* colours of HDF galaxies; we have hardly begun exploiting the 2-D resolved data which is the true benefit of using HST. I will conclude by illustrating the next step in this sense which is based on preliminary

† Actually the fit of Baugh et al. (1997) to the star formation history discussed by Madau (1997a) is not particularly good, but the observations are hardly completely determined at this point.

work done in Cambridge (Abraham et al. 1997, for a preliminary discussion see Abraham 1997).

Instead of using integrated 4-colour data to estimate redshifts, why not use those sources for which spectroscopic redshifts are available (Cowie et al. 1996) and analyse the resolved pixel-by-pixel 4-colour data in constraining the star formation history of each galaxy and its physical sub-components? As most of the faint galaxies are extraordinarily blue, the HDF colours are primarily sensitive to relative main sequence ages, modulo small uncertainties in dust, the IMF and metallicity. These effects would be worrisome if absolute ages were sought, but not if the goal is to examine the *relative spread* of burst ages ($\delta t/t$) across a galaxy in relation to its constituent spatial components. The technique can be tested for intermediate redshift spirals and ellipticals where sensible results emerge for the bulge and disk components and so, in Fig. 6, we illustrate how this method can be applied to the enigmatic high z irregulars and so-called 'chain galaxies.' Our resolved photometric analysis determines that these systems are consistent with star formation progressing in distinct bursts occurring several hundred Myr apart, as might be expected if physically-independent young components are slowly assembling into larger structures.

Clearly this technique has limitations in terms of the history accessible when using optical data at high z, but the benefit of extending this kind of study to include NICMOS K-band features sensitive to the mass of the incoming components would be considerable, as would the possibility of using GEMINI's integral field unit spectrographs to obtain the associated dynamical data. Clearly this is a tough observational project but I believe we should now complete the logical progression that has delineated this subject observationally over the past two decades: from integrated photometry (counts) and spectroscopy (redshift surveys) through to HST resolved imaging (MDS, HDF) to IFU-based spectroscopy. Such exciting datasets we can fully expect to come shortly. They will demand equal progress in theoretical modeling which will have to become more realistic to give physical insight into the evolutionary processes which are clearly occurring.

6. Conclusions

It's an exciting time to be doing cosmology! The last time we changed government in England was when the first deep photographic counts were published and Beatrice Tinsley suggested measuring the redshifts of a sample of galaxies to $B = 21$. She predicted a small fraction of the bluest sources might be high redshift primordial galaxies. After 15 years of ground-based work and only 4 years of post-refurbishment HST data, we have clearly come a long way. The acceleration of this subject in the past 2 years owes a great deal to HST and, within that context, to the HDF itself. There is much more data to come and much more physics to do. The much heralded 'synthesis' of theory and data is premature in my view. So far we are mainly surveying. The more fundamental task of understanding will take considerably longer. When we finally get there, I believe we will all recall that moment when we first saw the spectacular image of the Hubble Deep Field.

I thank the organisers for inviting me to give this opening review and for generous financial assistance. Full credit for the success of the HDF must go to Bob Williams and his talented team at ST ScI.

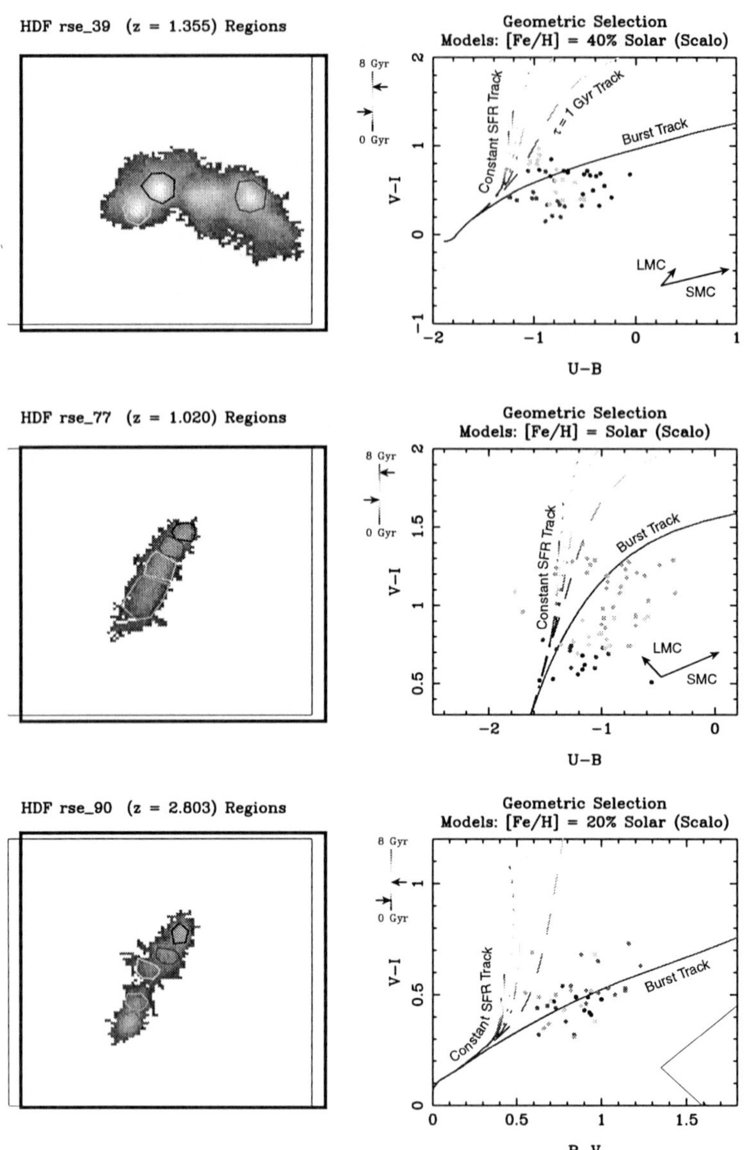

FIGURE 6. Resolved stellar populations for 3 high redshift HDF galaxies from the forthcoming study of Abraham et al. (for a preliminary discussion see Abraham (1997)). The left panels show F814W images for 3 galaxies with Keck spectroscopy courtesy of Len Cowie. The right panels show the pixel-by-pixel HDF colours grouped in physically-distinct regions within each image as indicated by the colour scheme in the left hand panels. Model tracks show evolutionary changes expected for various stellar histories according to the cosmic clock in the vertical scale. The analysis suggest these three galaxies assembled from physically and temporally-distinct stellar components.

REFERENCES

ABRAHAM, R. G. 1997 to appear in *The Ultraviolet Universe at Low and High Redshift*, in press.

ABRAHAM, R. G., TANVIR, N. R., SANTIAGO, B. X., ELLIS, R. S., GLAZEBROOK, K., VAN DEN BERGH, S. 1996a *MNRAS* **279**, L47.

ABRAHAM, R. G., VAN DEN BERGH, S., GLAZEBROOK, K., ELLIS, R. S., SANTIAGO, B. X. 1996b *ApJS* **101**, 1.

ABRAHAM, R. G., ELLIS, R. S. & FABIAN, A. C. 1997 in preparation.

ALLINGTON-SMITH, J. R., BREARE, J. M., ELLIS, R. S., GELLATLY, D. W., GLAZEBROOK, K., JORDEN, P. R., MACLEAN, J. ET AL. 1994 *PASP* **106**, 983.

BAUGH, C., COLE, S., FRENK, C. S. & LACEY, C. 1997 *ApJ* in press [**astro-ph/9703111**].

BAUM, W. A. 1962 in *Problems of Extragalactic Research, IAU Symposium 15*, (ed. G. McVittie). p. 390.

BRINCHMANN, J., SCHADE, D., TRESSE, L., ELLIS, R. S., LILLY, S. J., LEFÉVRE, O., COLLESS, M. & GLAZEBROOK, K. 1997 *ApJ* submitted.

BUTCHER, H. 1982 *Proc. S.P.I.E.* **331**, 296.

CONNOLLY, A. J., SZALAY, A. S., DICKINSON, M., SUBBARAO, M. U., BRUNNER, R.J. 1997 *ApJ* in press. [**astro-ph/9706255**].

COLLESS, M., ELLIS, R. S., TAYLOR, K. & HOOK, R. N. 1990 *MNRAS* **244**, 408.

COLLESS, M., ELLIS, R. S., TAYLOR, K., BROADHURST, T. J. & PETERSON, B. A. 1993 *MNRAS* **261**, 19.

COLLESS, M., SCHADE, D., BROADHURST, T. J., ELLIS, R. S. 1994 *MNRAS* **167**, 1108.

COLLEY, W., RHOADS, J. E., OSTRIKER, J. P., SPERGEL, D. N. 1996 *ApJL* **473**, L63.

COWIE, L., SONGAILA, A., HU, E. M., COHEN, J. L. 1996 *AJ* **112**, 839.

DRIVER, S. P., WINDHORST, R. A., OSTRANDER, J., KEEL, W. C., GRIFFITHS, R. E., RATNATUNGA, K. U. 1995 *ApJL* **449**, L23.

EBBELS, T., ELLIS, R. S., KNEIB, J-P., LE BORGNE, J-F., PELLÓ, R., SMAIL, I. & SANAHUJA, B. 1997 *MNRAS* in press [**astro-ph/9703169**].

ELLIS, R. S. 1996 in *The Early Universe with the VLT.* (ed. J. Bergeron). p. 65, Springer.

ELLIS, R. S. 1997 *Annual Reviews Astron. Astrophys. 35*, in press [**astro-ph/9704019**].

ELLIS, R. S., COLLESS, M., BROADHURST, T. J., HEYL, J. S. & GLAZEBROOK, K. 1996 *MNRAS* **280**, 235.

FALL, S. M., PEI, Y. C. & CHARLOT, S. 1996 *ApJL* **464**, L43.

GARDNER, J., COWIE, L. L. & WAINSCOAT, R. 1993 *ApJL* **415**, L9.

GIAVALISCO, M., STEIDEL, C. C. & MACHETTO, F. D. 1996 *ApJ* **470**, 189.

GIRAUD, E. 1992 *A&A* **257**, 501.

GLAZEBROOK, K., ELLIS, R. S., COLLESS, M., BROADHURST, T. J., ALLINGTON-SMITH, J. R. & TANVIR, N. R. 1995a *MNRAS* **275**, 157.

GLAZEBROOK, K., ELLIS, R. S., SANTIAGO, B. X. & GRIFFITHS, R. E. 1995b *MNRAS* **275**, L19.

GRIFFITHS, R. E., RATNATUNGA, K. U., NEUSCHAEFER, L. W., CASERTANO, S., IM, M. ET AL. 1994 *ApJ* **437**, 67.

GUHATHAKURTA, P., TYSON, A J. & MAJEWSKI, S. R. 1990 *ApJL* **357**, L9.

HILL, J., ANGEL, J. R. P., SCOTT, J. S., LINDLEY, D., HINTZEN, P. 1980 *ApJL* **242**, L69.

KNEIB, J-P., ELLIS, R. S., SMAIL, I. R., COUCH, W. J. & SHARPLES, R. M. 1996 *ApJ* **471**, 643.

KOO, D. C. 1981 Ph.D. thesis, University of California at Berkeley.

KOO, D. C. 1985 *AJ* **90**, 418.

KOO, D. C. & KRON, R. 1992 *Annual Reviews A&A* **30**, 613.

KRON, R. 1978 Ph.D. thesis, University of California at Berkeley.

LEFÉVRE, O. CRAMPTON, D., FELENBOK, P. & MONNET, G. 1994 *A&A* **282**, 325.

LILLY, S. J., COWIE, L. L. & GARDNER, J. P. 1991 *ApJ* **369**, 79.

LILLY, S. J., TRESSE, L., HAMMER, F., CRAMPTON, D., & LEFÉVRE, O. 1995 *ApJ* **455**, 108.

LOH, E. D. & SPILLAR, E. J. 1986 *ApJ* **303**, 154.

LOWENTHAL, J., KOO, D. C., GUZMAN, R., GALLAGHER, J., PHILLIPS, A. C. ET AL. 1997 *ApJ* **481**, 673.

MADAU, P. 1997a in *Star Formation Near & Far*, 7th Annual Astrophysics Conference, in press.

MADAU, P. 1997b in *7th International Origins Conference*, in press. PASP Conference Series.

MADAU, P., FERGUSON, H., DICKINSON, M., GIAVALISCO, M., STEIDEL, C. C. & FRUCHTER, A. 1996 *MNRAS* **283**, 1388.

METCALFE, N., FONG, R. & SHANKS, T. 1995 *MNRAS* **174**, 768.

MOUSTAKAS, L. A., DAVIS, M., GRAHAM, J. R., SILK, J., PETERSON, B. A. & YOSHII, Y. 1997 *ApJ* **475**, 445.

PARRY, I. & SHARPLES, R. M. 1988 in *Fiber Optics.* (ed. S. Barden). p. 93. PASP Conference Series.

PETERSON, B. A., ELLIS, R. S., KIBBLEWHITE, E. J., BRIDGELAND, M. T., HOOLEY, T. & HORNE, D. 1979 *ApJL* **233**, L109.

PETTINI, M., SMITH, L. J., HUNSTEAD, R. W., KING, D. L. 1994 *ApJ* **426**, 79.

ROCHE, N., RATNATUNGA, K. U., GRIFFITHS, R. E., IM, M., NEUSCHAEFER, L. W. 1996 *MNRAS* **282**, 1247.

SMAIL, I., HOGG, D. W., LIN, Y. & COHEN, J. G. 1995 *ApJL* **449**, L105.

STEIDEL, C. C. & HAMILTON, D. 1992 *AJ* **104**, 941.

STEIDEL, C. C., GIAVALISCO, M., PETTINI, M., DICKINSON, M., ADELBERGER, K. 1996a *ApJL* **462**, L17.

STEIDEL, C. C., GIAVALISCO, M., DICKINSON, M., ADELBERGER, K. 1996b *AJ* **112**, 352.

STORRIE-LOMBARDI, L., MCMAHON, R. G., IRWIN, M., HAZARD, C. 1996 *ApJ* **468**, 121.

TYSON, A. J. 1988 *AJ* **96**, 1.

TYSON, A. J. & JARVIS, X. 1979 *ApJL* **230**, L153.

WILLIAMS, R. E., BLACKER, B., DICKINSON, M., VAN DYKE DIXON, W., FERGUSON, H. ET AL. 1996 *AJ* **112**, 1335.

WINDHORST, R. A., DRIVER, S. P., OSTRANDER, E. J., MUTZ, S. B., SCHMIDTKE, P. C. ET AL. 1995 in *Galaxies in the Young Universe.* (ed. H. Hippelein). p. 265. Springer.

Kinematics of distant galaxies

By GARTH D. ILLINGWORTH†

UCO/Lick Observatory/Astronomy and Astrophysics Department, University of California, Santa Cruz, CA, 95064, USA

Characterizing and understanding galaxy evolution is ultimately dependent on establishing mass scales for galaxies as a function of redshift and environment. This requires high resolution, multi-band imaging from HST combined with spatially resolved spectroscopy from large ground-based telescopes. Such an approach has been adopted for the field by the Santa Cruz DEEP group and for clusters by a Netherlands-Santa Cruz collaboration. In the latter case, wide field, multi-pointing HST WFPC2 images are being obtained of several intermediate redshift clusters for detailed studies of the cluster population. Both these programs combine HST images with Keck multi-slit spectroscopy for kinematics. For the field project, spectra of 115 HDF galaxies to I $\sim$ 25 mag were obtained by the Santa Cruz DEEP group with the Keck LRIS multislit spectrograph and combined with detailed structural measurements from the HDF images to (i) derive evolution for a sample of disk galaxies at $z \sim 0.5$ using the Tully-Fisher relation, (ii) characterize, from size and velocity width data, the global properties to $z \sim 1$ of a sample of $\sim$60 "compact" galaxies, (iii) determine star formation rates in that "compact" sample, and (iv) derive structure and redshifts for a sample of U and B-band "dropout" objects, of which 11 have been confirmed to be at high redshift ($z = 2.2$–3.5). For the cluster project, velocity dispersions have been obtained for 53 early-type galaxies in a distant cluster Cl1358+62, and used to derive the fundamental plane relation at $z = 0.33$. This is one of the largest FP samples, at any redshift. Interestingly, this cluster program has also resulted in the discovery of the highest redshift galaxies reported to date, both of which are $z = 4.92$. One of these two objects is strongly gravitationally-lensed and so provides the clearest view yet of the structure of star-forming regions at early times. These results, from both programs, demonstrate that length scales, surface brightnesses and morphological data from HST imaging, when combined with Keck redshifts, velocity width and rotation curve data, provide an unprecedented opportunity to characterize the nature of distant galaxies, particularly with relations such as the fundamental plane and Tully-Fisher.

1. Background

Over the last few years we have witnessed a great advance in our understanding of distant galaxies, driven, as is often the case, by new observational capabilities. The remarkably detailed images from the refurbished HST, of which the HDF is the most striking and a great tribute to the foresight of Bob Williams, have been complemented by equally striking spectroscopic results from the Keck telescopes, the first of a whole new generation of large ground-based observatories with forefront instruments. Many groups are now using these new capabilities to delineate the properties of distant galaxies. The results reported here will be largely those from two major new projects, the Keck Deep project developed in collaboration with Sandra Faber and David Koo, and the HST intermediate redshift cluster environment project initiated by Marijn Franx.

A central guiding philosophy in both these programs is that establishing mass scales is key to characterizing and understanding the evolution of galaxies from intermediate to high redshifts. The very large variance in luminosities at a given mass, due to widely

† The work reported here is the result of collaborations with many inspiring colleagues, namely, Pieter van Dokkum, Sandra Faber, Marijn Franx, Jesus Gallego, Rafael Guzmán, Dan Kelson, David Koo, James Lowenthal, Dan Magee, Drew Phillips, Luc Simard, Kim-Vy Tran & Nicole Vogt. Any errors in summarising the results of the projects reported here are mine.

varying star formation rates and dust content and distribution, greatly complicates the interpretation of galaxy fluxes and colors. High spatial resolution images from HST for length scales and structural characteristics, combined with spectroscopy for kinematics from Keck (or other 8–10 m class telescopes in the future) are key to putting galaxy evolution onto a much more quantitative footing.

The DEEP Project (Deep Extragalactic Evolutionary Probe—see our home page at http://www.ucolick.org/~deep/home.html for more information) was developed initially with the Center for Particle Astrophysics at Berkeley with the goal of elucidating the physical processes and time scales of the formation and evolution of galaxies. The Keck telescopes and their multi-object spectrographs (LRIS—the Low Resolution Imaging Spectrograph Oke et al. 1995, and, when it becomes available in 1999, DEIMOS, the Deep Imaging Multi-Object Spectrograph) are central to this program, providing redshifts and velocity widths as well as absorption and emission line characteristics. Of equal importance is the high spatial resolution imaging capability provided by HST and its WFPC2 camera (and NICMOS in 1998, and later the Advanced Camera, the ACS, in 2000). These HST instruments provide the structural and morphological data that complements the 8–10 m class spectroscopic data. As the HDF has demonstrated, such HST-imaged regions for the study of galaxies at intermediate to high redshift are crucial for progress. Ground-based telescopes, even with adaptive optical systems, will not provide comparable capability for the foreseeable future. Broadly, the goal of DEEP is to obtain kinematical data on $\sim 10,000$ galaxies from several HST-imaged fields spread at ~ 6 hour intervals around the sky at declinations easily reachable with the Keck telescopes.

The second aspect of this report, the intermediate redshift cluster program, is complementary to the DEEP project in that it is focused on exploring and establishing the kinematic, structural and morphological properties of galaxies in rich clusters and their immediate neighborhood. It differs from other cluster programs at similar redshifts $z \sim 0.3$–1 in that the observations are made over a wide field (from many pointings with HST), and so provides results across a wide range of galaxy densities and environments. We are studying three clusters, Cl1358+62 at $z = 0.33$, MS2053-04 at $z = o.58$, and MS1054-03 at $z = 0.83$. These are all strong emitters in X-rays (they are all EMSS clusters). This program is focused primarily on early-type galaxies which preferentially lie in such clusters, and much of the attention of the program is being given to establishing the fundamental plane for the early-type galaxies in these clusters (see Kelson et al. 1997).

2. HDF and "Groth" strip results

For a specific example of the importance of HST for clarifying the nature of galaxies in the young universe, one needs only to think of the impact of the Hubble Deep Field (Williams et al. 1996), and of the many projects that the HDF images have precipitated. The DEEP team also focused on this field. The first results from the DEEP team's Keck LRIS observations of the HDF and its flanking fields have now been published in 4 papers. These results are summarized here, and are based on samples of galaxies selected from the HDF to meet a number of scientific goals. Structural parameters, colors, surface brightnesses and morphological attributes were determined for 140 galaxies from the HST HDF images. Of this sample, redshifts were ultimately determined for 115 galaxies, which were reported initially on the DEEP home page, and then in the above papers.

In addition to the HDF, the DEEP team has also been concentrating on a high latitude field that was observed by the WFPC GTO team. This field has become known as the "Groth strip" after the PI of that component of the WFPC GTO program. Together this

WFPC GTO field and the HDF have provided high spatial resolution imaging data that exemplifies the role of HST in understanding the formation and evolution of galaxies. Such images of distant galaxies will be a legacy of HST, and will lie at the heart of distant galaxy studies for many, many years into the future. The HST images have proven to be so crucial that it is now hard to imagine any work on the evolution of distant galaxies that does not involve HST images as a key component of the program.

Of particular interest to the DEEP team are the kinematical properties of distant galaxies, and so further measurements of rotation curves, velocity widths and velocity dispersions have been made or will be made on subsets of the galaxy samples both in the HDF and the "Groth strip." Such detailed spectroscopic data also lend themselves to the measurement of line strengths and ratios, providing further clues to the stellar populations, ionization conditions and star formation rates as well.

2.1. *Tully-Fisher at intermediate redshift*

In our first paper (Vogt et al. 1996), HST data from the "Groth" strip was used to demonstrate that it was possible to combine measurements from HST WFPC2 images with ground-based Keck multislit observations at moderate spectral resolution to derive rotational velocities for suitably-inclined spiral galaxies to redshifts $z \sim 1$ (see Figure 1). Based on a comparison of the velocity widths in these nine distant galaxies with the local Tully-Fisher (TF) relations, a limit of $\Delta M_B < 0.6$ mag could be set on the brightening at intermediate redshift. The more recent Tully-Fisher work on the disk/spiral galaxy sample in the HDF from Vogt et al. (1997) greatly improves the reliability of this estimate, first by adding a comparable number of galaxies, and second, adding galaxies of lower absolute luminosity. The distant galaxy TF relation is much more clearly defined, since it now spans 3 magnitudes in luminosity (see the paper by Vogt & Phillips in this volume for a figure of the current TF relation). These data provide strong evidence for only modest evolutionary brightening of $\Delta M_B \sim 0.4$ at $<z> \sim 0.5$ in disk galaxies comparable to disk galaxies today.

As noted in Figure 1, the estimates of brightening from these rotational velocities are conservative, in the sense that the most likely sense of any systematic error in the rotational velocities is towards underestimating the maximum velocity. Thus the derived brightening in these distant galaxies is an upper limit to the brightening and hence of the luminosity evolution. The discussion of a larger sample from both the HDF and "Groth strip" fields, as well as the details of the procedures used to derive rotational velocities, are given in Vogt et al. (1998).

2.2. *Nature of compact galaxies: I.*

One of the striking results of HST images of faint, distant field galaxies has been the prevalence of compact, relatively high surface brightness galaxies (with effective radii r_e typically 0.2″–0.3″). Our early work on CNELGs (compact narrow emission line galaxies—Koo et al. 1995; Guzmán et al. 1996) showed that these objects had kinematical, structural and photometric properties comparable to the more extreme examples at the current epoch of low mass, active star forming galaxies, the HII galaxies. These papers were amongst the first to demonstrate the power of kinematical measurements for discriminating between the evolutionary options that were being discussed for the high redshift, blue galaxy population. However, these objects are rather extreme in there properties, and so the question remained for the more typical compact objects seen in deep HST images—"what are the current-day analogues of the compact galaxies seen at redshifts $0.3 \lesssim z \lesssim 1$?" This question was addressed in two parts. A complementary study

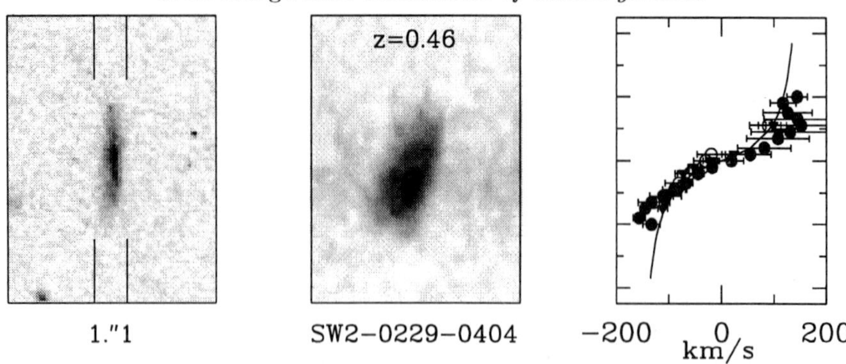

FIGURE 1. An example of a rotation curve for an intermediate redshift galaxy. The HST image of a galaxy at $z = 0.46$ is shown in the left panel with the Keck LRIS slit position overlaid, The two-dimensional spectrum of one of the observed emission lines ([OII]) is shown in the center, while the measured rotational velocities are shown in the right panel. The model rotation curve is also shown. Note that any underestimate of the rotational velocity will lead to an *overestimate* of the brightening! The Tully-Fisher relation derived from such measurements is shown in Vogt et al. (1996, 1997), as well as Vogt & Phillips paper in this volume.

by Rix et al. (1997) has also been carried out on a sample of objects at somewhat lower redshift ($z \sim 0.25$).

First, Phillips et al. (1997—see also Phillips, this volume) used the HST HDF flanking field images to derive sizes, magnitudes and morphologies for a sample of 61 faint ($I_{814} \sim 21$–24 mag), compact ($r_{1/2} \leq 0.5''$), high surface brightness galaxies ($\mu_{814} < 22.2$ mag). These data were then combined with the Keck redshifts to derive luminosities and length scales. The compact galaxies are luminous, with median $M_B \sim -20$ ($H_0 = 50$ km s^{-1} Mpc^{-1}), and have restframe colors typical of nearby star-forming galaxies (later than Sc, with many like NGC 4449 locally). There sizes and luminosities are similar to local luminous, active star-forming galaxies known as HII galaxies (Telles 1995). The similarity is suggestive, but a more quantitative comparison could be made using velocity widths. The Keck spectra yielded velocity widths (characterized by their gaussian σ) from emission lines for 45 of the 61 observed, allowing us to estimate their masses. The velocity widths ranged from $\sigma \sim 40$ km s^{-1} (the measurement limit) to 150 km s^{-1}. Combined, these data allowed us to compare and characterize the high redshift compact sample (see Figure 2). The clarity that kinematical data brings to understanding the nature of these objects is clear from the right panel in Figure 2. While a small fraction appear to be ellipticals, the vast majority are emission-line galaxies whose global properties resemble nearby HII galaxies, i.e., compact, late-type, star-forming galaxies. The comoving volume densities of these compact galaxies are very high compared to the local value and indicate a strongly evolving population.

2.3. *Nature of compact galaxies: II.*

The second aspect, the spectroscopic properties of the emission-line "compact" sample galaxies, is discussed in Guzmán et al. (1997). Star formation rates were derived from measurements of the [OII] equivalent widths; these ranged from 5 Å to 95 Å. The corresponding star formation rates are < 1–15 M$_\odot$ yr^{-1} (based on procedures and calibrations given in Guzmán et al.). Based on all the available data, namely velocity widths, excitations, Hβ luminosities, star formation rates, and M/L estimates, a substantial fraction of these "compact" emission-line galaxies ($\sim 60\%$) appear to be like local, low mass,

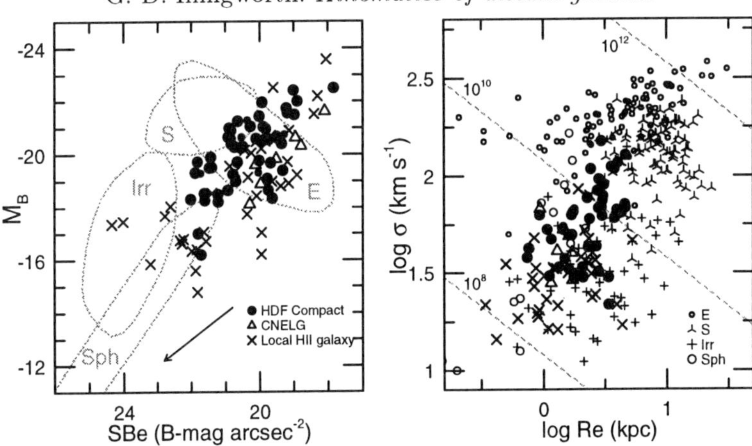

FIGURE 2. The luminosity-surface brightness relation for local samples of galaxies with the high redshift "compact" sample overplotted, on the left panel, from Phillips et al. (1997). The right panel again shows the relationship of the distant sample with local samples in the velocity width (σ)—effective radius plane. Typical masses are indicated. The "compact" objects are the filled circles. The properties of the high redshift galaxies overlap those of local, late-type, actively star-forming galaxies.

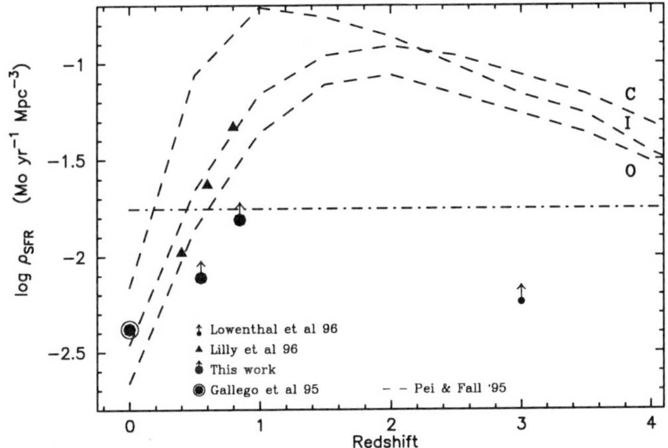

FIGURE 3. Estimates of the star formation rate per unit volume at several redshifts (from Guzmán et al. 1997). The sources are given on the figure. Models (closed, infall and outflow) from Pei and Fall (1995) are shown, as is the average at -1.75 derived from the local luminosity density (from Madau et al. 1996). Note that "compact" galaxies make a substantial contribution at intermediate redshifts, and that the SFR decline from $z \sim 1$ to the current epoch is large–a factor $\sim$10.

extreme star forming HII galaxies. The remainder are actively star forming, but appear to be a more heterogeneous sample of late-type galaxies.

An interesting result of this effort is that some additional constraints can be now placed on the evolution of the star formation rate per unit volume with redshift (see Figure 3 and the discussion in Guzmán et al. 1997). The data are consistent with a large decline in the SFR from $z \sim 1$ to the present epoch.

2.4. *Galaxies at very high redshift*

The use of U-band "dropout" galaxies to identify potential high redshift objects was beautifully exemplified by Steidel et al. (1996a) in their spectroscopic observations from

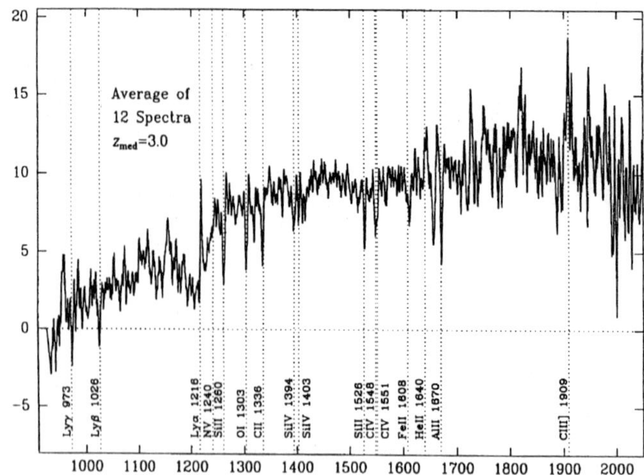

FIGURE 4. The average spectrum of the 11 high-redshift galaxies in the HDF from Lowenthal et al. (1997). The weak Lyα is particularly noteworthy. Other features to note are the strong continuum break and broad absorption line trough at Lyα, the weak emission lines of He II and C III], and the many (largely interstellar) absorption lines. The high ionization lines of C IV and Si IV are weak and narrow, (compared to those seen in spectra of high-metallicity hot stars), indicating that the metallicity in these objects is low.

Keck. This was followed by one of the very first HDF papers utilizing spectroscopic data (Steidel et al. 1996b); they found 5 high redshift ($z \sim 3$) galaxies. The "dropout" technique relies upon the large flux decrements that occur from line-of-sight and source absorption as both Lyα and the Lyman-limit pass through the filter bandpass, with the biggest effects being found at the Lyman-limit. Decrements in the U-band relative to other filters typically indicate sources at redshifts $z \gtrsim 2.5$, while decrements in the B-band suggest even higher redshift sources ($z \gtrsim 3$).

This technique, applied through the use of both U and B-band "dropout/decrement" objects, was the means by which we selected a sample of 24 galaxies for observation at Keck with LRIS. Of this sample 11 were confirmed to have redshifts in the range $2.2 < z < 3.4$. The remaining galaxies, with one exception, are consistent with being high redshift objects, but could not be confirmed spectroscopically. Interestingly, one faint source whose U-band measurement uncertainty was large, was found to be at low-redshift, consistent with improved U-band data (C. Steidel, private communication). Furthermore, two high redshift sources were found serendipitously in the Phillips et al. (1997) study of the HDF flanking fields; the redshifts for these objects were $z = 2.27$ and $z = 2.99$ (since multiple colors were not then available for the HDF flanking fields, the "dropout" technique could be applied there).

The high redshift objects in the HDF are quite small, but luminous ($1.8 < r_{\frac{1}{2}} < 6.5h_{50}^{-1}$ kpc and $-21.5 < M_B < -23$), and show a wide diversity of forms. Interestingly, they are not as morphologically uniform as the brighter sample studied by Giavalisco et al. (1996). Just as for the Steidel et al. (1996a, 1996b) samples the spectra showed a great diversity of characteristics, with L_α ranging from quite strong to non-existent, and a variety of stellar or interstellar absorption lines. The mean spectrum for the sample of 11 galaxies is shown in Figure 4, and is distinguished by quite weak Lyα emission, unlike many objects which have been detected through other searches (some of which are constrained to only detect strong Lyα sources). Given the externally imposed selection criteria (line-of-sight absorption), this indicates that the spectrum in Figure 4 is likely

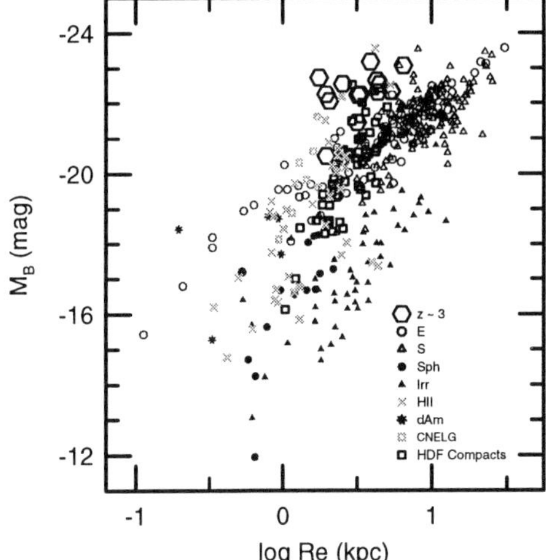

FIGURE 5. The location of the high redshift sample in the M_B–R_e plane, compared to local galaxies (Lowenthal et al. 1997). Note that R_e is equivalent to $R_{1/2}$. The large symbols are the $z \sim 3$ galaxies. Even though they are very luminous, the high redshift galaxies do overlap with the local, bright, star forming HII galaxies. A key issue is whether the high redshift objects have comparable velocity widths (and hence masses) to the HII objects, or whether they are more massive and similar to present-day bulges.

to be very typical of the spectral characteristics of $z \sim 3$ galaxies, and that any searches based on Lyα detection will be heavily biased towards "unusual" objects.

Combining these new data with the 5 in the HDF from Steidel et al. (1996b), we have derived co-moving volume densities and star formation rates (Lowenthal et al. 1997). It is striking that these HDF data already indicate that the volume density is comparable to the local volume density of all galaxies brighter than L* (and 3–4× higher than the first estimate of Steidel et al. (1996a) based on a brighter sample of "dropout" candidates. If the remaining objects in the original "dropout" sample prove to be at high redshift, the co-moving volume density could be at least two times higher still. Star formation rates have been estimated from the rest-frame UV continuum fluxes. They are found to be typical of local HII galaxies and many spiral galaxies (at 6–27h_{50}^{-2} $M_\odot$ yr^{-1} for $q_0 = 0.5$, or 2–9h_{50}^{-2} $M_\odot$ yr^{-1} for $q_0 = 0.05$). The relationship of the $z \sim 3$ galaxies in the Lowenthal et al. (1997) sample to nearby galaxies is shown in the M_B–$R_{1/2}$ plane in Figure 5 (R_e is equivalent to $R_{1/2}$). Since such strongly star-forming objects as these high redshift objects are quite likely to be dusty, the star formation rate (or the mass involved in the burst) is likely to be significantly higher when IR data (J, H and K images) are used to add longer wavelength measures of the flux from these galaxies.

Just as for the lower redshift sample of Phillips et al. (1997) and Guzman et al. (1997), velocity widths could be used to determine the masses of these objects. This would provide a much clearer understanding of the type of object. Are they really relatively low mass as the association with luminous HII galaxies in Figure 5 would suggest, or are they already part of a substantial potential well—one that might end up being a current-day elliptical, or a luminous spiral or S0 galaxy? Unfortunately, interpreting the observed widths of UV spectral lines as velocities characteristic of the gravitationally potential in which the object sits is a risky step. As noted by Conti et al. (1996), and discussed in this context by Lowenthal et al. (1997), such lines can be broadened by winds and shocks from supernovae in these actively star forming regions. Thus a simple direct estimate of the characteristic masses of these high redshift objects is unlikely to be forthcoming from the observed line widths. This is discussed further below in the section on the $z = 4.92$ lensed galaxies in Cl1358+62.

2.5. *Early-type galaxies in clusters*

Elliptical and S0 galaxies have been shown to lie in a plane in the three-space of effective radii, velocity dispersion and surface brightness. The plane, known as the fundamental plane since the early work of Dressler et al. (1987) and Djorgovski and Davis (1987), is defined by $r_e \propto \sigma^{1.24} I_e^{-0.82}$ for Coma (Jorgensen et al. 1993). The scatter about this relation in Coma is only 17%. The implication of these results if is that the M/L ratio in early-type galaxies is nearly constant, being only a weak function of mass, i.e., $M/L_V \propto r_e^{0.22}\sigma^{0.49} \propto M^{0.24} r_e^{-0.02}$. The work on nearby clusters was first extended to higher redshift clusters by Franx (see, e.g., Franx 1995 and van Dokkum and Franx 1996). These results for clusters at $z = 0.18$ and $z = 0.39$ have recently been complemented by the first Keck and HST results from the clusters Cl1358+62 at $z = 0.33$ and MS2053-04 at $z = 0.58$ by Kelson et al. (1997). A unique feature of these studies is the use of multiple HST pointings in two bands (F606W and F814W—roughly V and I) to cover large areas in and around the cluster. In the case of Cl1358+62, the data consisted of 12 pointings each of 3600 s integration in these two filters in a total of 18 CVZ orbits (equivalent to 36 "normal" orbits since an orbit in the continuous viewing zone leads to roughly twice the on-target integration time).

Kelson et al. (1997) used the Keck LRIS to measure velocity dispersions σ for a sample of galaxies in both clusters, and combined them with measurements of effective radius r_e and surface brightness I_e from HST images of the clusters. The fundamental planes developed from these observations, along with that of Cl0024+16 at $z = 0.39$, can be used to address the question of the degree of evolution of the old population and hence its formation epoch. These fundamental planes are shown in Figure 6. As for Coma, the scatter in these higher redshift fundamental planes are low, around 17% in r_e, suggesting that the structure of galaxies like present-day S0s and ellipticals is largely unchanged. An interesting aside is that galaxies that are in the present day sample may not be present in these distant clusters with lookback times of 3–5 Gyr ($H_0 = 65$). Galaxies that are (or have recently been) star forming at the cluster epoch would be too blue to fall in the sample, yet such objects would redden and fall into early-type galaxy samples at the present day (Franx 1995). Kelson et al. (1997) exemplify this effect with their discussion of the two E+A galaxies in Cl1358+62 that were measured as part of the fundamental plane study (see Figure 6).

Given the offsets in Figure 6, corrected appropriately for cosmological dimming, the M/L_V ratios of early-type galaxies in these four clusters can be plotted as a function of redshift. The expected brightening of the stellar population is seen, and can be compared with suitable evolutionary models as in Figure 7. The data suggest that the last major epoch of star formation in these early-type galaxies (the sample includes both S0 and E galaxies) was at a redshift $z > 2$. The E+A galaxies, shown in Figures 6 and 7 by crosses, show a much larger decrement in M/L, ~ 3, consistent with a time interval since the last burst of about one Gyr (as expected from the strong "A" star component to their spectra). These results show that the fundamental plane can be used as a sensitive diagnostic of the evolutionary history of galaxies.

The future of fundamental plane work is bright. Much larger samples of the kinematic data needed to derive velocity dispersions can be obtained with 8–10 m class telescopes. The primary requirement is for a large-field, high-throughput multi-slit spectrograph with FWHM resolution $\lambda/\Delta\lambda \gtrsim 3000$. An excellent example of the potential of such programs is given by Kelson et al. (1998). In his thesis and the subsequent papers,

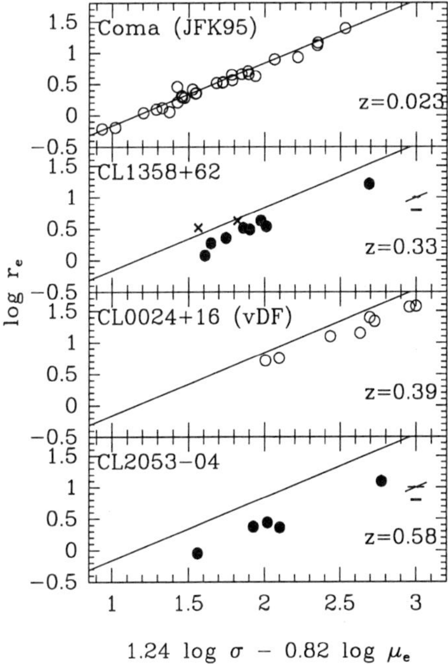

FIGURE 6. Fundamental planes for the clusters Cl0024+16 from van Dokkum and Franx (1996—open circles), Cl1358+62 and MS2053-04 from Kelson et al. (1997—filled circles), and Coma from Jorgensen et al. (1996—open circles). The Coma mean relation is plotted on each panel. The two E+A galaxies in Cl1358+62, indicated by crosses, are brightened, as expected, indicative of their brighter, younger age component. The offset in these relations is due to a combination of brightening from the evolution of the galaxy's stellar population in the past, offset by $(1+z)^4$ cosmological dimming.

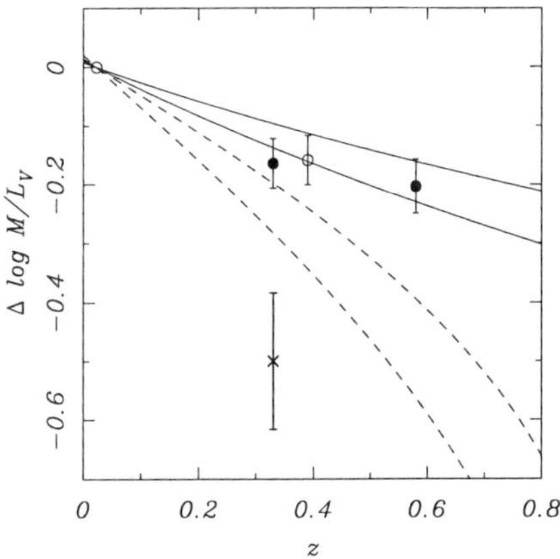

FIGURE 7. Mean M/L_V offsets from Coma as a function of redshift for the 3 intermediate redshift clusters of Figure 5. These are for $q_0 = 0.05$. The region bounded by the solid lines corresponds to single-burst models with $z_{form} >> 1$, and a range of IMFs (see, e.g., van Dokkum and Franx 1996). The region marked by the dashed lines corresponds to similar models, but with $z_{form} = 1$. The Coma and Cl0024+16 results are shown as open circles, with the Cl1358+62 and MS2053-04 results as filled circles. The cross indicates the mean M/L_V for the two E+A galaxies in Cl1358+62, with an offset as expected for their younger age since the last major burst. Systematic and random errors have been added in quadrature for this figure.

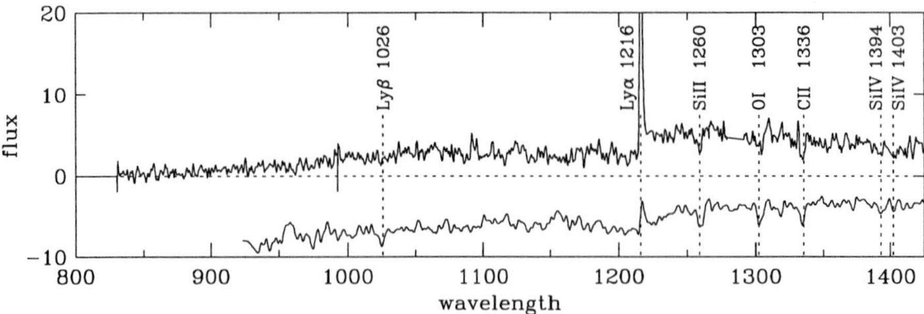

FIGURE 8. The spectrum of G1 in Cl1358+62, averaged along the slit. Comparison is made with the average spectrum of 12 $z = 3$ galaxies from Lowenthal et al. (1997—plotted beneath the G1 spectrum). The Lyα line, the sharp break at Lyα due to intervening absorption by the Lyα forest, the drop to zero flux (within the noise) at the Lyman continuum, and a number of interstellar absorption features in the galaxy are all consistent with a redshift $z = 4.92$ object. The smaller-than-expected decrement at Lyα is due to a weak contribution from an elliptical galaxy in the $z = 0.33$ cluster that is near to the arc.

Kelson extends the fundamental plane sample in Cl1358+62 to 53 galaxies, making it the largest, homogeneous fundamental plane sample of any cluster, at any redshift. It is clear also that the Keck and 8 m class telescopes can extend these results to redshifts $z \sim 1$. Already much more extensive data from Keck is available for MS2053-04 at $z = 0.58$, and preliminary data is available for MS1054-03 at $z = 0.83$ (van Dokkum et al. 1998). The biggest impediment to progress may well be the difficulty of getting adequate time on HST to carry out such systematic studies, given the huge competition for WFPC2 time for high redshift studies of galaxies and cosmology. HST data is essential, given the small scale sizes of distant galaxies.

2.6. *Galaxies at $z \sim 5$*

One of the remarkable aspects of acquiring new data with HST or large telescopes like Keck are the serendipitous discoveries that arise. This was certainly the case for the Cl1358+62 HST image. While checking the image soon after the HST data were taken, Franx noted that the cluster contained a strongly-lensed object but one with an unusual characteristic. The arc was quite red in (F606W-F814W). This was sufficiently unusual that it was included in our Keck observational program of the fundamental plane galaxies. The first results from the spectroscopic observations were tantalizing, since it was clear that this arc was potentially a very high redshift object if the single emission line seen at 7204 Å was Lyα. The redshift would then be $z = 4.92$. Subsequent higher S/N data allowed us to confirm this result, as reported in detail in Franx et al. (1997) The spectrum of one of the two high redshift galaxies is shown in Figure 8. By a remarkable coincidence, the serendipitous discovery of the arc in the HST image was matched by another serendipitous discovery of a second $z = 4.92$ galaxy. This galaxy just happened to fall on one of the slitlets in the multi-slit mask that was being used to measure velocity dispersions for the cluster members. This second object is $150h_{65}^{-1}$ kpc from the first, and differs in redshift by only 450 km s^{-1}!

Steidel and his collaborators (e.g., Steidel et al. 1996a, 1996b; Giavalisco et al. 1996) have made the detection and acquisition of redshifts at $z \sim 3$ a remarkably routine activity—so much so that it is hard to recall the dark ages of just a few years ago when such galaxies were very few, and unusual in their properties, being mostly radio galaxies. While the $z = 4.92$ galaxies, G1 and G2, in Cl1358+62 are distinguished, momentarily,

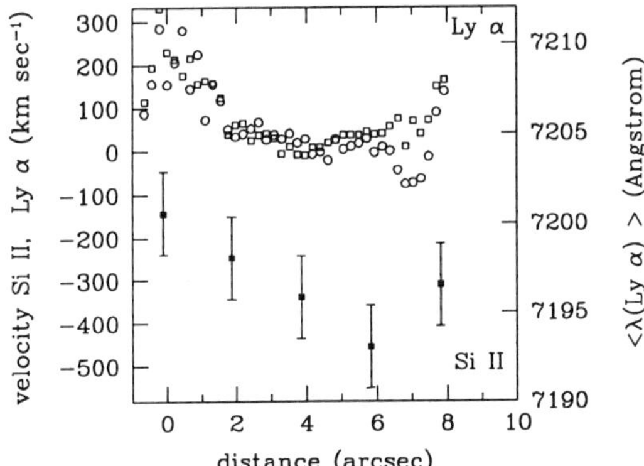

FIGURE 9. The velocity structure in the emission and absorption lines in Cl1358+62 G1. This velocity structure was measured along the arc using the long-slit mode of LRIS. The open symbols are measurements of the Lyα emission, while the filled symbols are the interstellar absorption feature Si II 1260 Å. The absorption line is sytematically blueshifted with respect to the Lyα emission. Such effects are seen in gas outflows where the blue wing of Lyα is absorbed, resulting in a "blueshift" of the Si II absorption line. The asymmetry of the Lyα profile, with a sharp blue edge and a red "wing" is also indicative of such effects.

by being the highest redshift galaxies, and the first galaxies which have exceeded the redshift of all know QSOs since the discovery of QSOs, what is most valuable about the strongly-lensed object is that the spatial resolution on the reconstructed source galaxy of G1 is better than 20 milliarcseconds in some parts of the image. This corresponds to about $100h_{65}^{-1}$ pc for $q_0 = 0.5$. The reconstructed images are shown in Franx et al. (1997). The high magnification in this object allows one to study the distribution of star formation in a way that is not possible for virtually all other high redshift galaxies (the number of strongly-lensed sources at $z \sim 3$ or greater is still very small).

For example, it is striking that a very substantial fraction (~50%) of the star formation in G2 is taking place within a single "knot" or "complex" that is only $200h_{65}^{-1}$ pc in size (FWHM). The optical data suggest a star formation rate of 36 $M_\odot\ yr^{-1}$ for G1; any dust would significantly raise this value (see Soifer et al. 1998 for a discussion of the effects of dust and for a revised number). As Franx et al. note, even this modest rate allows for the formation of 20–40% of the Galactic bulge in 10^8 yr, a period that is about 20% of a Hubble time at $z \sim 5$.

The high magnification allows us to also investigate one of the other key issues surrounding these high redshift galaxies. A central question that has been asked a number of times in connection with these objects, as noted above in the discussion of the Lowenthal et al. (1997) paper is: "what are their masses?" The high magnification allows one to reconsider this issue since the arc is resolved for ground-based observations (the total length is about 8″). Using the long-slit mode of LRIS, we obtained data that could be useful for such purposes. However, in endeavoring to interpret the results it became clear that the "motions" we were seeing were again likely to be due to turbulent motions resulting from the large energy input from supernovae and winds from massive, young stars. The velocity structure was most naturally due to outflows and not gravity. The velocity variations that we found are shown in Figure 9, where the velocities from Lyα are compared to those from the interstellar absorption line of Si II 1260 Å. Such velocity differences are seen locally in starburst galaxies (e.g., Heckman, Armus and Miley 1990), as well as other high redshift galaxies (see e.g., Lowenthal et al. 1997; Trager et al. 1997).

These results thus make it difficult to use the observed velocity structure in high redshift galaxies to determine the masses of these very actively star forming galaxies, as Steidel et al. (1996a), as well as others, including us, hoped to do. While it is likely that a variety of techniques will be developed that will place constraints on the masses of high redshift galaxies, it is not going to be as easy as we hoped!

2.7. *The future—kinematics of distant galaxies*

The recent Keck data from LRIS (and even HIRES—see Koo et al. 1995) have demonstrated very clearly the value of kinematical data on high redshift galaxies. The high star formation rates in typical high redshift galaxies make it very hard to estimate masses from stellar population models, especially given the ubiquitous presence of dust in starbursting systems. Even with multi-slit spectrographs, acquiring adequate kinematical data on large samples of galaxies is a slow process. Thus we can look forward to a substantial increase in the number of galaxies for which such data can be obtained as the new generation 6.5 m—8 m class telescopes become operational in the next few years. Along with these expanded effort, I can only encourage all those studying such galaxies from the ground to make every effort to acquire the HST images that have now become an essential part of such programs. As we all know, length scales are as important for determining masses as velocities, and the small characteristic sizes of high redshift galaxies are one of HST's many discoveries.

The HDF dramatically demonstrated the value of HST images for studies of distant galaxies. The complementary, more extensive areal surveys, such as those discussed here (the "Groth strip," the multiple-pointing cluster fields like Cl1358+62) also play a central role in studies of distant galaxies. Those of us working in this field look forward to many more such surveys with HST over the next decade. Such archived surveys would be a legacy of HST that would extend well beyond even HST's long life.

My collaborators and I are grateful for funding from a number of sources and programs, namely NSF grant AST–91–20005 through the CfPA (Center for Particle Astrophysics) at Berkeley, NSF grant AST–9529098, and ST ScI/NASA grants for HST programs AR–6337.08–94A, AR–6337.21–94A, GO–5994.01–94A, AR–5801.01–94A, AR5798.01–94A,GO5991.01–94A, GO5989.01–94A, and AR–6402.01–95A. I would like to thank Bob Williams for his foresight in choosing to carry out the HDF program, and then to follow it up with the southern HDF. The study of distant galaxies has exploded over the last couple of years, due in large part to the impact of his decision to carry out the HDF survey. I am certain that I am not alone in hoping that the next Director will continue the tradition instituted by Bob!

REFERENCES

CONTI, P. S., LEITHERER, C., & SONGAILA, A. 1996 *ApJ* **110**, 1576.

DJORGOVSKI, S., & DAVIS, M. 1987 *ApJ* **313**, 59.

DRESSLER, A., ET AL. 1987 *ApJ* **313**, 42.

FRANX, M. 1995 in *IAU 164, Stellar Populations* (ed. P. C. van der Kruit & G. Gilmore). p. 269. Kluwer.

FRANX, M., ET AL. 1997 *ApJL* **486**, L75.

GALLEGO, J. ET AL. 1995 *ApJL* **455**, L1.

GIAVALISCO, M. ET AL. 1996 *ApJ* **470**, 189.

GUZMÁN, R., ET AL. 1996 *ApJL* **460**, L5.

GUZMÁN, R., ET AL. 1997 *ApJ* **489**, 559.

Heckman, T. M., Armus, L., & Miley, G. K. 1990 *ApJS* **74**, 833.

Jorgensen, I., Franx, M., & Kjaergaard, P. 1993 *ApJ* **411**, 34.

Kelson, D. D., et al. 1997 *ApJL* **478**, L13.

Koo, D. C. et al. 1995 *ApJL* **440**, L49.

Lilly, S. J. et al. 1996 *ApJL* **460**, L1.

Lowenthal, J. D. et al. 1997 *ApJ*, submitted.

Madau, P., et al. 1996 *MNRAS* **283**, 1388.

Oke, J. B., et al. 1995 *PASP* **107**, 375.

Pei, Y. C., & Fall, S. M. 1995 *ApJ* **464**, 69.

Phillips, A. C., et al. 1997 *ApJ* **489**, 543.

Rix, H.-W., et al. 1997 *MNRAS* **285**, 779.

Steidel, C. C., et al. 1996a *AJ* **112**, 352.

Steidel, C. C., et al. 1996b *ApJL* **462**, L17.

Telles, E. 1995 thesis, Univ. Cambridge.

Trager, S. C., et al. 1997 *ApJ* **485**, 92.

van Dokkum, P. G., & Franx, M. 1996 *MNRAS* **281**, 985.

Vogt, N. P., et al. 1996 *ApJL* **465**, L15.

Vogt, N. P., et al. 1997 *ApJL* **479**, L121.

Vogt, N. P., et al. 1998, in preparation.

Williams, R. E., et al. 1996 *AJ* **112**, 1335.

Redshift clustering in the Hubble Deep Field†

By JUDITH G. COHEN

Palomar Observatory, Mail Stop 105-24, California Institute of Technology

We present initial results from a redshift survey carried out with the Low Resolution Imaging Spectrograph on the 10–m W. M. Keck Telescope in the Hubble Deep Field. Merging the published 1996 observations of Caltech, the University of California, and the University of Hawaii groups with our extensive as yet unpublished and not completely reduced 1997 data set produces a sample of about 440 objects in the HDF and the surrounding flanking fields with measured redshifts. The strong clustering noted earlier persists in this larger sample and extends out to redshifts in excess of 1. It is expected that at the end of this observing season, when the observational effort will finish, there will be in $\approx$ 550 objects with measured redshifts from LRIS/Keck within a radius of 4 arc-min from the center of the HDF.

The evidence continues to support (or at least not to contradict) the hypothesis that the redshift peaks correspond to "walls" which are the boundaries of giant ($\sim$50 – 100 Mpc) voids as seen in the local universe. There are also regions in redshift that are completely devoid of any galaxies, as well as evidence for the presence of groups of galaxies, but not rich galaxy clusters, in the sample.

The spectroscopic analysis has barely commenced, but one new result of interest is that the emission line ratios in the brighter objects with $z < 0.6$ resemble those of gas photoionized by starlight, not those of AGNs.

---

1. The Redshift Survey underway at the Keck Observatory in the HDF

The Hubble Deep Field (HDF hereafter; Williams et al. 1996) is the deepest image ever taken and this, combined with the superb spatial resolution of the HST, has excited the interest of the entire astronomical community. Follow up activities in every possible spectral region and in every possible mode are underway world wide. The unique contribution that can be made by the Keck Observatory given the huge collecting area at optical wavelengths of a 10–m telescope is to provide redshifts for a large sample of very faint galaxies in the HDF.

I first describe the redshift survey underway in the HDF using the LRIS spectrograph (Oke et al. 1995) at the Keck Telescope. I am collaborating with Roger Blandford and David Hogg of Caltech in this effort. Our redshift survey focuses on a field 8 arc-min in diameter centered on the HDF. We are attempting to observe every object with $R < 24$ mag in the HDF itself and every object with $R < 23$ mag in the flanking fields within this area. The sample is culled from a catalog of objects detected in a 5–m Hale Telescope image taken with Cosmic (Dressler 1993) through a R filter and analyzed with FOCAS (Tyson & Jarvis 1979, Valdez 1982) by Charles Steidel. Our sample includes about 470 objects.

In 1996 I coordinated a campaign among the 3 major groups using the LRIS at Keck, the Lick group, the University of Hawaii group, and the Caltech group. The goal was to avoid duplication in our observing efforts by circulating among ourselves the lists of objects a group planned to observe before each run and the list of objects for which the

† Based in large part on observations obtained at the W. M. Keck Observatory, which is operated jointly by the California Institute of Technology and the University of California.

group believed they had determined redshifts after each Keck run during the winter and spring of 1996.

This worked extremely well. The Caltech and Hawaii groups merged their efforts and collectively published a paper (Cohen et al. 1996a) that included redshifts for 140 extragalactic objects and identified 12 galactic stars in the HDF and the surrounding flanking fields. About half of the objects are in the HDF itself. The Lick group, which also participated in this effort to avoid duplication, has published their data in two papers, Lowenthal et al. (1997) and Phillips et al. (1997), the first dealing with 11 objects with $z \approx 3$, while the second presented redshifts for 105 objects in this area.

Thus by the end of the 1996 observing season there were about 250 objects with redshifts from LRIS/Keck in the HDF and the surrounding flanking fields. Continuation of this coordination was not possible during the 1997 observing season as the objects in the HDF without redshifts by the end of 1996 were so faint that careful reduction and adding up of multiple spectra were necessary to determine a redshift. This meant that immediate notification to other groups of a list of objects successfully observed after each Keck run was not possible.

In 1997 I continued this survey, while as far as I am aware, the Lick and Hawaii groups have concentrated on other things this observing season. As of May 29, 1997, the last time I completely merged all the data known to me, there were a total of 440 objects in the HDF and surrounding flanking fields with redshifts. Of these 120 are in the HDF itself. This includes my unpublished data from 1997, the sources given above, 6 objects from Steidel et al. (1996), 10 unpublished redshifts for objects in the HDF communicated to me in 1997 by Len Cowie, and one object that I had not already observed from the WWW page of Moustakas, Zepf & Davis (1997).

I had a very successful observing season at Keck during the winter and spring of 1997 with 9 dark nights part or all of which could be used for this effort. I have more than 350 spectra taken during that period, and at the present time 5 multi-slit masks containing about 150 spectra in the HDF area from my last spring observing run are still completely unreduced.

Figure 1 illustrates the spatial distribution of the sample of objects in the HDF and its flanking fields with known redshifts, with the outline of the HDF marked. The more intense observational effort in the HDF itself is clear from the increased areal density of objects with known redshifts.

2. Redshift clustering in the HDF sample

Our major scientific thrust thus far with the sample has been to study the clustering properties. As expected from our earlier work in the HDF (Cohen et al. 1996a) and our survey in the Caltech 0 Hour Deep Field (Cohen et al. 1996b) and from the pioneering work of Broadhurst et al. (1990), the galaxies over the entire region are concentrated in many strong peaks in redshift space. There are also ranges in redshift that are empty of galaxies. There is no question now given the large sample in hand that this result is statistically valid. We believe this to be a manifestation at high redshift and in the young universe of the same sort of large scale structure seen locally and best exemplified locally by the Great Wall and the voids around it, as originally described by de Lapparent et al. (1986) for the CFA survey, and for example recently found by Landy et al. (1996) in their analysis of the Las Campanas Redshift Survey, by Bellanger & de Lapparent (1995) for the ESO Slice Survey, and by LeFèvre et al. (1996) for the CFRS.

Figure 2 shows the redshift histogram for objects with $z < 1.3$ in our sample. Objects in the HDF itself are denoted by solid filled area, while those in the flanking fields are

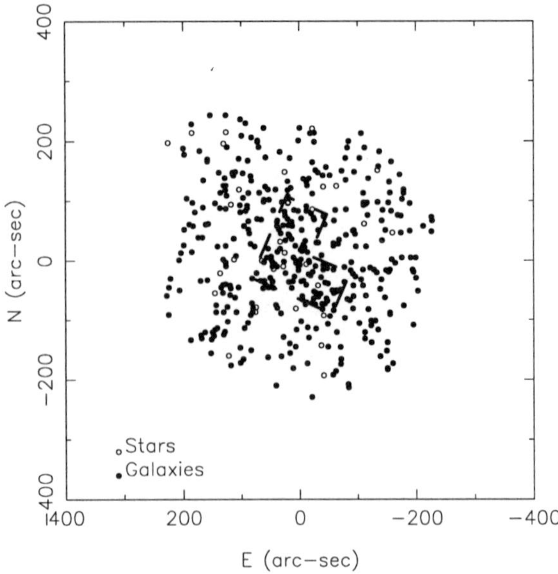

FIGURE 1. The spatial distribution for the stars (open circles) and extragalactic objects (filled circles) in the HDF and surrounding fields from the merged sample as of May 1997. The outline of the HDF itself is marked.

denoted by unfilled area. The arrows mark the five redshift peaks we identified based on our 1996 sample (Cohen et al. 1996a). In the present sample, which is now roughly three times larger, these peaks are still prominent. However, there are also several large peaks we missed in our earlier effort, particularly at high redshift ($z > 0.8$); the initial observations were biased towards the brighter objects and hence could not delineate structure at high z with any statistical validity.

The discrete nature of the redshift histogram is illustrated in another way in Figure 3, where redshift is plotted as a function of R magnitude. The location of the major redshift peaks from our original paper (Cohen et al. 1996a) are indicated as dashed vertical lines. As in Figure 2, the filled symbols denote objects in the HDF, while the open symbols denote objects in the flanking fields. Clustering in this figure appears as vertical bands, and is apparent even at $z > 1$.

Once the 1997 observations are completely reduced, sometime this fall, there are many issues to be explored. For example, we will examine the velocity dispersion within each clump, and whether this is changing with magnitude, i.e., are the less luminous and presumably less massive galaxies still falling into the structures? (Of course, this is closely related to the issue whether the uncertainties in redshift are independent of magnitude.) Is there any difference between the spectroscopic properties of galaxies within the structures and outside the structures? We will also produce a robust and statistically valid method of defining the redshift peaks in this larger sample of galaxies in contrast to the hand-waving arguments I have advanced thus far. (Our published work does include a careful statistical treatment for finding and establishing the validity of the peaks in the redshift distributions.)

Given the large area of our survey (large, for such a deep and complete redshift survey), we will of course explore the spatial structure of the galaxies within each of the redshift peaks, looking for clues to the nature of the redshift structures in physical space. Thus

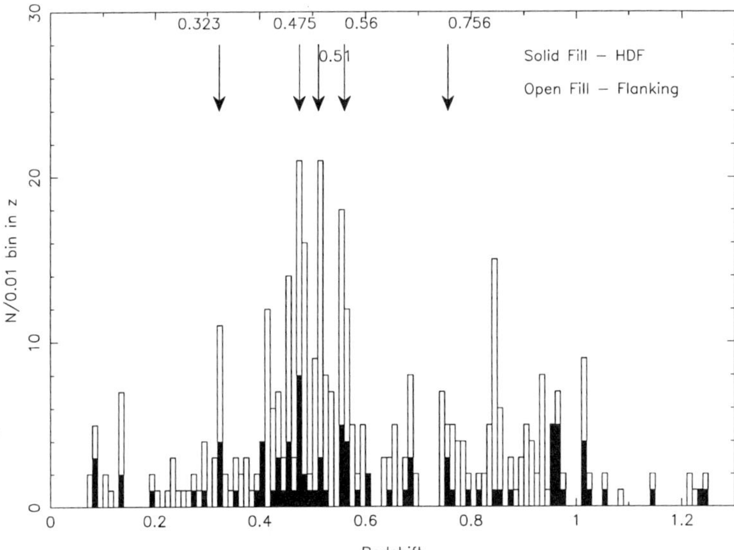

FIGURE 2. The histogram in redshift for $z < 1.3$ of the merged sample is shown. Galactic stars are excluded. The solid fill denotes objects in the HDF, the open fill represents objects in the flanking fields. The five strong redshift peaks from Cohen et al. (1996a) are indicated by vertical arrows.

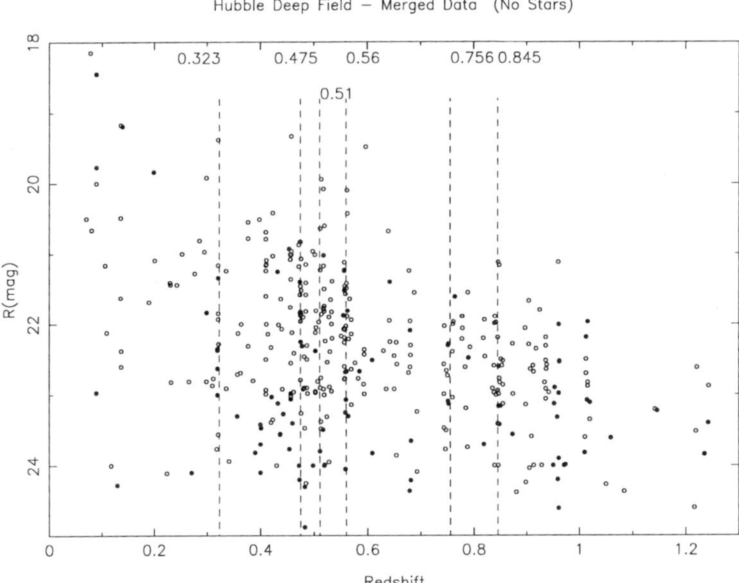

FIGURE 3. Redshift as a function of R magnitude for $z < 1.3$ is shown for the merged sample. Galactic stars are excluded. The filled circles denotes objects in the HDF, the open circles represents objects in the flanking fields. The five strong redshift peaks from Cohen et al. (1996a) plus one at $z = 0.845$ are indicated by dashed vertical lines.

far there is no evidence for centrally concentrated organized structures such as would be expected for normal rich galaxy clusters. The spatial distributions appear consistent with galaxies semi-randomly distributed throughout the entire area on the sky. There are, however, clearly a lot of small groups here with several members of a group sometimes represented in the sample.

The areal density on the sky in the most prominent of the redshift peaks has been calculated (extrapolating down to 0.1L*) and is about 1/2 that of the Virgo cluster within a radius of 6° determined by Kraan-Korteweg (1981) and by Bingelli, Sandage, & Tammann (1985).

To explore the issue of large scale structure over a larger spatial baseline, we have begun probing fields located between 30 and 45 arc-min from the center of the HDF. These fields reach separations (depending on redshift of interest and cosmology) of 10 to 20 Mpc from the center of the HDF. Our current sample in two such fields located west of the HDF is about 80 galaxies.

The galaxy luminosity function derived from this sample is the subject of David Hogg's thesis, which should be completed this fall.

3. The magnitude—redshift relation for galaxies in the HDF

Figure 4 presents z versus R magnitude for the full range of the available data, folding in the high redshift end. Also shown here are predictions for the behavior of non-evolving elliptical (solid line) and irregular (dashed line) galaxies with L = 2L*. An Einstein-de Sitter universe is assumed with $M_R(*) = -20.2$ mag. The sample is nicely bounded by the curve representing the behavior of an irregular galaxy. It is clear from the spectra that at high redshift most of these galaxies have considerable star formation, and hence this is not a surprise. Figure 4 does show that these galaxies probably do not have very high luminosities, more than 2L* is quite rare.

We will shortly able to do much better, as we now have obtained at Palomar (Pahre et al. 1997) a mosaic at 2.2μ covering the entire HDF and its flanking fields, i.e., the entire area of our spectroscopic sample. We already have optical images in several filters covering the entire HDF and flanking fields, and hence can now compute galaxy luminosities directly by integrating (with some kind of fit) the observed flux from 0.35μ to 2.2μ. This bypasses the issue of the difference in the change of spectral energy distribution with redshift for various types of galaxies.

4. Spectroscopic analysis

Beyond determining redshifts, the spectra have not as yet been analyzed carefully. The spectral resolution is 10 Å, and the wavelength range covered is basically the entire optical window from 0.4 to 0.9μ. The signal-to-noise ratio for the brighter galaxies in our sample is sufficiently high that measurement of a large number of emission and absorption line features for the brighter half of the sample will be possible. For the fainter half, only broad continuum parameters and the strength of the 3727 Å [OII] emission line will be measured. Based on my examination of the spectra thus far, it is clear that, as recognized by many other groups (see for example Cowie et al. 1996), galaxies at $z > 0.7$ are dominated by those with emission line spectra. While there are a few galaxies that show the absorption lines characteristic of $z \approx 0$ ellipticals, they are rare. There are many galaxies in our sample that show E+A spectra, but objects that show broad emission lines similar to those characteristic of various types of AGNs such as Seyfert 1 galaxies, broad-lined radio galaxies or QSOs are very rare.

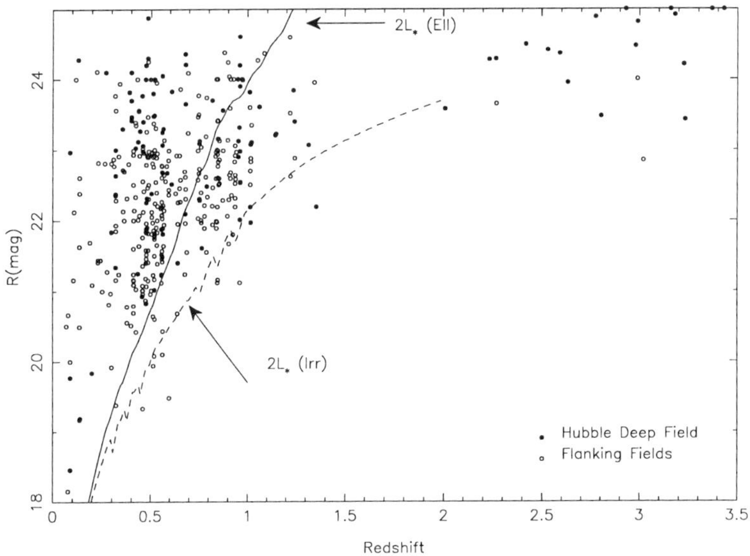

FIGURE 4. Redshift as a function of R magnitude is shown for the entire merged sample. Galactic stars are excluded. The filled circles denotes objects in the HDF, the open circles represents objects in the flanking fields. Curves indicating the predicted behavior of a non-evolving elliptical galaxy (solid line) and of a non-evolving irregular galaxy (dashed line) with $L = 2L*$ are indicated.

Our most important new spectroscopic result to date is that there is little/no sign of AGN ionization in these spectra. Figure 5 illustrates the conventional Osterbrock (1989) line ratio diagram of $5007/H_\beta$ versus 6584 $[\mathrm{NII}]/H_\alpha$ diagnostic line ratios. The areas occupied by AGNs (power law ionization spectrum) and by normal galaxies whose gas is ionized by starlight are shown. It is possible to measure the required line ratios in the brighter galaxies out to $z \approx 0.6$. The emission line ratios measured are consistent with photoionization by starlight. Even though these are measurements over a large fraction of the light from the galaxy, while line ratios measured in local AGNs are those of a small region around the nucleus, study of local AGNs shows that the nucleus dominates the integrated light of the galaxy, and the nuclear line ratios prevail throughout most of the spatial extent of the galaxy (Tadhunter, Metz & Robinson 1994 and Storchi-Bergmann et al. 1996).

5. Future work

There is a lot of work to be done before this sample can be published. I still need to do a second pass through all the spectra to work harder on the absorption line redshifts and in particular to concentrate as much possible on the objects which I suspect lie in the difficult zone with $1 < z < 2$, where there are no strong emission lines falling within the optical bandpass. This is the region where people desperately need a training set of spectroscopically confirmed objects as calibrators for various photometric redshift schemes. I need to measure emission line strengths, absorption feature strengths, and continuum shape indices for all of this data. As well as collecting more redshift data in the outlying fields, we still have a lot of work to do and thinking to do before we can

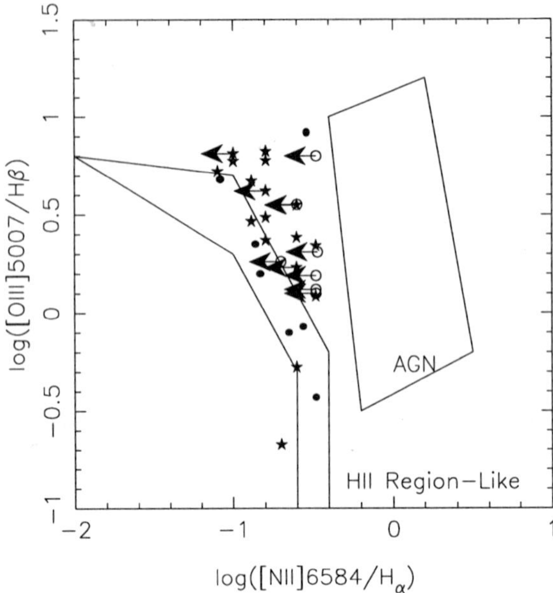

FIGURE 5. The Osterbrock (1989) diagnostic emission line ratios are shown for bright galaxies with $z < 0.6$. Open circles denote objects in the Caltech 0 Hour Deep Field where only an upper limit to the 6584 Å [NII] line is available, while filled circles denote galaxies in this field for which all relevant emission lines were detected. Objects in the HDF and flanking fields are denoted by asterisks.

claim to understand the clustering in redshift space, how it relates to the distribution of objects in physical space, and what this all means in terms of large scale structure.

We thank the Hubble Deep Field team, led by Bob Williams, for planning, taking, reducing, and making public the HDF images. We are grateful to Gerry Neugebauer, Mike Pahre and Jim Westphal for helpful conversations. We thank Charles Steidel for providing access to his Cosmic images and the catalogs derived from them of the HDF and surrounding fields. The entire Keck user community owes a huge debt to Bev Oke, Jerry Nelson, Gerry Smith, and many other people who have worked to make the Keck Telescope a reality. We are grateful to the W. M. Keck Foundation, and particularly its late president, Howard Keck, for the vision to fund the construction of the W. M. Keck Observatory. Support by ST ScI/NASA grant AR-06337.12-94A is greatly appreciated.

REFERENCES

BELLANGER, C. & DE LAPPARENT, V. 1995, *ApJ* **455**, L103.

BINGELLI, B., SANDAGE, A. & TAMMANN, G. A. 1985, *AJ* **90**, 1681.

BROADHURST, T., ELLIS, R., KOO, D. & SZALAY, A. 1990, *Nature* **343**, 726.

COHEN, J. G., COWIE, L. L., HOGG, D. W., SONGAILA, A., BLANDFORD, R., HU, E. M. & SHOPBELL, P. 1996, *ApJ* **471**, L5.

COHEN, J. G., HOGG, D. W., PAHRE, M. A. & BLANDFORD, R. 1996, *ApJ* **462**, L9.

COWIE, L. L., SONGAILA, A., HU, E. M. & COHEN, J. G. 1996, *AJ* in press.

DRESSLER, A. 1993, in Neugebauer, G., ed., *Palomar Observatory Annual Report 1993* **2**.

KRAAN-KORTEWEG, R. 1981, *A&A* **104**, 280

DE LAPPARENT, V., GELLER, M. & HUCHRA, J. P. 1986, *ApJ* **302**, L1.

LANDY, S. D., SHECTMAN, S. A., LIN, H., KIRSHNER, R. P., OEMLER, A. A. & TUCKER, D. 1996, *ApJ* **456**, L1.

LEFÈVRE, O., HUDON, D., LILLY, S. J., CRAMPTON, D., HAMMER, F. & TRESSE, L. 1996, *ApJ* **461**, 534.

LOWENTHAL, J. D., KOO, D. C., GUZMAN, R., GALLEGO, J., PHILLIPS, A. C., FABER, S. M., VOGT, N. P., ILLINGWORTH, G. D. AND GRONWALL, C. 1997, *ApJ* **481**, 673.

MOUSTAKAS, L., ZEPF, S. & DAVIS, M. 1997, http://astro.berkeley.edu/davisgrp/HDF.

OKE, J. B., COHEN, J. G., CARR, M., CROMER, J., DINGIZIAN, A., HARRIS, F., LABRECQUE, W., LUCINIO, R., SCHAAL, W., EPPS, H. & MILLER, J. 1995, *PASP* **107**, 307.

OSTERBROCK, D. E. 1989, *Astrophysics of Gaseous Nebulae and Active Galactic Nuclei*, University Science Books, Mill Valley, Ca.

PAHRE, M., NEUGEBAUER, G., ET AL. 1997, manuscript in preparation.

PHILLIPS, A. C., GUZMAN, R., GALLEGO, J., KOO, D. C., LOWENTHAL, J. D., VOGT, N. P., FABER, S. M. AND ILLINGWORTH, G. D. 1997, *ApJ* in press.

STEIDEL, C. C., GIAVALISCO, M., DICKINSON, M. & ADELBERGER, K. L. 1996, *AJ* **112**, 352.

TADHUNTER, C. N., METZ, S. & ROBINSON, A. 1994, *MNRAS* **268**, 989.

TYSON, J. A. & JARVIS, J. F. 1979, *ApJ* **230**, L153.

VALDEZ, F. 1982, *Proc. SPIE*, **331**, 465

WILLIAMS, R. E., BLACKER, B., DICKENSON, M., DIXON, W. V., FERGUSON, H. C., FRUCHTER, A. S., GIAVALISCO, M., GILLILAND, R. L., HEYER, I., KATSANIS, R., LEVAY, Z., LUCAS, R. A., MCELROY, D. B., PETRO, L. & POSTMAN, M. 1996, *AJ* **112**, 1335.

Radio observations of the Hubble Deep Field

By K. I. KELLERMANN[1] AND E. A. RICHARDS[1,2]

[1]National Radio Astronomy Observatory, 520 Edgemont Rd., Charlottesville, VA 22903

[2]University of Virginia, Charlottesville, VA

We have used the NRAO Very Large Array at 8.5 GHz to image the Hubble Deep Field and surrounding flanking fields. Twenty-eight sources were catalogued above a limiting flux density ranging from 9.0 μJy at the center of the VLA primary beam to 112 μJy at 4.6 arc minutes from the field center. Seven of these sources lie within the boundaries of the Hubble Deep Field. All are identified with galaxies brighter than $I_{AB} = 21.2$. An additional six galaxies, all brighter than $I_{AB} = 23.1$ in the HDF have radio emission somewhat below the completeness level of our radio catalogue. All but one of the 21 sources laying outside the HDF, are covered by sensitive ground based optical imaging. Sixteen of these are identified with galaxies brighter than R = 25.5. There are no identifications with stars or quasars. The optical counterparts of the microjansky radio sources are a mixture of early type galaxies and merging or interacting galaxies undergoing intense star-formation, along with AGN, primarily in elliptical galaxies but also possibly associated with a few spirals (LINERS and Seyferts). The optical counterparts have redshifts spread over the range from zero to two, with radio luminosities ranging between 10^{19} W/Hz and 10^{24} Watts/Hz. The slope of the integral radio source count for the 28 sources in the 6×10^{-6} sr region included in our primary beam is -1.0 ± 0.2, comparable to that found in other deep surveys to this level.

---

1. Introduction

One of the classical problems facing radio astronomers is to acquire optical follow-up material for radio source surveys. Often it may take years of work to obtain the optical identifications and redshifts needed to find the optical counterparts of the radio sources and to determine their redshift and luminosity. The availability of the Hubble Deep Field (Williams et al. 1996) with its aggressive ground based optical follow-up as well as auxiliary space based observations in other wavelength bands offers a unique opportunity to investigate the nature of the radio emission from normal or nearly normal galaxies at cosmologically interesting distances. In this paper we report on preliminary results of Very Large Array observations of the radio counterparts of galaxies in the Hubble Deep Field.

2. Classification of extragalactic radio sources

Extragalactic radio sources cover a wide range of luminosity extending from 10^{19} Watts/Hz for normal spiral galaxies to more than 10^{28} Watts/Hz for the powerful FRII radio galaxies and radio loud quasars. Intermediate in luminosity are the less powerful FRI radio galaxies, the radio quiet quasars, and galaxies with active star formation.

With few exceptions the observed radio emission is non thermal synchrotron radiation. The nature of individual radio sources, e.g., normal galaxies, starbursts, AGNs, FRI or FRII radio galaxies, may be distinguished by their luminosity, the observed radio morphology, the radio spectral index, variability, the optical counterpart and spectral features, and observed characteristics in other wavelength bands, particularly x-ray and infrared.

In the powerful radio galaxies and quasars, the source of energy is thought to lie in a massive central engine, whereas in many of the sources of intermediate luminosity the energy source appears to lie in regions of active star formation and supernovae activity which accelerate relativistic electrons into the interstellar medium. This starforming activity may be found in a population of faint blue galaxies (Windhorst 1985, Thuan and Condon 1987) or associated with the strong IR galaxies detected in the IRAS survey at 60 microns (Danese et al. 1987, Franceschini et al. 1988). Because of the close association of 60 micron and radio wavelength emission (Helou, Soifer, & Rowan-Robinson 1985), both believed to be closely linked to star forming activity, deep radio observations are sensitive to star formation at early epochs, unaffected by the obscuration which plagues optical and infrared observations. Star forming rates may be estimated from the observed radio flux density (Condon 1992), but only if the contribution to the observed radio flux density from a massive central engine can be determined from the radio or optical data.

Typically, both the FRI and FRII radio galaxies, as well as radio loud and radio quiet quasars have linear dimensions of the order of a few hundred kiloparsecs with structure characterized by radio lobes which are well separated from the optical counterpart, but often joined to the optical counterpart by a thin jet. FRI and FRII radio galaxies are optically thin with radio spectral indices, $\alpha \sim -0.8$, but may have a self absorbed flat spectrum core. The compact radio cores of quasars and AGN typically have angular dimensions much less than an arcsecond, and due to self absorption have flat radio spectra with observed spectral indices near zero. These very compact radio sources are often variable, and are identified with galaxies showing broad emission line spectra in their nuclei.

Radio emission from galaxies with active star formation is typically confined to dimensions comparable to that of the galactic disk. Optical counterparts are often unusually bright at far infrared wavelengths as a result of the absorption of uv emission from young massive stars by dust and its subsequent thermal reradiation.

Radio source surveys made over a wide range of wavelengths and flux density have catalogued about two million discrete radio sources. Figure 1 shows the normalized differential radio source counts compiled from surveys made at 1.4, 5, and 8.4 GHz as presented by Windhorst et al. (1993). Optical identification and spectroscopic redshifts show that most catalogued radio sources stronger than a few millijanskys are relatively distant powerful radio galaxies or quasars with radio luminosities greater than 10^{25} Watts/Hz, or nearby normal or nearly normal galaxies with much weaker radio emission. The space density of powerful radio galaxies and quasars quickly converges beyond redshifts of unity, (e.g., Condon 1984, 1989) so that nearly all of these powerful sources are included in the radio source counts above one millijansky.

At submillijansky levels, the normalized radio source count begins to turn up, corresponding to a new population of radio sources (e.g., Windhorst et al. 1985, Fomalont et al. 1991, Windhorst et al. 1993). Identification of these sub-millijansky radio sources are with a mixture of faint starforming galaxies and low luminosity AGN which are often found in pairs or small groups (e.g., Weistrop et al. 1987, Benn et al. 1993, Windhorst et al. 1995, Hammer et al. 1995). These previous programs of optical identification of weak radio sources, which have been done with 4-m class telescopes and with HST prior to refurbishment, typically reach a limiting R magnitude of about 25. The radio observations of the HDF offers the possibility of examining the optical counterparts of mirojansky radio sources down to 29–30 magnitude with angular resolution up to an order of magnitude better than previous work. Moreover, the extensive ground-based follow-up work has been important in obtaining spectroscopic classification of the optical

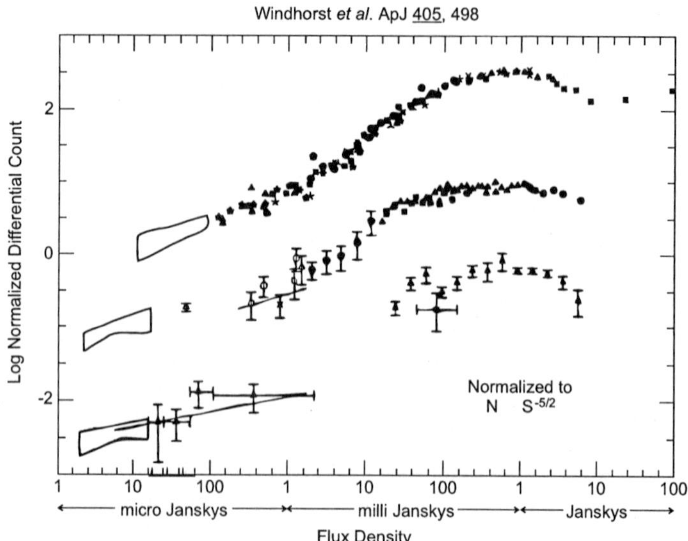

FIGURE 1. Differential number flux density relation at 1.4, 5, and 8.4 GHz compiled by Windhorst et al. (1993). Data are shown multiplied by $S^{2.5}$ so that the expected count, in a uniformly filled static universe with Euclidean geometry, is represented by a horizontal line.

counterparts, as well as for determining redshifts needed to constrain the radio luminosity function.

3. Radio emission from the Hubble Deep Field

We have used the NRAO VLA to observe the Hubble Deep Field for a total of 152 hours at a frequency of 8.5 GHz. The VLA field of view at 8.5 GHz has a diameter of 9.2 arc minutes down to the eight percent power point of the VLA primary beam covering an area of 6×10^{-6} sr. This includes the HDF with nearly full sensitivity as well as all of the surrounding flanking field images with reduced sensitivity. The observations were divided among the A, C, and D configurations of the 27-element array to examine the radio emission from the HDF on angular scales up to about 100 arc seconds with resolutions ranging from 0.1 to 10 arcseconds, corresponding to linear scales of 1 kpc to 1 Mpc for moderate redshifts. We observed with both LHC and RHC polarizations with an effective bandwidth of 200 MHz. The rms noise level using all of the data was 1.8 microjanskys with an effective resolution of about 4 arc seconds.

At 8.5 GHz, we catalogued a total of 28 sources within the 9.2 arc minute diameter field of view stronger than 9.0 μJy (snr = 5). Of these, seven sources are found within the HDF. All seven are identified with galaxies of various types, and five have been detected at 6.7 or 15 microns by ISO (Goldschmidt et al. 1997). We also find weaker radio emission from six other relatively bright galaxies below the 9 μJy completeness level but stronger than 6.5 μJy ($3 < \text{snr} < 5$). Figure 2 shows the radio contours at 8.5 Ghz superimposed on the HDF optical image.

We also used the VLA for about 50 hours in its A configuration at 1.4 GHz in order to obtain information on the radio spectra of galaxies in the HDF. At 1.4 GHz, the resolution of the VLA is limited to about 2 arc seconds. Complementary observations made with the MERLIN array by Muxlow et al. (in preparation) give greatly improved

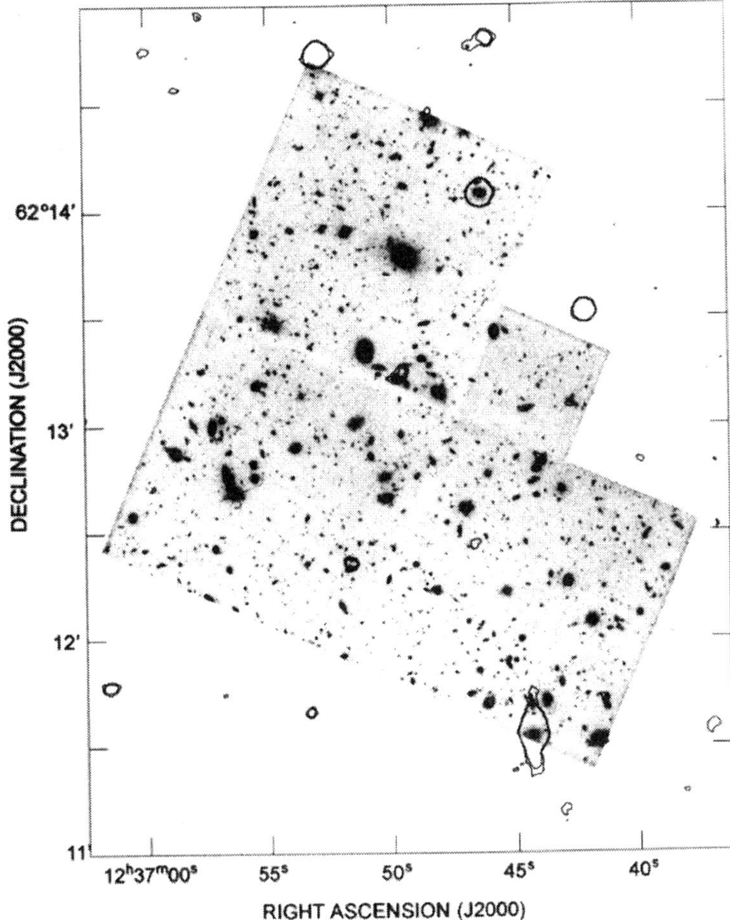

FIGURE 2. Radio contours superimposed on HDF image. Contour levels are 3.5 and 5 times the rms noise or 6.3 and 9.0 μJy which corresponds to the lower limits of our supplementary and complete lists respectively.

resolution. The combination of the VLA and MERLIN data will give sensitive radio images of the HDF with an angular resolution of 0.2 arc sec. This will allow a detailed comparison of the optical and radio emission.

4. Optical counterparts of radio sources in the HDF

Figure 3 shows the distribution of I_{AB} magnitudes of the 13 radio sources in the HDF. Of the seven sources found in the complete sample, all have optical counterparts brighter than $I_{AB} = 21.2$. At least half of the optical identifications are with spiral or irregular galaxies, many of which appear to be merging or are in small groups. The remaining optical counterparts are composed of nearby field spirals, red ellipticals, and possibly a few late type AGN. There are no identifications with quasars or with galactic stars.

Spectroscopic or photometric redshifts are available for all of the galaxies in the HDF with detected radio emission. All have redshifts less than 2.01 and the seven identified sources in the HDF complete sample have redshifts less than 1.01 (Figure 4). Although these microjansky radio sources are a million times weaker than Jansky level sources,

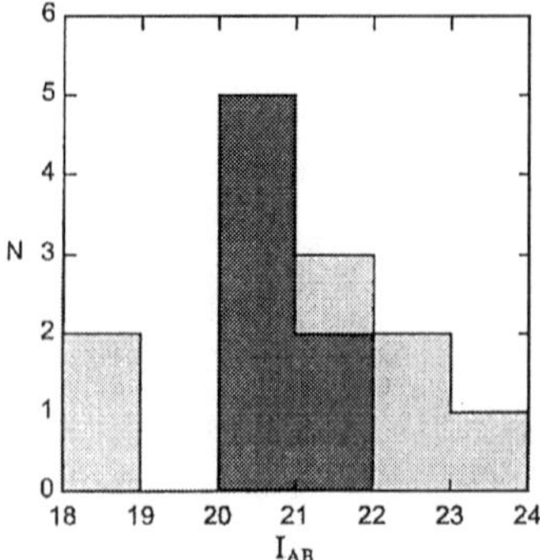

FIGURE 3. Magnitude distribution for radio sources in the HDF. The black areas refer to the complete sample of seven sources and the lightly shaded areas the six additional radio sources not found in the complete sample. I_{AB} magnitudes are taken from Williams et al. 1996.

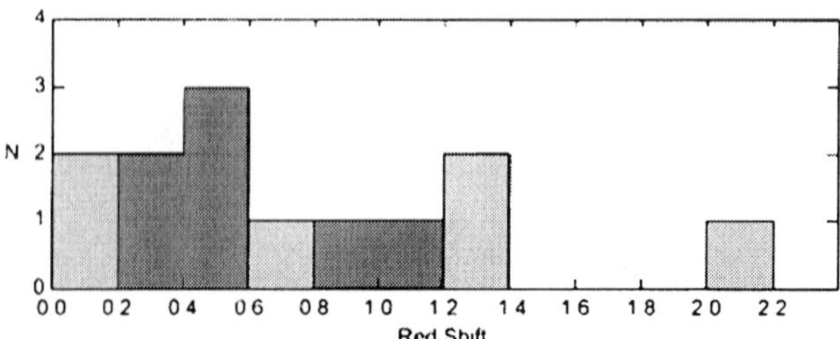

FIGURE 4. The redshift distribution for radio sources in the HDF. The black areas refer to the complete sample of seven sources and the open areas the six additional radio sources not found in the complete sample which are identified with galaxies having measured redshifts. Redshifts are taken from the Keck/HDF Consortium (Moustakis et al. 1997).

such as found in the all sky 3CR or Parkes surveys, it is important to note that the redshift distribution of the microjansky sources does not differ appreciably from that of the strong radio source population.

This is because the radio luminosity function is relatively steep. Unlike the optical galaxy counts which reach to successively more distant galaxies at faint magnitudes, the radio counts, at microjansky levels, sample roughly the same part of redshift space as the strong source samples, but reflect the distribution of a much lower luminosity group of radio sources.

None of the galaxies in the HDF are strong FRII radio galaxies. As shown in Figure 5, the radio luminosity of all of the galaxies in the HDF is less than 6×10^{24} W/Hz at 8.5 GHz, typical of FRI radio galaxies, weak AGN found in some elliptical and Seyfert galaxies, and other galaxies with active star formation. The weakest source in the field, HDF 3648+1427 (we denote radio sources in the HDF and HFFs by their coordinates, truncating the 12 hours and +62 degrees), which is identified with a bright ($I_{AB} = 18.4$) relatively nearby ($z \sim 0.02$) galaxy which has a luminosity of only 2×10^{19} W/Hz, comparable with that of normal spiral galaxies such as M31.

Only one source in the Hubble Deep Field (HDF 3646+1404) shows clear evidence of variability over the approximately 18 month period covered by our observations. It is a relatively powerful radio source with a luminosity at 8.5 GHz of 3×10^{24} W/Hz and has an unresolved radio core less than 0.1 arc seconds in diameter and a flat radio spectrum. It is identified with a face on Sb galaxy at a redshift of 0.96 which is very red in color

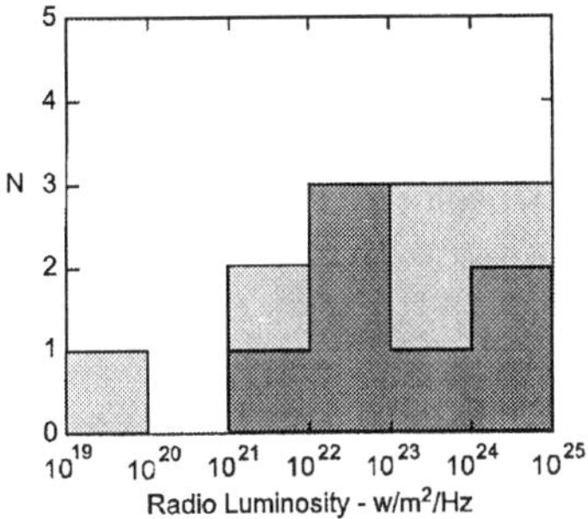

FIGURE 5. The luminosity distribution of radio sources in the HDF calculated assuming H_o = 50 km/sec/Mpc and q_o = 1/2. The dark areas refer to the complete sample of seven sources and the open areas to the six additional radio sources not in the complete sample which have measured redshifts.

and contains a compact optical nucleus with broad emission lines (Phillips et al. 1997). It is bright in the infrared H and K bands (Cowie et al. 1997). Rowan-Robinson (1997) interpret the ISO 6.7 micron flux density of 52 μJy as evidence for a massive starburst with a SFR of about 200 M_o/yr. However, the observed radio variability combined with its flat radio spectrum and small size suggests that the radio emission in HDF 3644+1404 originates in an AGN, and is not primarily due to star formation.

The strongest radio source (S = 600 μJy) in the Hubble Deep Field, HDF 3644+1133, is identified with a bright ($M_B = -23$) red elliptical galaxy at a redshift of 1.01. The radio luminosity is 6×10^{24} W/Hz, typical of the stronger FRI radio galaxies. It has a strong flat spectrum radio core which is unresolved at our highest angular resolution and is less than 0.1 arc seconds in diameter. A double lobe structure reaches some 15 arc seconds to the north and south of the unresolved core. HDF 3644+1133 has also been observed by ISO and has a 6.7 micron flux density of 50 μJy (Goldschmidt et al. 1997). Curiously, there is a very blue apparently elongated chain of galaxies, or possibly merging system, which lies within the northern radio lobe of HDF 3644+1133 and is at essentially the same redshift. This feature is similar in appearance to the "chain" galaxies described by Cowie et al. (1995), although it appears much larger in extent than the galaxies discussed by Cowie et al. It is not clear what relation exists, if any, between HDF 3644+1133 and the "chain" galaxy which has multiple apparently unresolved "hot spots." Possibly, the "chain" galaxy is undergoing active star formation induced by the jet emerging from the red elliptical. We have detected radio emission from another "chain" galaxy, HDF 3652+1354, but which is not in our complete radio sample.

Interestingly, there is one source which appears just below our completeness level with a snr of 4.7 which is also seen in our 20 cm data with a snr of 3.5, but which has no optical counterpart down to the limiting magnitude of $I_{AB} = 27.6$ of the HDF. If real, it is unlikely that this unidentified source is an intrinsically faint galaxy at moderate redshift as no other radio sources are known to be identified with such faint galaxies. The observed radio emission could be the displaced radio lobe of an extended source with asymmetric structure and no detectable radio emission from the parent galaxy, but we consider this also unlikely as other microjansky sources all appear coincident with their optical counterpart. More likely, it may be a very high redshift ($z > 6$) *I dropout* galaxy, in which case H and J band observations with NICMOS may be able to detect the parent galaxy.

5. Radio sources in region surrounding the HDF

Of the 28 sources in our complete radio catalogue, 21 are located outside the region of the Hubble Deep Field. Twenty of sources are covered by a Palomar 5-m image (Steidel, private communication) which allows identifications down to R = 25.5, and 12 of these are included in one of the Hubble flanking fields. We were able to identify 16 of the 20 sources with optical counterparts between R magnitudes 18 and 24 on the Palomar image, five of which appear to have ISO counterparts at 6.7 microns (Goldschmidt et al. 1997). The optical counterparts of the identified sources in the region surrounding the HDF are a mixture of galaxy types, and includes one relatively bright (R = 18.1) nearby (z = 0.08) face-on spiral, HFF 3637+1135, with a radio luminosity of about 10^{21} W/Hz. Two of the radio sources in the region surrounding the HDF are variable, suggesting the presence of an AGN. One is unidentified on the Palomar image and has R > 25.

6. Evolution of the microjansky population

The best fitting slope to the integral source count for the 28 radio sources contained in our complete sample is -1.0 ± 0.2, comparable to that found by Windhorst et al. (1993) and Fomalont et al. (1991) for other microjansky samples. As discussed earlier, redshifts of the microjansky radio sources extend up to two or more, thus in the absence of any evolution, the source count would be expected to rapidly converge. The steepening of the observed count below one millijansky implies an increase in the density and/or luminosity of the source population. However, as reported by Windhorst et al. (1993), below about one microjansky, the count must converge rather rapidly, otherwise the integrated contribution of the weaker sources would distort the microwave background spectrum which is known to be within a part in 10^4 of a perfect black body. Thus, the space density radio sources which contribute to the microjansky counts appears to peak near redshifts of unity and evolve in much the same manner as strong radio galaxies, optically selected quasars (e.g., Schmidt et al. 1991) or the population of faint starforming galaxies reported by Madau (1997).

7. Summary

Radio sources in the Hubble Deep Field and surrounding region are identified with a variety of galaxy counterparts, mostly spiral and irregular galaxies in small groups, but including a few late type systems. Redshifts of HDF radio sources are comparable to those of radio galaxies identified with much stronger radio source samples, but the microjansky radio sources are much less luminous. Many of the galaxies with microjansky radio emission have also been detected by ISO at 6.7 or 15 microns. Measurements of the radio luminosity, morphology, variability, and spectral index suggest an emission mechanism which is powered in part by AGN and in part by starforming activity. Our estimates of star forming rates estimated from the observed radio emission of HDF galaxies is typically within a factor of two of that given by Rowan-Robinson (1997) from ISO observations at 6.7 microns. One radio source appears to have no optical counterpart in the HDF and may be an *I dropout* galaxy at a redshift greater than six.

The results reported here are based on work done in collaboration with E. Fomalont, R. Windhorst, and B. Partridge. A more detailed account will be given by Richards et al. (1997). Optical magnitudes and redshifts quoted in this paper are taken from data made available by J. Cohen et al. (1996), C. Steidel, L. Moustakas et al., (1997), A. Phillips et al. (1997), L. Cowie, (1997) and others. We are grateful to these colleagues for making the follow-up ground based optical data available. Part of this work was supported by NASA through grant AR-06337.02-94A from the Space Telescope Science Institute, which is operated by the Association of Universities for Research in Astronomy, Inc., under NASA contract NAS5-2655. EAR acknowledges support of a Sigma Xi Grant-in-Aid-of-Research. The National Radio Astronomy Observatory is a facility of the National Science Foundation, operated under cooperative agreement by Associated Universities, Inc.

REFERENCES

BENN, C. R., ROWAN-ROBINSON, M., MCMAHON, R.G., BROADHURST, T. J. & LAWRENCE, A. 1993 *MNRAS* **276**, 1085.

COHEN, J. G., COWIE, L. L., HOGG, D. W., SONGAILA, A. 1996 *ApJ* **471**, L5.

CONDON, J. J. 1984 *ApJ* **287**, 461.

CONDON, J. J. 1989 *ApJ* **338**, 13.

CONDON, J. J. 1992 *ARAA* **30**, 575.

COWIE, L. I., HU, E. M., & SONGAILA, A. 1995 *AJ* **110**, 1576.

COWIE, L. ET AL. 1997 *(http://www.ifa.hawaii.edu/~cowie/tts.html)*, and in preparation.

DANESE, L., DE ZOTTI, G., FRANCESCHINI, A., & TOFFOLATTI, L. 1987 *ApJ* **318**, L15.

HAMMER, F., CRAMPTON, D., LILLEY, S., LE FEVRE, O., & KENET, T. 1995 *MNRAS* **276**, 1085.

MADAU, P. ET AL. 1997 *MNRAS* **289**, 465.

HELOU, G., SOIFER, B. T., & ROWAN-ROBINSON, M. 1985 *ApJ* **298**, L7.

FOMALONT, E. B., WINDHORST, R. A., KRISTIAN, J. A., & KELLERMANN, K.I. 1991 *AJ* **102**, 1258.

FRANCESCHINI, A., TOFFOLATTI, L., DANESE, L., & DE ZOTTI, G. 1989 *ApJ* **344**, 35.

MADAU, P. 1997, this volume.

MOUSTAKAS, L., ET AL. 1997 *(http:/astro.berkeley.edu/davisgrp/HDF/)*.

PHILLIPS, A. C., ET AL. 1997 *ApJ*, in press.

RICHARDS, E. A., KELLERMANNN, K. I., FOMALONT, E. B., WINDHORST, R. A., & PARTRIDGE, R. B. 1997, submitted to *AJ*.

ROWAN-ROBINSON, M., ET AL. 1997 *MNRAS* **289**, 490.

SCHMIDT, M., SCHNEIDER, D. P., & GUNN, J. E. 1991 in *The Space Distribution of Quasars* (ed. D. Crampton). p. 109. Astr.

THUAN, T. X., & CONDON, J. J. 1987 *ApJ* **322**, L9.

WEISTROP, D., WALL, J. V., FOMALONT, E. B., & KELLERMANN, K. I. 1987 *AJ* **93**, 803.

WILLIAMS, R. ET AL. 1996 *AJ* **112**, 1335.

WINDHORST, R., MILEY, G. K., OWEN, F. N., KRON, R. G., & KOO, D. C. 1985 *ApJ* **289**, 494.

WINDHORST, R., FOMALONT, E. B., PARTIDGE, R. B., & LOWENTHAL, J. D. 1993 *ApJ* **405**, 498.

WINDHORST, R. A., FOMALONT, E. B., KELLERMANN, K. I., PARTRIDGE, B., RICHARDS, E. A., FRANKLIN, B. E., PASCARELLE, S. M., & GRIFFITHS, R. E. 1995 *Nature* **375**, 471.

The ISO Survey of the Hubble Deep Field

By MICHAEL ROWAN-ROBINSON, S. J. OLIVER, R. G. MANN, S. SERJEANT, P. GOLDSCHMIDT, A. EFSTATHIOU, N. EATON, B. MOBASHER, T. J. SUMNER, L. DANESE, D. ELBAZ, A. FRANCESCHINI, E. EGAMI, M. KONTIZAS, A. LAWRENCE, R. MCMAHON, H. U. NORGAARD-NIELSEN, I. PEREZ-FOURNON, J. I. GONZALEZ-SERRANO

Blackett Laboratory, Imperial College, Prince Consort Road, London SW7 2BZ

We present the results of deep imaging at 6.7μm and 15μm from the CAM instrument on the Infrared Space Observatory, centred on the Hubble Deep Field. These are the deepest integrations made to date at these wavelengths in any region of sky.

We have constructed algorithmically selected "complete" flux-limited samples of 19 sources in the 15μm image, and 7 sources in the 6.7μm image. The typical flux limit at 15μm is $\sim$ 0.2 mJy and at 6.7μm is $\sim$ 0.04 mJy. We discuss the completeness and reliability of the connected pixel source detection algorithm used, by comparing the intrinsic and estimated properties of simulated data. The most pessimistic estimate of the number of spurious sources in the "complete" samples is 1 at 15μm and 2 at 6.7μm.

Both source counts and a $P(D)$ analysis suggest we have reached the ISO confusion limit at 15μm and may be close to it at 6.7μm. The counts appear steeper than expected from models of far infrared emission from different galaxy populations which fit the IRAS 60μ counts.

We associate the detected ISO sources with objects in existing optical and near-ir HDF catalogues using the likelihood ratio method. We find 15 ISO sources to be reliably associated with 13 bright $[I_{814}(AB) < 23]$ galaxies in the HDF. Amongst optically bright HDF galaxies, ISO tends to detect luminous star-forming galaxies at moderate redshift and with disturbed morphologies, in preference to nearby ellipticals.

We have modelled the spectral energy distributions of the 13 ISO-HDF galaxies. For two, the emission detected by ISO is consistent with being starlight or the infrared 'cirrus' in the galaxies. For the remaining 11 galaxies there is a clear mid-infrared excess, which we interpret as emission from dust associated with a strong starburst.

We derive star formation rates for these galaxies of 8–1000 $\phi M_\odot$ per yr, where ϕ takes account of the uncertainty in the IMF. The HDF galaxies detected by ISO are clearly forming stars at a prodigious rate compared with nearby normal galaxies. We discuss the implications of our detections for the history of star and heavy element formation in the universe. Although uncertainties in the calibration, reliability of source detection, associations, and starburst models remain, it is clear that dust plays an important role in star formation out to redshift 1 at least.

---

1. Introduction

Because of its great depth, high resolution, and the intensive follow-up which has been carried out in it, the Hubble Deep Field (HDF) is an exceptional resource for cosmological studies. The central area of the HDF consists of 5 square arcmin of sky. It was imaged by the Hubble Space Telescope on 150 orbits in December 1995 and reaches to at least 29th magnitude in I (800nm), V (600nm) and B (450nm), and to 27th magnitude in U (300nm) (Williams et al. 1996). We were successful in bidding for Director's Time on the Infrared Space Observatory (ISO) and were awarded a total of 12.5 hours to map the HDF with ISO-CAM in the LW2 (6.7μm) and LW3 (15μm) filters. The observations were carried out in July 1996 and have been described by Serjeant et al. (1997). The

Parameter	LW-2 6.7μm	LW-3 15μm
Pixel field of view	3″	6″
M, N steps	8, 8	8, 8
M, N step size	5″, 5″	9″, 9″
$T_{\rm int}$	10 sec	5 sec
$N_{\rm stab}$	80	100
$N_{\rm obs}$	10	20

TABLE 1. Summary of our CAM01 astronomical observation templates

images have been searched for point sources by Goldschmidt et al. (1997): a total of 15 sources were found in the central HDF area at 6.7μm, and 5 at 15μm, of which 6 and 4, respectively, are from complete and reliable sub-samples. A further 27 sources were found in the flanking fields around the HDF. The resulting source-counts have been discussed by Oliver et al. (1997) and shown to be consistent with the strongly evolving starburst models previously used to model the 60μm and 1.4 GHz counts (Franceschini et al. 1994, Rowan-Robinson et al. 1993, Pearson and Rowan-Robinson 1996). Associations for the 17 ISO sources in the central HDF area (2 were detected at both 6.7 and 15μm) were sought with HDF galaxies using a likelihood method (Mann et al. 1997) and 13 credible associations were found. We discuss the spectral energy distribution of these 13 galaxies and consider the implications for star formation rates and the overall history of star formation in the universe.

A striking feature of the fainter HDF galaxies is their blue colours, indicative of high redshift galaxies undergoing bursts of star formation. This is confirmed both by systematic analyses of the colours of the HDF galaxies (Mobasher et al. 1996) and by studies of the morphologies of the galaxies (Abraham et al. 1996), which show a high proportion of interacting and merging systems. In both respects the HDF galaxies look like a higher redshift version of the starburst galaxies found by IRAS. This was one of the strong motivations for seeking observing time with ISO. The fact that HDF galaxies have been detected by ISO is sufficient to demonstrate that this analogy with IRAS galaxies is highly relevant.

An Einstein de Sitter model ($\Omega_0 = 1$), with a Hubble constant $H_0 = 50$ km/s/Mpc, has been used throughout the paper.

2. Data acquisition and analysis

2.1. *Observation strategy*

The observing strategy was designed to reach the predicted confusion limit of ISOCAM. The observations were performed in two bands, 6.7μm and 15μm (the LW-2 and LW-3 filters respectively). Typically more than one HDF galaxy falls within the Airy disk, even at 6.7μm, so sub-pixel offsets were needed to maximise the available spatial information.

We made three 8×8 rasters in microscanning mode (CAM01) analogous to the "dithering" in the HDF, each centred on one of the HDF WF frames, at both 6.7μm and 15μm, using in total $\sim$ 41.9 ks of time including overheads. Our choice of 3″ (6″) pixel sizes at 6.7μm (15μm) and raster step sizes of 5″ (9″) yields optimal flat fielding accuracy and spatial resolution. Cosmic ray transients (discussed below) ruled out readout integration times longer than 10 seconds, and the signal-to-noise vs. time predictions from the CAM simulator implied diminishing returns for more than 10 (20) readouts per raster position at 7μm (15μm). The Astronomical Observation Template (AOT) parameters are summarised in Table 1; for a more detailed discussion of these parameters see the ISOCAM Observers Manual.

2.2. *Dark subtraction, deglitching and flat fielding*

The default dark frame in the April 1996 CIA version was subtracted from the data.

Cosmic ray events were easily identified in the readout histories of each pixel as $> 4\sigma$ rises followed (one or two readouts later) by $> 4\sigma$ falls. A similar algorithm was used to find readout troughs. These events were masked out in the mosaicing discussed below.

However, a minority of cosmic rays appear to cause transients in subsequent readouts with roughly exponential decays persisting over a few readouts. These glitch transients are in general difficult to model. No attempt was made to identify and remove them; instead, they are effectively removed by median filtering in the mosaicing below.

Use of the ESA-supplied flat field gave very unsatisfactory results. Instead, we created our own sky flat by noting that each detector pixel samples 64 different sky positions during the raster. For each detector pixel, we examined the histogram of (unmasked) readouts and fitted Gaussians to find a mean value. To eliminate both sources and glitch transients, $> 5\sigma$ outliers from the mean were eliminated and the fit was iterated.

2.3. *Shift-and-add and drizzled maps*

The rasters were mosaiced together using two competing algorithms: drizzling, and variants on shift-and-add. The latter are expected to have higher signal-to-noise but at the expense of spatial resolution.

In the shift-and-added frames, we began by defining an image with pixel size one-sixth that of the CAM detector pixels (i.e., one-sixth of $3''$ at 7μm, and of $6''$ at 15μm), encompassing the area surveyed by all three rasters. Each position in this fine-gridded image may have been observed several times by the CAM detector array, so we compiled a list of such CAM pixel readouts for each fine-gridded image position. The data at a given sky position could then (for example) be median filtered or mean averaged to produce a final image.

The glitch transients discussed above make a large contribution to the noise in final mosaics made by simple mean averages of readouts. An obvious alternative is median filtering, though this has a signal-to-noise penalty (about $\sqrt{2}$). The glitch transients have timescales of order or less than the duration of a pointing, so a possible compromise is to mean average over pointings, and median filter the means. This may also have the advantage of identifying affected readouts more efficiently.

Several mosaics were therefore created from the readout arrays: (a) using the median readout at each position, to eliminate glitches; (b) using the mean readout at each position, with the $\pm 5\sigma$ outliers eliminated and iterated to eliminate glitches; (c) using the mean readout within each raster position, followed by the median of these means; (d) using a clipped, iterated mean of pointing means. We found the best 6.7μm map to be the simple median filtered image, but at 15μm the optimal map is the median-of-means, presumably reflecting a greater sensitivity to glitch transients in the final mosaic.

The drizzled images were produced with the same code used to produce the optical HST images. Briefly, instead of superimposing overlapping pixels, the drizzle algorithm allows the user to shrink the input pixel sizes (the "footprint") before superimposing. At one limit the drizzle algorithm is equivalent to interlacing; at the other it is similar to shift-and-add above. Since the ISO HDF images were taken with fractional pixel spacings between them maps with increased resolution can be constructed. The footprint was chosen to produce roughly the highest resolution image without leaving gaps in the output image.

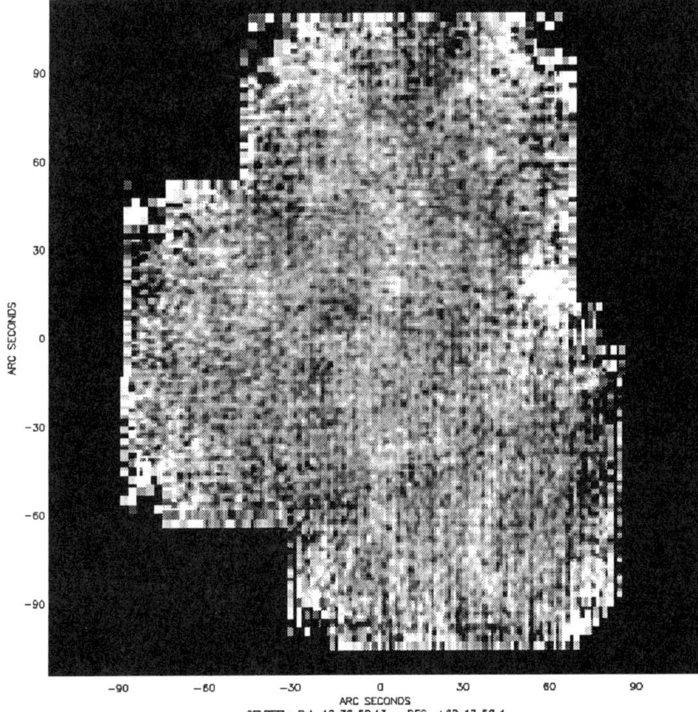

FIGURE 1. The 6.7μm drizzled ISO map of the HDF.

Glitch transients were first removed from the input images using the following median filtering scheme, noting that at 6.7μm the raster step offsets correspond to $1\frac{2}{3}$ pixels and at 15μm the offsets are $1\frac{1}{2}$ pixels. The 64 individual pointings were grouped into sets whose members have integer pixel offsets (9 groups for LW-2 and 4 for LW-3), and for each of these groups the medians of the data at each pixel position (allowing for the shifts) was put into an image and the variance was computed. These 9 images at 7μm and 4 images at 15μm served as input to the drizzling algorithm.

Flux calibration assumes the standard conversion in the ISO-CAM handbook. This gives good agreement between our measured backgrounds and a zodiacal background model.

Figures 1 and 2 show the results of the drizzling algorithm. The correspondence between the drizzling and the shift-and-add mosaics (not shown) is excellent suggesting that neither algorithm is introducing artifacts. Note the poorer coverage at the edges, which causes the poorer signal-to-noise at the corresponding edges of the mosaics.

Several sources are clearly seen in the maps. However, we note here that a visual inspection clearly shows that at 15μm we are approaching the confusion limit of one source per 40 beam sizes.

3. Detected objects and their estimated fluxes

We used the object detection algorithm, PISA, which counts contiguous pixels above a user-supplied threshold to search for sources in the ISO maps of the HDF. If the number of contiguous pixels exceeds a user-supplied minimum, then these pixels are defined as

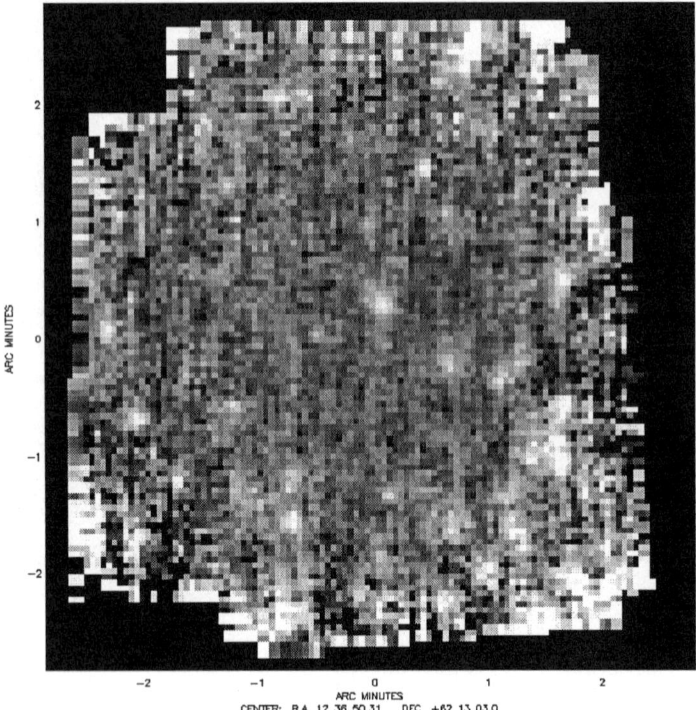

FIGURE 2. The 15μm drizzled ISO map of the HDF.

being an object. The minimum number of contiguous pixels that a detected object must have is set by considering the number of pixels in the point spread function and also from tests with simulated data.

PISA uses a constant threshold per pixel across each image and this threshold depends on the noise, therefore before running PISA we split the image up into sub-images, according to the noise level.

The detection criteria were as follows: each pixel had to have a flux greater than 2σ above the modal sky value in that region, where σ was estimated from the variance in the sky counts and also from the pixel history. The two estimates of σ are in good agreement with each other. The minimum number of contiguous pixels in a detected object was set to be 8. Running PISA produced 19 objects in the 15μm image and 7 objects in the 6.7μm image.

To estimate the fluxes of the detected objects, we measured the local sky, using a concentric annulus around the adopted point-source aperture. At 15μm the adopted aperture was of radius 12 arcseconds, and the sky annulus had an inner radius of 15 arcseconds and an outer radius of 24 arcseconds. At 6.7μm the corresponding values were 6, 8 and 13 arcseconds.

To estimate the errors at each wavelength, we created "sky maps" by convolving the sky annulus with each image to give the estimated sky value at each pixel. We then subtracted this from the original image and convolved the resulting image with the object aperture. This image has, at each pixel, an estimate of the "object-minus-sky" flux that a detected object would have if it were centred on that image. The variance in this resulting image therefore gives the variance in the fluxes of the estimated objects. We find that at 15μm the 1 sigma error is ~ 0.1 mJy and at 6.7μm it is 0.02 mJy.

In addition, a number of objects were selected by PISA with slightly less stringent criteria. This meant that a higher number of spurious objects were selected. A slightly lower threshold was used in the selection, and the resulting objects were "eyeballed" to see if they looked genuine. This is because PISA chooses connected pixels regardless of shape, i.e., a chain of connected pixels above the threshold will be selected as an object in the same way as a circle of pixels. Eyeballing these selected objects therefore was designed to select the "round" objects that appeared to have the similar sort of smooth profiles that one would expect a genuine object convolved with the point-spread function would have. These objects will be referred to as the supplementary objects. Mann et al. (1997) discusses the likelihood of these objects being genuine by matching them up to the optical HDF image. Note that these objects are *not* selected with any reference to the optical images.

The reliability of the detected sources was tested by using the same source detection algorithm on simulated data.

We estimate that, at worst, 1 object at 15μm and 2 objects at 6.7μm in the "complete" lists are spurious detections, and at best, less than 1 object at each wavelength is spurious. A pessimistic estimate is that 1 object at 15μm and 5 objects at 6.7μm in the supplementary lists are spurious.

We have performed a number of independent checks on the reality of the ISO sources, particularly those at 6.7μm: (1) clippping the data at $\pm 2\sigma$ yields maps almost identical to those of Figs. 1 and 2, (2) when the data is split in half most sources can be seen on each separate map, (3) the time lines on the central pixel in each source have been examined to see if the sources can be seen in the time domain. 9/10 sources in Table 2 detected at 6.7μm are confirmed by at least one of these tests, and 6/10 are confirmed in all 3.

4. Source counts at 6.7 and 15μm, comparison with models

To convert these source lists into source counts requires an estimate of the area within which a source of observed flux (S_ν) could have been detected. This requires a systematic study of the ISO Point Spread Function, which is discussed by Oliver et al. (1997).

The faintest flux limit that can provide useful information is defined by the area of the survey to that flux limit. We choose the smallest useful area to be 200 beams, which translates to 38.6 and 255μJy respectively. These flux limits include 6 of the 7 6.7μm sources and 17 of the 19 15μm sources. Allowing a smaller area of 40 beams would have suggested a flux limit of 30.4 and 161μJy in which case the faintest 15μm source would be excluded and the faintest 6.7μm source would be at the flux limit.

We consider two models for far infrared galaxy populations. Pearson and Rowan-Robinson (1996) have described a galaxy population model involving five populations: normal galaxies; star-burst galaxies; hyper-luminous galaxies; Seyfert 1 galaxies and Seyfert 2 galaxies. We have predicted the counts at 6.7μm, 12μm and 15μm using these models ignoring the hyper-luminous population which will have negligible contribution.

A second model is due to Franceschini et al. (1994). The total counts are modelled as the sum of 5 populations: AGN; star-burst galaxies; spiral/irregular galaxies; S0 and elliptical galaxies. The late-type systems (Spirals, Irr and star-bursts) evolve as: $L(z) = L(0)e^{2\tau(z)}$. For the early-type systems (Ellipticals, S0) Franceschini et al. (1994) assume that a bright phase of active star-formation at $z \sim 2-4$ is obscured by dust quickly produced by the first stellar generations. This same model accounts in some

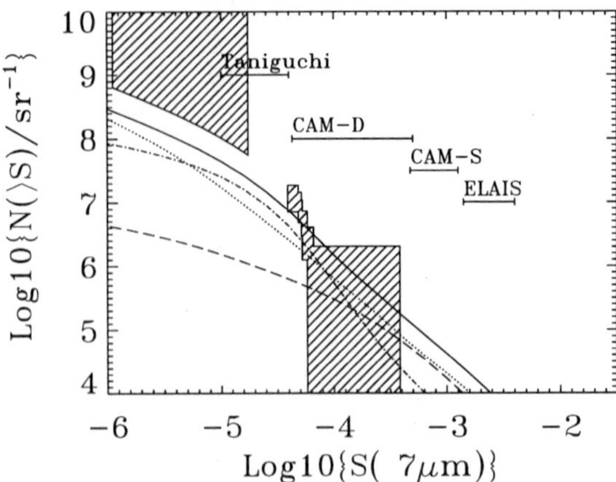

FIGURE 3. 6.7μm source counts from HDF. Models based on Franceschini et al. (1994): all components solid; spiral galaxies dotted; E+s0 dot-dash.

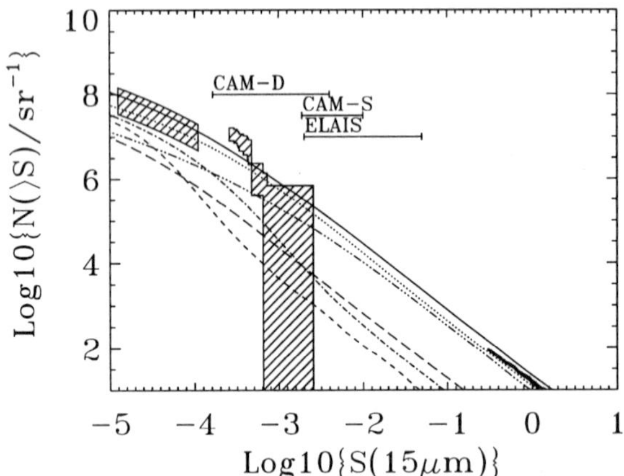

FIGURE 4. 15μm source counts from HDF with IRAS 12μm counts at bright end, from this paper and Rush et al. 1993 (asterisks) (IRAS data shifted to 15μm using cirrus spectrum). Models based on Franceschini et al. (1994): all components solid; spiral galaxies dotted; star-bursts dash-dot-dot-dot; s0 dot-dash; AGN-dash Elliptical galaxies-short dashes.

detail for the sub-mm background as estimated by Puget et al. (1996). The AGN number density is set by the local IRAS samples at 12μm (see Rush et al. 1993). The evolution is calibrated so as to produce the hard X-ray background with a suitable distribution of the dust/gas absorbing column densities. The integral counts predicted by this model are shown in Figures 3 and 4.

The first impression from Figs. 3 and 4 is that the agreement between the models and the data is surprisingly good, considering that in the case of the 15μm data the predictions were made from data 3 decades brighter, while at 6.7μm there were no previous data with which to normalise the models. The slope of the observed counts at both wavelengths is slightly steeper than either model would predict. Either models with stronger evolution are required to fit these data, or revisions to the K-corrections using improved SEDs may be required.

5. ISO source associations in the Hubble Deep Field

There are several basic problems which complicate the association of ISO-HDF sources with HDF galaxies in existing optical and near-infrared catalogues. The most obvious of these is the poor match between the resolution of ISOCAM and the high source density of galaxies in the HDF: the radius of the Airy disk is 2.8 arcsec at 6.7μm and 6.0 arcsec at 15μm, while there are several hundred galaxies per square arcminute detected to $I_{814} \sim 29$, so we expect several galaxies per randomly placed Airy disk at both 6.7μm and 15μm. Thus, not only is there a high likelihood of chance associations with optical galaxies, but any given ISOCAM beam may be integrating over more than one source, and significant flux may be contributed by more than one galaxy: this latter is exacerbated both because many luminous mid-infrared sources are likely to be interacting or merging galaxy systems (e.g., Lawrence et al. 1989), and by the fact that our ISO maps appear to be at, or close to, the confusion limit in both bands.

We have performed the likelihood ratio procedure of on the I_{814} images of Williams et al. (1996), which is both the reddest and deepest band, as well as the only one available in the Hubble Flanking Fields (HFF).

In Fig. 3 of Mann et al. (1997) we show the immediate surroundings (in an I_{814} band mosaic) of the 13 ISO-HDF sources reliably associated with HDF galaxies.

We detect 10 of the 44 brightest I_{814} band objects in the Williams et al. (1996) catalogue (i.e., those with $I_{814} < 22.04$): 8 of these 44 objects are stars, which we discuss no further. Of the 36 galaxies, we detect 13 percent (2 out of 15) of the ellipticals, 30 percent (6/18) of the spirals and 67 percent (2/3) of the irregulars/mergers. We find that, amongst bright HDF galaxies, ISO has a tendency to detect luminous, star-forming galaxies at moderate redshift and with disturbed morphologies, in preference to nearby ellipticals.

6. Spectral energy distributions

For the 13 galaxies reliably detected by ISO in the HDF we have modelled their spectral energy distributions (seds) from 0.3 to 15μm (Rowan-Robinson et al. 1997). Spectroscopic redshifts are available for 9 of the galaxies (Cowie 1996, Cohen et al. 1996, Phillips et al. 1997). For the remaining 4 we have used photometric redshifts determined by the method of Mobasher et al. (1996).

The U, B, V, I data (AB magnitudes) from HST and the J, H/K data of Cowie (1996) have been fitted with galaxy models from the library of Bruzual and Charlot (1993). Excellent fits to the U to K data for our galaxies can be obtained using models with starburst of duration 10^9 yrs, viewed at a range of subsequent times $\tau = 1$ to 2.4 Gyr. In the case of the galaxies for which we have only photometric redshifts, the good fits of the models to the data provide support for the photometric redshifts. No allowance is made for reddening at this stage. For one galaxy (12 36 43.9 +62 11 30), the 6.7μm emission can be accounted for almost completely by starlight. For the remaining 12 galaxies, there is a clear excess of infrared radiation. We have considered first the possibility that we are seeing the infrared 'cirrus' in the galaxies, emission from starlight in the galaxies absorbed by interstellar grains and reemitted in the infrared. The cirrus models of Rowan-Robinson (1992) have been revised to incorporate very small grains and PAHs correctly (Efstathiou et al. 1997). For 12 of the galaxies, there was no plausible cirrus model, because the resulting far infrared luminosity was always at least 3 times the total optical-uv luminosity of the galaxy. A typical value of L_{fir}/L_{opt-uv} for cirrus emission is 0.2–0.3 (Rowan-Robinson et al. 1987, Rowan-Robinson 1992). For one galaxy (12 36

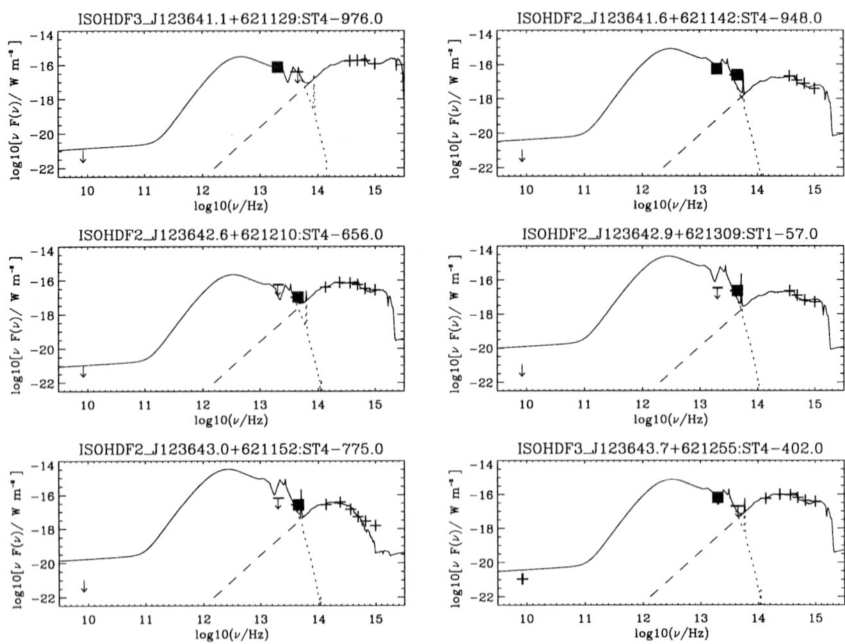

FIGURE 5. Spectral energy distributions ($\nu F(\nu)$ in Wm^{-2}) for galaxies of Table 2, compared with Bruzual and Charlot (1993) models (optical and near ir) and Efstathiou et al. (1997) cirrus and starburst models (mid and far ir). Crosses: HST, IFA and VLA data, filled squares: ISO data.

48.1 +62 14 32), in which the cirrus model gave a far infrared luminosity comparable to that seen in the optical and uv, we accepted the cirrus model fit. Figure 5 shows fits of the standard starburst model of Efstathiou et al. (1997) to the infrared spectral energy distributions for 6 of the remaining 11 ISO-detected HDF galaxies. Although in most cases we have only a single mid-infrared data point, it is satisfactory that in the case where we have detections at both 6.7 and 15μm, the model fits both data points well. Where only upper limits are available at one of the ISO wavelengths, these are generally consistent with the predictions of the model. Even more impressive is the fit for 3 galaxies to the VLA detections by Fomalont et al. (1997), for which we have assumed in the model the standard radio-far ir correlation (S(60μm)/S(1.4 GHz) = 90) and a radio spectral index of 0.8. This supports the idea that we really are seeing dust emission from starbursts with ISO. In most other cases the models are consistent with the 1.4 GHz upper limit, taken as 12.2 μJy (Fomalont et al. 1997) (but for objects 4, 5 and 10, the radio limits lie significantly below the predictions of the starburst model).

From Table 2 it can be seen that $L_{60}/L_{0.8}$ ranges from 2 to 200, and $L_{60}/L_{0.3}$ ranges from 3 to 1000. Thus the implication of the ISO detections is that, at least for the detected galaxies, the bulk of the bolometric luminosity of the galaxies is emitted at far infrared wavelengths. The interpretation of this is similar to that for the starburst galaxies found in IRAS surveys: most massive star formation takes place in dense molecular clouds and is shrouded from view by a substantial optical depth in dust. What is seen in the optical and uv represents stars formed near the edges of clouds, so that the light from these stars can escape directly.

Galaxy (ISOHDF)	redshift	$L_{0.3}/L_{\odot}$	$L_{0.8}/L_{\odot}$	$L_{15}/L_{\odot}$	$L_{60}/L_{\odot}$	$(\dot{M}_{*,all}/ \times M_{\odot})\phi^{-1}$	notes
12 36 41.1 +62 11 29	(0.047)	2.8×10^8	4.0×10^8	2.1×10^8	7.6×10^8	0.20	S sb c
12 36 41.6 +62 11 42	0.585	2.8×10^9	7.3×10^9	1.1×10^{11}	3.9×10^{11}	101	I sb b,d,e
12 36 42.6 +62 12 10	0.454	6.9×10^9	1.8×10^{10}	1.7×10^{10}	6.0×10^{10}	16	S sb b
12 36 42.9 +62 13 09	(0.74)	5.5×10^9	1.4×10^{10}	5.3×10^{11}	1.9×10^{12}	495	S sb b
12 36 43.0 +62 11 52	(0.82)	3.4×10^9	3.0×10^{10}	8.9×10^{11}	3.2×10^{12}	840	S sb a
12 36 43.7 +62 12 55	0.558	1.4×10^{10}	3.6×10^{10}	8.6×10^{10}	3.1×10^{11}	80	I sb c,e
12 36 43.9 +62 11 30	1.01	1.75×10^{10}	1.35×10^{11}	2.0×10^{10}	1.4×10^{11}	—	E sl b,e
12 36 46.4 +62 14 06	0.960	2.2×10^{10}	8.1×10^{10}	1.1×10^{12}	3.9×10^{12}	1010	E sb a,e
12 36 48.1 +62 14 32	(0.023)	6.0×10^7	1.5×10^8	1.8×10^7	1.26×10^8	—	E cirr a,c,e
12 36 48.4 +62 12 15	(0.778)	4.5×10^9	1.15×10^{10}	6.8×10^{11}	2.45×10^{12}	640	S sb a
12 36 49.7 +62 13 15	0.475	1.75×10^9	1.26×10^{10}	9.5×10^{10}	3.4×10^{11}	88	S sb a,c,e
12 36 51.5 +62 13 57	0.557	4.7×10^9	2.7×10^{10}	5.2×10^{10}	1.86×10^{11}	48	S sb d
12 36 58.9 +62 12 48	0.320	2.4×10^9	6.0×10^9	9.0×10^9	3.2×10^{10}	8	S sb a

S - spiral, E - elliptical, I - interacting pair
sb - sed fitted with starburst model
cirr - sed fitted with cirrus model
sl - sed fitted with starlight
a - detected at 6.7 μm, from reliable and complete sub-sample
b - detected at 6.7 μm, from supplementary list
c - detected at 15 μm, from reliable and complete sub-sample
d - detected at 15 μm, from supplementary list
e - detected at 1.4 GHz (Fomalont et al. 1997)

TABLE 2. 60μm luminosities and star formation rates for ISO-HDF galaxies.

7. Star formation rates for ISO-HDF galaxies

A number of authors have discussed how the star formation rate in a galaxy can be inferred from its optical, ultraviolet or far infrared luminosity (Scoville and Young 1983, Thronson and Telesco 1986, Madau et al. 1996).

To convert from 60μm luminosity to star formation rate, we assume that a fraction ϵ ($\simeq 1$) of the optical and uv energy emitted in a starburst is absorbed by dust and emitted in the far infrared, so that

$$L_{bol,fir} = \epsilon L_{bol,opt-uv} \quad . \tag{7.1}$$

Rowan-Robinson et al. (1997) then find:

$$\dot{M}_{*,all}/[L_{60}/L_{\odot}] = 2.6\phi/\epsilon \times 10^{-10} \quad . \tag{7.2}$$

$\dot{M}_{*,all}/[L_{15}/L_{\odot}] = 9.3\phi/\epsilon \times 10^{-10}$ where the factor ϕ incorporates (1) the correction from a Salpeter IMF to the true IMF ($\times 3.3$ if the Miller-Scalo IMF is the correct one), (2) a correction if the starburst event is forming only massive stars ($\times 1/3.1$ if only O,B,A stars, $> 1.6 M_o$, are being formed).

Table 2 gives the inferred 15 and 60μm luminosities, and star formation rates based on eq. (7.2), for the 11 HDF starburst galaxies detected by ISO. The star formation rates range (with one exception) from 8–1000 $\phi M_{\odot}$ per yr. The galaxies detected by ISO are forming stars at a prodigious rate compared with nearby normal spirals. Although star formation rates based only on 6.7 and 15μm detections are bound to be rather uncertain, because most of the energy is emitted at much longer wavelengths, it is clear that the star formation rates deduced from the uv fluxes detected by HST are a severe underestimate for these galaxies.

8. Infrared luminosity-density and the history of star formation

Madau et al. (1996) have used the Canada-France Redshift Survey (Lilly et al. 1996) and HDF data to calculate the history of star formation and heavy element generation, under the assumption that the uv gives a complete view of the star formation that is occurring. Integrating over the star formation density as a function of redshift, they conclude that all the heavy elements associated with the visible matter in galaxies can be generated. However if their calculated baryonic density is converted to a value for Ω, a value of 0.0035 is obtained, only 7% of the baryonic density of 0.05, for an assumed $H_o = 50$, derived from cosmological nucleosynthesis of the light elements (Walker et al. 1991). It is not unreasonable to assume that some of the remaining 93% or baryons has participated in star formation and heavy element production.

We first estimate the far infrared luminosity density for the luminosity function derived from IRAS 60μm data. Oliver et al. (1997) have shown that the 6.7 and 15μm source-counts are consistent with strongly (luminosity-)evolving starburst models.

The solid curve in Fig. 6 shows the luminosity-density at 60μm as a function of redshift. For $z < 0.3$ this is directly derived from IRAS galaxy redshift surveys (the luminosity function given in line (23) of Table 3 of Saunders et al. 1990). The extrapolation to higher redshift is the luminosity evolution model used in Pearson and Rowan-Robinson (1996), Rowan-Robinson et al. (1993), and Oliver et al. (1992) to fit the deep 60μm and 1.4 GHz source-counts, for which

$$L_*(z) = (1+z)^{3.1}, z < 2 \ , \qquad (8.3)$$
$$L_*(z) = 3^{3.1}, 2 < z < 5 \ .$$

Using eq. (7.2) we can convert this to a density of star formation, and integrate to derive a total mass-density in stars or in heavy elements. We find

$$\Omega_* = 0.008\ h_{50}^{-2}\ \phi\ , \quad \Omega_Z = 0.00019\ h_{50}^{-2}\ . \qquad (8.4)$$

These values are not unreasonable. They require that twice as much star formation as has been inferred by Madau et al. (1996) from the uv integrated light has taken place shrouded by dust. The total fraction of baryonic matter that has participated in star formation would be of order 20%, with about 1/3rd of the resulting heavy elements now residing in the luminous parts of galaxies. The remainder could be in baryonic objects in the halos of galaxies or in intergalactic gas (including the hot X-ray emitting gas in clusters). In fact evolutionary rates appreciably steeper than that assumed in eq. (8.3) can probably not be ruled out at this stage. If the star-forming galaxies we have detected with ISO are typical of the fainter HDF galaxies, then we may require that more than 50% of baryons have participated in star-formation and heavy element production, presumably with a truncated IMF so that most of the baryons now reside in dark remnants. Similar conclusions are reached if we use the model for evolution of infrared galaxies of Franceschini et al. (1994, 1997 in preparation), shown as a broken line in Fig. 6.

We have also estimated the contribution to the 60μm luminosity-density implied directly by the ISO-HDF starburst galaxies. There are 5 starburst galaxies in the redshift bin 0.4–0.7 and 3 in the redshift bin 0.7–1.0 (omitted the galaxy with broad lines). Estimating the volume of the universe sampled by the HDF survey we find contributions of $6.0 \pm 2.0 \times 10^8$ and $2.6 \pm 1.5 \times 10^9 L_o$ Mpc^{-3} for the redshift ranges 0.4–0.7 and 0.7–1.0 respectively. These estimates take no account of sources fainter than the ISO limit and they are subject to any uncertainty in the ISO calibration (probably a factor of 50% either way), as well as the considerable uncertainty associated with extrapolating from 6.7 and 15μm to 60μm, so must be seen as very preliminary. They appear to be

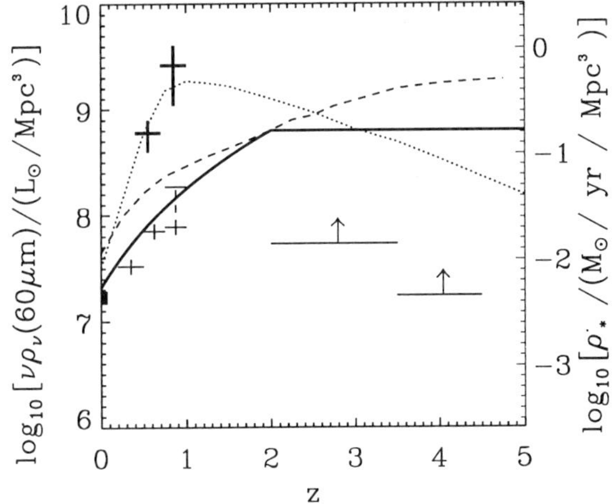

FIGURE 6. The luminosity density at $60\mu m$ as a function of redshift (left-hand scale) and the star-formation rate (right-hand scale, from eq. (7.2) assuming $\phi = 1$). The solid curve is derived from the IRAS $60\mu m$ luminosity function of Saunders et al. (1990, assuming evolution of the form (3). The broken curve is the model of Franceschini et al. (1994, 1997) and the dotted curve is the infall model of Pei and Fall (1995) with $\Omega_g = 8 \times 10^{-3}\ h^{-1}$. The thick crosses are the estimates of luminosity density derived directly from the ISO data for the galaxies of Table 2. Also shown (thin crosses and lower limits) are the star formation rates derived from the ultraviolet (2800 Å) luminosity density by Madau et al. (1996).

significantly higher than the predictions for evolution of the form (3) (the solid line in Fig. 6) by factors of 5 and 10 respectively. This may imply that the evolution of starburst galaxies is steeper than assumed in eq. (8.3) for $0 < z < 1$. Alternatively our models may overestimate the $60\mu m$ luminosities for at least some of the galaxies, for example because the 6.7 and $15\mu m$ radiation comes from dust tori around AGN.

We have also shown in Fig. 6 one of the more extreme of the models of Pei and Fall (1995, dotted line). This illustrates that the luminosity density estimated from the ISO-HDF galaxies is not at odds with current data on the number-density of quasar absorption-line clouds or the observed heavy element abundance at high redshift. However for this model $\Omega_* = 0.032 h_{50}^{-1}\phi$, $\Omega_Z = 0.00076 h_{50}^{-1}$, which would probably imply that only higher mass stars were being formed in the ISO-HDF galaxies. Support for the idea that stronger evolution than (3) is required for the starburst galaxy population comes from the fit to the ISO counts by Oliver et al. (1997). It will also be interesting to observe the HDF galaxies detected by ISO at submillimetre wavelengths, for example with SCUBA on the JCMT.

9. Conclusions

We have modelled the spectral energy distributions of the 13 galaxies reliably associated with ISO sources detected at 6.7 and/or $15\mu m$. For 2 galaxies the emission detected by ISO is consistent with being starlight or normal infrared 'cirrus' in the galaxies. For the remaining 11 galaxies there is a strong mid-infrared excess, which we interpret as emission from dust associated with a strong starburst. Inferred rest-frame luminosities (νL_ν) at 0.3, 0.8, 15 and $60\mu m$ are given and $L_{60}/L_{0.3}$ ranges from 3 to 1000 for the 11 galaxies. Thus most of the bolometric luminosity of the galaxies is predicted to emerge at far infrared wavelengths.

We derive star formation rates of 8–1000 ϕM_o per yr, where ϕ takes account of the uncertainty in the initial mass function (= 1 for Salpeter IMF). The HDF galaxies detected by ISO are clearly forming stars at a prodigious rate compared with nearby normal galaxies. We discuss the implications of our detections, and of the IRAS 60μm luminosity function and evolution, for the history of star and heavy element formation in the universe. We conclude that at least 20% of baryons must have participated in star formation.

Further details can be found on our web page: http://artemis.ph.ic.ac.uk/hdf/.

The work was supported by PPARC (grant number GR/K98728) and by the EC TMR Network programme (FMRX-CT96-0068).

REFERENCES

ABRAHAM, R. G., ET AL. 1996 *MNRAS* **279**, L47.

ACOSTA-PULIDO, J. A., ET AL. 1996 *AA* **315**, L121.

BRUZUAL, A. G., & CHARLOT, S. 1993 *ApJ* **405**, 538.

CESARSKY, C., ET AL. 1996 preprint, to appear in *A&A*.

COHEN, J. G., ET AL. 1996 *ApJ* **471**, L5.

COWIE, L. L. 1996 http:/www.ifa.hawaii.edu/cowie/hdf.html

FOMALONT, E. B., ET AL. 1997 *ApJ* **475**, L5.

FRANCESCHINI, A., MAZZEI, P., DE ZOTTI, G., DANESE, L. 1994 *ApJ* **427**, 140.

GOLDSCHMIDT, P., ET AL. 1997 *MNRAS* in press.

LILLY, S. J., LE FEVRE, O., HAMMER, F., CRAMPTON, D. 1996 *ApJ* **460**, L1.

MADAU, P., ET AL. 1996 *MNRAS* **283**, 1388.

MANN, R. G. ET AL. 1997 MNRAS in press

MOBASHER, B., ROWAN-ROBINSON, M., GEORGAKAKIS, A., EATON, N. 1996 *MNRAS* **282**, 7L.

OLIVER, S. J., ROWAN-ROBINSON M., SAUNDERS W. 1992 *MNRAS* **256**, 15.

OLIVER, S., ET AL. 1997 *MNRAS* in press.

PEARSON, C., ROWAN-ROBINSON, M. 1996 *MNRAS* **283**, 174.

PEI, Y. C., & FALL, S. M. 1995 *ApJ* **454**, 69.

PHILLIPS, A. C., ET AL. 1997 *ApJ* (in press).

PUGET, J.-L., ET AL. 1966 *AA* **308**, 5.

ROWAN-ROBINSON, M., HELOU, G., WALKER, D. 1987 *MNRAS* **227**, 589.

ROWAN-ROBINSON, M. 1992 *MNRAS* **258**, 787.

ROWAN-ROBINSON M. 1995 *MNRAS* **272**, 737.

ROWAN-ROBINSON, M., EFSTATHIOU, A. 1993 *MNRAS* **263**, 675.

ROWAN-ROBINSON M., ET AL. 1993 *MNRAS* **263**, 123.

ROWAN-ROBINSON, M., ET AL. 1997 *MNRAS* in press.

RUSH B., MALKAN M. A., SPINOGLIO L. 1993 *ApJS* **89**, 1.

SAUNDERS, W., ET AL. 1990 *MNRAS* **242**, 318.

SCOVILLE, N. Z., YOUNG, J. S. 1983, *ApJ* **265**, 148.

SERJEANT, S. ET AL. 1997, *MNRAS*.

THRONSON, H., TELESCO, C. 1986 *ApJ* **311**, 98.

WALKER, T. P., STEIGMAN, G., SCHRAMM, D. N., OLIVE, K. A., KANG, H.-S. 1991 *ApJ* **376**, 51.

WILLIAMS, R. E., ET AL. 1996 *AJ* **112**, 1335.

Galaxy counts vs. type for $19^m \lesssim B \lesssim 29^m$, and galaxy formation from sub-galactic clumps

By R. WINDHORST,[1] S. PASCARELLE,[1] S. ODEWAHN,[1] S. COHEN,[1] C. BURG,[1] W. KEEL[2] AND S. DRIVER[3]

[1]Dept. of Physics & Astronomy, Arizona State University, Tempe, AZ 85287

[2]Dept. of Physics & Astronomy, University of Alabama, Tuscaloosa, AL 35487

[3]School of Physics, University of New South Wales, Sydney, NSW 2052, Australia

We review the results from recent parallel and other deep HST surveys. Deep HST/WFPC2 images in U,B,V,I were analyzed using artificial neural network (ANN) classifiers, which are based on galaxy surface brightness and light profiles, and which distinguish quite well between E/S0's, Sabc's, and Sd/Irr for $B \lesssim 27$ mag. We discuss effects from the cosmological SB-dimming and from the redshifted UV-morphology on the classifications, and correct for the latter. The median scale-length at $B \simeq 27$ mag is $r_{hl} \simeq 0''.25$–$0''.3$ (1–2 kpc at $z \simeq 1$–2). Early and late-type galaxies are fairly well-separated in BVI color-magnitude diagrams for $B \lesssim 27$ mag, with E/S0's being the reddest and Sd/Irr's generally blue. We present the B-band galaxy counts for ~ 36 WFPC2 fields as a function of morphological type for $B \lesssim 27$ mag. E/S0's are only marginally above the no-evolution predictions, and Sabc's are at most 0.5 dex above the non-evolving models for $B \gtrsim 24$ mag. The faint blue galaxy counts in the B-band are dominated by Sd/Irr's, and are explained by a combination of a *moderately* steep local luminosity function undergoing strong luminosity evolution plus low-luminosity lower-redshift dwarf galaxies.

Deep WFPC2 images in the medium-band filter F410M (Lyα at $z \simeq 2.4$) yielded 18 faint, compact objects surrounding the radio galaxy 53W002 at $z \simeq 2.390$. These objects appear to be star-forming spheroids smaller than the bulge of a spiral galaxy. They are much smaller ($r_{hl} \simeq 0''.1 \simeq 0.5$–1 kpc) and fainter ($M_V$ $(z = 0) = -21 \rightarrow -17$) than typical galaxies today, and may the building blocks from which many of the luminous nearby galaxies were formed through repeated hierarchical mergers. Parallel F410M images of two other random fields yielded 15 more $z \simeq 2.4$ candidates—as faint and compact as the $z \simeq 2.4$ candidates in the 53W002 field—confirming that there may exist a widespread population of compact, faint Lyα emitting galaxy building blocks at $z \simeq 2.4$, and presumably anywhere in the range $z \sim 1$–4. The difference in the number of objects from each F410M parallel field can be explained if these subgalactic clumps exhibit some level of clustering at $z \simeq 2.4$ on Mpc scales, and suggests that the group around 53W002 is no more than a $\sim 2\sigma$ fluctuation in the density distribution. Deep KPNO 4m F410M imaging and spectroscopy shows that the that the 53W002 cluster stretches over $7'(\sim 5$ Mpc), and may be part of some larger-scale structure at $z \simeq 2.39$.

Deep *HST/PC* images at $\sim 0''.06$ FWHM resolution in BVI—as well as in redshifted Lyα—of the weak radio source 53W002 show several morphological components for this compact narrow-line galaxy at $z = 2.390$: (1) a blue point source (central AGN, $\lesssim 500$ pc) with $\lesssim 20$–25% of the total continuum light; (2) a small, concentrated and somewhat redder inner core; (3) an $r^{1/4}$-like light distribution with colors indicating an overall stellar population age ~ 0.4 Gyr and no large color or age gradient out to $r \sim 9$ kpc; and (4) two small blue clouds roughly aligned with the radio axis and the main stellar population, and separated by a red linear feature, possibly a "dust" lane. The size and shape of the Lyα clouds suggests reflected AGN continuum-light shining through a cone (plus re-radiated Lyα in emission). A recent OVRO interferometric CO-detection on *both* sides of 53W002—and in the same direction as the continuum clouds *and* the radio jet—suggest a star-bursting region induced by the radio jet. Hence, both mechanisms likely play a role in explaining the "alignment effect." We discuss the formation and evolution scenario for 53W002 in context of its surrounding sub-galactic objects.

---

1. Introduction

1.1. *Faint galaxy classifications with HST*

The majority of explanations for the faint blue galaxy (FBG) excess observed in deep ground-based images (e.g., Kron 1982, K82—hereafter references are abbreviated; see bibliography—T88, B92, L93, D94, NW95a) involve Irregular/dwarf populations, or objects with $M_B \gtrsim -17.0$ mag (using $H_0 = 75$, $q_0 = 0.1$ throughout, unless indicated otherwise). The ~0''.1 FWHM resolution provided by HST's Wide Field Planetary Camera 2 (WFPC2) allows the determination of the sub-kpc morphology of distant galaxies—and hence of the population fractions of different galaxy types—over a wide range of epochs. Gr94, D95a, D95b, G95a, C95, O96, A96, and R97 used WFPC2 to study the morphological properties of faint galaxies. D95a showed that the FBG counts for $I_{814} \lesssim 24.5$ mag are dominated by late-type/Irregulars, and suggest that neither a steep local luminosity function (LF), nor strong evolution alone can explain the high observed counts (O96). A96 used a 2-parameter space to assign types for $I_{814} \lesssim 25$ and reached similar conclusions as D95a. In view of these HST results, the development of a rigorous morphological classification scheme is important to study galaxy formation and evolution in an unbiased manner. One promising approach maps multivariate information to morphological types using an Artificial Neural Network (ANN; cf. O92, O95, N95). This technique was applied to other available WFPC2 fields in B_{450} ($\simeq B$), since in B the galaxy counts are steepest and the FBG excess will be most pronounced (O96 & §2).

1.2. *Clues to galaxy formation with HST*

Among the many theories of galaxy formation, the idea that galaxies may form by the accumulation of smaller star-forming subsystems has recently received much attention (NW94; P96b, Low97). Two major lines of evidence lend support to this 'bottom-up' scheme. First, deep HST images and the angular size–redshift (or Θ–z) relation of luminous early-type galaxies (E/S0) and mid-type spiral galaxies (Sabc) indicate that these objects were assembled largely before $z \gtrsim 1$, and that they have been passively evolving since $z \lesssim 1$ (D95a, Mu94). In addition, there is evidence from both HST and ground-based work that the galaxy-merger rate was higher in the past, roughly increasing with redshift in proportion to $(1+z)^m$, with m $\simeq$ 2–3 (B94; C94a; YE95). Taken together, these two points suggest that the luminous galaxies seen today could have been assembled from the merging of smaller systems sometime before $z \gtrsim 1$. The second piece of evidence comes from the excess of blue objects at faint magnitudes which has been found in many deep imaging studies (e.g., T88, NW95a, M95). Although recent spectroscopic studies (E96; G95b; L95) have found many of these FBG's to be at modest redshifts ($z \sim 0.5$), the existence of a significant high-redshift 'tail' to the distribution (CHS95a) raises the possibility that a non-negligible fraction lies at $z \gtrsim 2$. O96 noted that for $B \gtrsim 25$ mag, the FBGs are extremely compact ($\lesssim$ 0''.2–0''.3). This (likely) higher-redshift FBG population, which is dominated by late-type or irregular galaxies (D95a), that have undergone substantial evolution since $z \lesssim 1$, could provide a reservoir of building blocks from which the luminous, present-day galaxies were made at $z \gtrsim 1$. Indeed, to produce the number of giant galaxies seen today, many more objects at $z \gtrsim 2$ are required by most merger-dependent models of galaxy evolution (CHS95a; B92) than we see for $z \lesssim 1$. Recent findings, such as a population of faint, compact ($B \gtrsim 24$ mag) blue galaxies at $1 \lesssim z \lesssim 3.5$ with rather unusual morphologies which suggest dynamical formation processes or mergers (P96b, S96a, CHS95), seem to support this hypothesis.

1.3. *Where and why are the primeval galaxies hiding?*

To resolve the issue of where and why are the long sought primeval galaxies hiding, direct observational evidence is critical. The difficulty is in finding large numbers of faint, and presumably distant, galaxies at or near their time of formation. Many theories of galaxy formation predict that there should exist a widespread population of primeval galaxies at a cosmic epoch somewhere before $z \gtrsim 2$. These primeval galaxies, undergoing an intense initial burst of star formation, should be detectable in several emission lines (e.g., Lyα, Ha, [O-III], etc.). Exhaustive ground-based Lyα searches have been made over the last 15 years for young galaxies with strong continuous star-formation at high redshift (e.g., TDT95), but have not yielded a large number of candidates for several possible reasons. First, even small amounts of dust, if properly distributed, can effectively quench Lyα and UV continuum emission (CK92, K96). Second, these objects may primarily occur at much higher redshifts ($z \gtrsim 5$–10) than have thus far been explored. Third, they might exist in protoclusters or larger-scale structures at high redshifts (although the clustering amplitude is predicted to be lower at high redshifts by CDM theories), causing most narrow-band searches to look in between any such structures (see §3.4). And/or fourth, in a hierarchical formation scenario—in which galaxies were assembled from pieces over a long time interval—the long-sought Lyα emitting primeval galaxies may be out there, but in a large number of small pieces (P96b; see also §3), which would be largely beyond the ground-based flux detection limits. Finding these elusive objects, or a similar population of very compact star-forming systems, may help us to improve our understanding of how exactly galaxies formed.

1.4. *Confrontation of recent HST results with CDM models*

The dominant dissipationless component of dark matter may accumulate by gravitational instability into clumps, the virialized parts of which would become dark matter halos of galaxies-to-be. Galaxies would then form by the cooling and condensation of gas within these halos. Hierarchical clustering currently appears to be the most successful at reproducing many of the properties of the real universe (RHS97), while remaining within the constraints produced by the COBE results. In the typical CDM model, structure grows from the gravitational collapse of small initial density perturbations, from which systems of progressively larger mass merge and collapse to form newly virialized systems. Observational evidence suggestive of hierarchical galaxy formation was presented by P96b, who found a significant number of sub-galactic sized star-forming objects to be gravitationally bound at $z \simeq 2.39$. P96b suggested that these "galaxy building blocks" are actually part of a widespread population existing throughout the redshift range $z \simeq 2$–5. P97 present a Cycle 6 HST project to image four random fields using the same F410M filter as P96b, to see if the faint Lyα emitting candidates exist at $z \simeq 2.4$ in other fields in the sky (§3).

1.5. *How much clustering is allowed at high redshifts and on what scales?*

It is becoming clear that galaxies over a wide range in redshift are more inclined to be in groups or clusters than isolated, giving the term "field galaxy" less meaning than it once had. As such, it is perhaps not surprising to find that in many "pencil-beam" galaxy redshift surveys, often several galaxies are found in a single narrow redshift bin (e.g., the HDF redshift distribution reported by C96 and Cohen et al. this volume). On a slightly larger scale, "spikes" have been found in the redshift distribution of several more extensive galaxy surveys out to $z \lesssim 1$ (B90; LF94), indicative of possibly some sort of larger structure at earlier epochs. What is questionable, according to CDM models, is whether or not any such structure could have existed at much higher redshifts than $z \sim 1$. RHS97 are able to reproduce the observational characteristics of QSO absorption systems at $z \sim 3$ over a wide range of column densities with a hydrodynamical model consisting of subgalactic clumps embedded in "sheet"-like structures and often lying along "ribbons" or "filaments" (see also Ostriker, this volume). Their simulations indicate that their ~ 20 clumps will merge to produce three L^* galaxies by $z = 0$, which agrees remarkably well with the independent observations of P96b, who conclude that their 18 $z \simeq 2.4$ candidates could produce a few L^* galaxies by $z = 0$ (§3). Is it possible that groups, clusters, or even some larger-scale structure existed to some extent at earlier epochs, although at much lower amplitude than that seen at lower redshifts? The random deep Cycle 6 WFPC2 parallel fields in the F450W and F410M filters presented by P97 are a first attempt to address this question (§3.4).

1.6. *Clues from the formation of a weak radio galaxy*

After reviewing some of the specific discoveries by HST on galaxy evolution (§2) and galaxy formation through sub-galactic clumps (§3), it will be instructive to review the formation and evolution of a specific high redshift (weak radio) galaxy (§4). Radio galaxies have played an important role in the study of galaxy evolution, since these were for a long time the only galaxies that could be identified easily at high redshifts. Jet-induced star formation or non-thermal radiation scattered in a reflection cone are the most probable radiation processes in ultraluminous high redshift 3CR and 1 Jy radio galaxies (C90, M93, B97). It is not clear that these processes are universal, and their role needs to be clarified at high resolution for $\sim$ 30–100$\times$ weaker radio galaxies (for $\alpha_{0.178}^{1.41} \gtrsim 1.2$), which may be more representative of what ordinary young galaxies would look like at those redshifts. To test these conjectures, deep multicolor high-resolution HST/PC images (at 0.″0455 pixels) were obtained of the faint, compact, radio galaxy LBDS 53W002, which has *narrow* emission lines at $z = 2.390$ (W91). Its line ratios suggest a weak Seyfert-like AGN, and constrain its non-stellar component to $\sim 35 \pm 15\%$ of its total restframe UV-continuum (W91; W92). Ground-based and pre-refurbished HST *continuum* images (W92) showed some alignment with the radio source axis on 0.″5–1.″0 scales (4.5–9 kpc), which itself is aligned with the much larger ground-based Lyα cloud ($\sim 25 \times 45$ kpc; W91). The PC images were obtained to constrain the relative contributions from 53W002's AGN and its young stellar population, and to examine the relations between these components and its dynamics during the galaxy collapse—whether this occurred as a global halo collapse (cf. ELS62), through the rapid merging of many sub-galactic sized objects (e.g., SZ78, P96b), through jet-induced star-formation (e.g., C90), or some combination thereof.

2. Automated morphological classification in deep HST fields

2.1. *WFPC2 observations, reduction, and ANN classifications*

2.1.1. *Primary and parallel observations in B, V, I, and/or U*

In Cycle 4–5, a single deep 63-orbit dithered WFPC2 field (hereafter the "W02" field) was imaged surrounding the radio galaxy 53W002 at $z = 2.39$, with 12 orbits in both V_{606} & I_{814} (Driver et al. 1995; D95a), plus 24 orbits in B_{450} (Odewahn et al. 1996; O96), and 15 orbits in the medium-band filter F410M ($\lambda_{eff} \simeq 4090$ Å or Lyα at $z \simeq 2.36 \pm 0.06$; Pascarelle et al. 1996b; P96b). The 48-orbit stack of BVI images is shown as a color Plate in W97. Calibration followed D95a, D95b & O96, using the best available WFPC2 super-sky flats. These deep WFPC2 images provided a 5σ point source detection limit for B_{450}, $V_{606} \lesssim 28.4$, & $I_{814} \lesssim 27.6$ mag, and a 1σ surface brightness (SB) sensitivity of $B, V \lesssim 27.5$ & $I \lesssim 26.7$ mag arcsec^{-2} (see W97 & D95a). O96 compare the galaxy counts in the 53W002 field to those from *UBVI* images in the HDF, which reach $\sim$ 0.8–1.2 mag deeper. In Cycle 6, 29 shallower B_{450} fields were collected in the WFPC2 *B*-band Parallel Survey (S. Cohen et al. 1997, Co97), as well as from several other Archival WFPC2 B_{450} and I_{814}-band fields from the Medium Deep Survey (R97), and from other Archival sources to get better statistics at the bright-end ($19 \lesssim B \lesssim 25.5$ mag, see Fig. 1–3).

2.1.2. *Data processing and catalog generation*

All images in each filter were spatially registered, CR-clipped, and averaged to create high S/N composites (see W94a, W97), from which object catalogs were computed (cf. D95a, O96). Potential objects were located and measured using aperture magnitudes grown-to-total (cf. P96b, O96). These $(U), B, V, I$ catalogs were used to produce a final catalog by selecting any object present in *any* filter with $S/N \geq 3$. Hence, very red or very blue objects are not excluded. Surface photometry was carried out with the automated image analysis package MORPHO (O95), giving ellipse fits over a range of isophotal levels, elliptically averaged SB-profiles, and a set of type-dependent photometric parameters (O96).

2.1.3. *ANN classifications*

O96 assigned a morphological type to each galaxy in several independent ways. First, 173 galaxies were classified by eye with $U_{360} \lesssim 26$ mag, 372 with $B_{450} \lesssim 26.5$ mag, and 542 with V_{606}, $I_{814} \lesssim 26$ mag, obtaining results mutually consistent within ± 2 (rms) Hubble classes, and scale errors $\lesssim 10\%$ on the 16-step Hubble scale. All objects with $B_{450} \lesssim 27.5$ mag were classified using an ANN analysis of the photometric parameters (O92, O95). ANN's are systems of weight vectors, whose component values are established through an iterative learning algorithm such as back-propagation. They take as input a linear set of patterns, and produce as output a numerical pattern encoding an object classification. O95 show that this pattern can be used to assign a confidence value to the estimated galaxy type. Fundamental to the development of such pattern classifiers is the existence of a large sample of examples (or "training set"). A set of ANN classifiers was developed for the I_{814} and V_{606} images using the morphological I_{814} eye-ball classifications as a training set, as described by D95b, D95b, & O96. Each classifier inspected both the digital image and the SB profile for each galaxy in each filter. Hence, some profile information was incorporated in these eyeball estimates. With the mean visual types from our samples in U, B, V and/or I, ANN galaxy classifiers were developed using six primary parameters: the SB at the 25% and 75% quartile radii, the mean SB within the effective radius and within an isophotal radius, as well as the slope and intercept of a linear fit to the SB-profile (in $r^{1/4}$ space). These ANN's are based on the observed galaxy SB-profile, but *not* on color, *nor* on scale-length. ANN classifiers were also developed to classify galaxies in their appropriate *rest-frame UBV* filters.

2.1.4. *Tests of the classification systems as a function of wavelength and redshift*

Since for faint galaxies ($B \gtrsim 24$ mag) the expected median redshift is $z \gtrsim 0.7$ (KK92; C96), the B_{450} filter could look back into the rest-frame mid-UV. At higher redshifts (and fainter fluxes), V_{606} and even I_{814} may have the same problem. One therefore has to be concerned how the cosmological $(1+z)^4$ SB-dimming and the wavelength-dependence of the rest-frame (UV-)morphology may affect the ANN classifications. As discussed by O96, little or *no* systematic change was observed in the eyeball morphological types between the U, B, V & I filters (see Fig. 3 here and Fig. 1 of Burg et al., this volume). This suggests that the properties on which our human classifiers rely (morphology and light profile) are more uniform with spectral bandpass than we would expect (cf. dJ94). Alternatively, it could suggest that the median redshift for the faint galaxy sample is somewhat smaller ($z \sim 1.0$ for $B \gtrsim 25$) and the redshift distribution rather much wider than expected (SLY97), so that cosmological effects possibly do not yet differ greatly between the $B, V,$ & I filters. To investigate this further, a set of ANN classifiers was designed that is valid for *rest-frame* filters. For this, the following steps were used: (1) the likely redshift range was estimated for each B_{450} interval using the B redshift distributions of KK92 and E96, and the models for Gunn g $\lesssim$ 26 mag of NW95a; (2) using the known $UBVI$ filter responses, these redshift ranges were used to estimate the most likely *rest-frame* central wavelength for each galaxy image; (3) a series of *rest-frame* U, B, and/or V images was selected, and the corresponding rest-frame ANN's were derived. For galaxies with $B_{450} \lesssim 26$ mag, these *rest-frame* ANN types were compared to the $UBVI$ eyeball classifications, as well as to the types predicted by the single-filter ANN classifiers. The *rest-frame* predictions for $B \lesssim 27$ (heavy black lines in Fig. 3) do not differ significantly from the mean eyeball estimates (color-coded or shaded lines), nor from the single-filter ANN estimates (open data points). The four-color counts in Fig. 3 show *no* obvious evidence that the shorter wavelength filters result in a larger fraction of late-type galaxies. The W02 and HDF samples yield consistent classes as a function of flux (see Fig. 3), yet have S/N that differs by 1.2–1.5 mag at a given flux. For $B \lesssim 27$ mag both methods of ANN classification are therefore reliable, but for $B \gtrsim 27$ significant discrepancies may exist between the I_{814} or V_{606} ANN's and the *rest-frame* ANN's, indicating that z_{med} may be $\gtrsim 2$ and that the uncertain far-UV morphology becomes important. Further tests of systematic errors are discussed by O96. About 15% of all galaxies with $B_{450} \lesssim 25.5$ mag and classified by the ANN systems as late-type spirals or irregulars were classified as merging systems by the human classifiers. For $B_{450} \gtrsim 25.5$ mag, the number of merging systems increases to 35% of those ANN-classed Sd/Irr galaxies. Hence, a good fraction of "late-type" systems as classified by the ANN are actually merger morphologies, so this class is designated as Sd/Irr+M systems.

2.2. *WFPC2 B-band results and discussion*

2.2.1. *The scale-length–magnitude relation*

Fig. 1 shows the scale-length r_{hl} (averaged over the BVI filters in which the object was detected to increase S/N) versus B-magnitude for both the W02 and the HDF galaxies. The three main galaxy classes resulting from the *rest-frame* ANN are indicated with different colors or shades, as are the formal 50% completeness limits for both fields computed for nearby RC3 galaxies of the same types (see O96). Also plotted are the best fit models to the *local* r_{hl} vs. M_B relations for RC3 mid-type spirals and ellipticals for a median $r_{hl} = 4.7$ and 3.2 kpc, respectively, assuming model redshift distributions (cf. CC92, NW95a) and K-corrections (BC93, D95a & b) as a function of galaxy type (using $q_0 = 0.5$; but in one case also for $q_0 = 0.01$). Generally, r_{hl} is smallest at a given flux for E/S0's. The median scale-length at $B \simeq 27$ mag is $r_{hl} \simeq 0''.25$–$0''.3$, which corresponds to $\sim$ 1.2–2.5 kpc for the redshift range $z \sim 0.5$–2.5. These values appear to be smaller than the characteristic scale-lengths of mid to late-type galaxies measured locally, as well as those measured with HST for $z \lesssim 0.8$ (or $B \lesssim 23$ mag; cf. Mu94). A possible explanation is that the faint galaxy population becomes progressively more dominated by lower luminosity and therefore smaller late-type objects at fainter fluxes (cf. D94, D95a, D95b & §3).

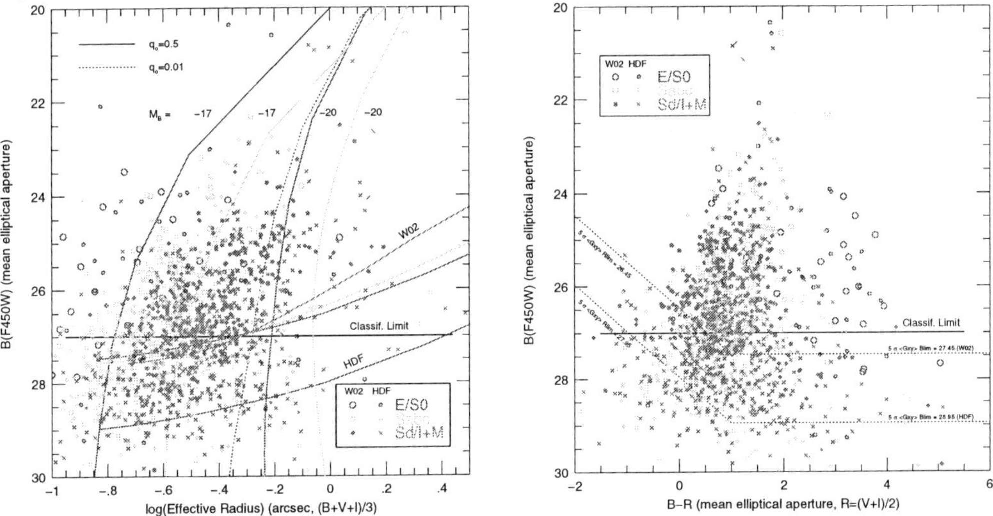

FIGURE 1 AND 2. (1) [Left] The B_{450} magnitude vs. half-light radius r_{hl} for all classified galaxies in the W02 field and the HDF. Symbols indicate membership in the categories E/S0, Sabc, & Sd/Irr+M classified by the *rest-frame* ANN (§2.1.4). The solid almost-horizontal curves indicates the B_{450}-detection limits. The almost-vertical curves indicate how the median scale-length of RC3 galaxies of given Hubble type and M_B decline towards fainter magnitude (see O96). Galaxies classified as E/S0 have on average smaller observed scale-lengths than Sabc's, which are generally smaller than Sd/Irr galaxies. The observed scale-lengths reach a median of r_{hl} $\simeq$ 0.″25–0.″3 at $B \simeq 27$ mag; (2) [Right] The $(B - R)$ vs. B_{450} color-magnitude diagram for all classified galaxies, as in Fig. 1. The two (slanted) dashed lines indicate the completeness limits in (R and) B for the two fields. The same ANN galaxy classes are indicated as in Fig. 1, which are reliable for $B \lesssim 27$ mag. The reddest objects at $B \lesssim 27$ mag are mostly classified as E/S0's and Sabc's, and do not increase as rapidly towards the completeness limits, suggesting that the formation of early-type galaxies was largely complete by $z \sim 1$. The late-type+merging population galaxies increases rapidly to the detection and classification limit ($B \sim 27$ mag; see Fig. 3d). A color version of Figs. 1–3 is given by O96.

2.2.2. *The color–magnitude diagram*

The total B_{450}-magnitude versus $(B - R)$ color is shown for ANN each type in Fig. 2, where "wide R" $\equiv (V + I)/2$ was used to increase S/N in the colors. Fig. 2 becomes increasingly incomplete at $B \gtrsim 26.0$ mag for the *bluest* W02 objects, because of the *red* detection limit at $R \simeq 26.5$ mag, as indicated by the slanted dashed lines (the HDF limits are ~ 1.5 mag fainter due to its finer substepping and the r_{hl} vs. B-mag relation of Fig. 1). Only objects detected in all three filters are plotted here, but the fraction of objects seen only in B_{450} (at $B \lesssim 27$ mag) and not in V_{606} and/or I_{814}—as well as those seen only in V_{606} & I_{814} ($\lesssim 26$ mag) and not in B_{450}—is $\lesssim 5\%$. Of significance is the clear segregation between the early and the late type galaxies for $B \lesssim 27$ mag, even though *no* color information was used by the ANN classifier. E/S0's are almost without exception the reddest galaxies at any flux level, and Sd/Irr+M's are generally blue, at least down to the formal detection limit. Mid-type spirals (Sabc's) have colors with large dispersion and about as blue as Sd/Irr's, as expected from the (slightly) different colors measured for these populations locally and the K-corrections for galaxies with ongoing star formation at $z \simeq 1$–2 (BC93). The increase in the relative fraction of late-type and merging galaxies (Sd/Irr+M) in Fig. 3 towards fainter B_{450} is quite remarkable. They exist essentially down to the very detection limit, despite their larger relative scale lengths (Fig. 1). The redder objects—mostly classified as E/S0 galaxies—do not increase

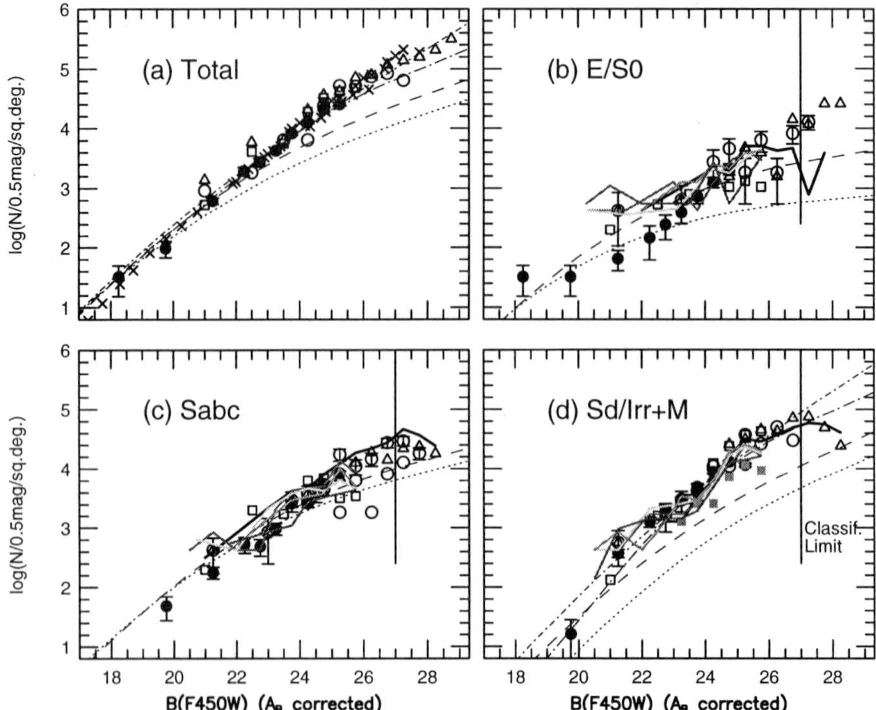

FIGURE 3. (a) B-band counts from the W02 field (open circles), the HDF (open triangles), the WFPC2 B-band parallel survey (filled circles), and other Archival B_{450} fields (open squares), as well as previous ground-based B_J-band counts (crosses). Panels show: (a) all galaxies, (b) E/S0's, (c) early-type spirals (Sabc's), and (d) late-type/Irregulars Sd/Irr+M's. The Merger (M) fraction is given by filled blue or shaded squares. Plotted are mean types from: (1) our *visually* classified images (color-coded or shaded lines); (2) the *single-filter* ANN types (open data points), and (3) from the *rest-frame* ANN types (**heavy** lines and filled circles). All models are plotted as thin curves (see §2.2.3). A color version of Fig. 3 is shown in O96.

as rapidly towards the completeness limits ($B \lesssim 27$ mag) as the late-types. Since early-type galaxies do not generally have the largest scale-lengths (Fig. 1) nor the lowest SB, this is probably not due to large differences in their redshift distribution—and therefore due to a larger cosmological SB-dimming. Instead, it may suggest that the formation of early-type galaxies was largely complete by the median redshift corresponding to $B \simeq 26$–27 mag (probably $z \gtrsim 1$, see NW95a), so that their counts should indeed be converging for $B \gtrsim 26$ mag (see O96 and Fig. 3b).

2.2.3. *The galaxy counts versus type for $B \lesssim 27$ mag*

Fig. 3 gives the differential B number counts for the different morphological types (E/S0's, Sabc's, Sd/Irr+M's) in the ~36 WFPC2 B_{450}-fields discussed in §2.1.1. The plotted counts are based on mean types from the *visually* classified $U, B, V,$ & I images (lines coded by colors or shades), from the I_{814} $+V_{606}$ *single-filter* ANN types (data points), and the *rest-frame* ANN types (heavy solid lines). Following D95a, D95b & O96, the B number counts were modeled for the three main morphological types separately. The best fit non-evolving models in Figs. 3b–3d are based on the local LF's of Ma94 (long-dashed curve) and Lo92 (dotted curve). Fig. 3b shows that the B-band counts of E/S0's follow the predictions for passively or mildly evolving models for $B \lesssim 24$ mag, and for $B \gtrsim 24$ mag the observed counts are at most 0.3 dex higher than these models.

Fig. 3c shows that Sabc's are consistent with these models for $B \lesssim 24$ mag, and are at most 0.5 dex higher for $24 \lesssim B \lesssim 27$ mag. Hence, a scenario invoking strong luminosity evolution is *not* required for the early-type galaxies out to $B \lesssim 27$ mag ($I_{814} \lesssim 25$ mag or $z \sim 1$), suggesting that their formation was largely complete by $z \sim 1$. From a spectroscopic survey for $z \lesssim 1$, L95 deduce a similar lack of evolution for early-type galaxies as found in D95a and O96. The morphological HST studies can push this work now another 3 magnitudes fainter than can be done spectroscopically from the ground.

Fig. 3d shows that *non*-evolving models are inadequate to explain the high WFPC2 B number counts for the Sd/Irr+M population. The fraction of visually classified Mergers are plotted as blue or filled squares. The fraction of mergers rises strongly for $B_{450} \gtrsim 23$ mag, with up to one third of the late-type systems possibly being mergers. For $B_{450} \gtrsim 22$ mag, the number counts of the *classical* Sd/Irr galaxies from both the ANN and visual types are well above the nonevolving models. It is clear that the FBG excess is dominated by galaxies with late-types, with a non-negligible fraction of merging morphologies. A *non-evolving* dwarf-rich population fits the FBG counts quite well [steep LF with $\alpha = -2.0$ (short-dash-dot) in Fig. 3d], but is inconsistent with the redshift distributions observed for $B \lesssim 24.5$ mag (cf. PD95), and therefore cannot be the only explanation. Consistent with the I_{814}-band counts of D95a, the Sd/Irr+M B_{450}-counts in Fig. 3d is best described by an *evolving* Sd/Irr+M population with a *moderately* steep local LF [starburst with $\Delta Lum = 1.5$ mag at $z \approx 0.5$ and a local Marzke LF (long-dash-dot), with the high normalization of D95b], that underwent significant luminosity evolution since $z \lesssim 1$ (cf. PD95). The late-type galaxy population therefore must have evolved strongly and/or be dwarf-dominated, and causes most of FBG excess.

In such a scenario, a non-negligible fraction of the FBG's with $25 \lesssim B \lesssim 29$ could be at $z \simeq 1$–3. In §3, we argue that these objects may have been the reservoir of building blocks from which the luminous early-type galaxies were formed through repeated merging—a process that must have been largely completed by $z \simeq 1$ for the early-type galaxies (Fig. 3b & c), with perhaps some residual epoch-dependent merger rate (cf. B94). The fact that 20–35% of the galaxies classified as late-type by the ANN appear to be merging systems—with that fraction increasing at fainter magnitudes—tends to support this possibility. With the completion of ~ 100 B-band parallel fields in Cycle 6–7, we hope to extend the B_{450} counts with sufficient statistics at the bright end, covering the entire range $18 \lesssim B \lesssim 28$ mag, thereby significantly constraining the "LF normalization problem" (Ma94) and the evolution for each population *separately*.

3. Subgalactic clumps at $z \simeq 2.4$ & implications for galaxy formation

3.1. *Deep F410M surveys for faint Lyα emitting candidates at $z \simeq 2.4$*

Pascarelle et al. (1996b; P96b) report evidence in support of the 'bottom-up' hypothesis, in the form of the detection of 18 significant compact emitters in deep HST images in the medium band filter F410M—likely Lyα at $z \simeq 2.4$ (see §3.2)—with spectroscopic confirmation thus far for eight of these subgalactic-sized clumps at $z \simeq 2.391 \pm 0.004$ with the MMT and the KPNO 4 m. Deeper Keck spectroscopic follow-up of the faintest $z \simeq 2.4$ candidates is presented by Armus et al. (1997; A97). These may be the building blocks from which present-day galaxies were made. In this section, we provide further evidence that galaxies, or the small star-forming subsystems from which they formed, tended to exist in small groups or proto-structures in the early Universe.

The initial discovery of a group of $z \simeq 2.4$ candidates was made from ground-based photometry with a medium-band (150 Å-wide) filter centered at 4130 Å (Lyα at $z \simeq 2.39$)

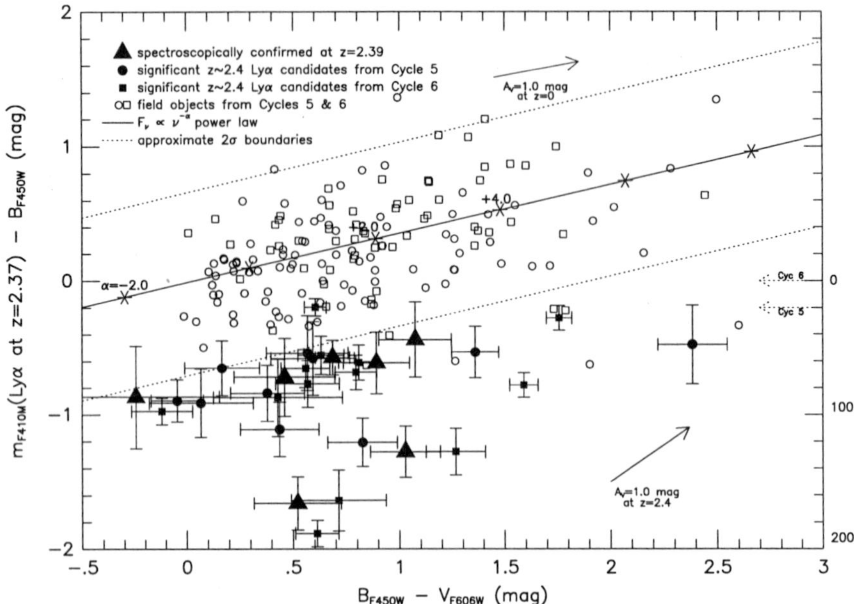

FIGURE 4. (F410M–B_{450}) versus (B_{450}–V_{606}) color-color diagram from 51 WFPC2 orbits on both the Cycle 5 53W002 field (circles, P96b) and the two Cycle 6 parallel F410M fields (squares, P97). The solid line shows the expected relation for a featureless power-law ($F_\nu \propto \nu^{-\alpha}$) labelled with values of α, around which most of the general field objects (open symbols) are distributed. Approximate 2σ error boundaries for the WFPC2 photometry are indicated by the dotted lines. Error bars are plotted only for objects $\gtrsim 2\sigma$ below the power-law line (filled symbols), which are significant candidate Lyα emitters at $z \simeq 2.39$. The five large filled triangles are spectroscopically confirmation of P96b at $z \simeq 2.39$. Approximate implied rest-frame Lyα equivalent widths are indicated on the right-hand vertical axis. Most $z \simeq 2.4$ candidates have $W_{Ly\alpha} < 100$ Å.

in the field surrounding the weak radio galaxy 53W002 at $z = 2.390$ (W91), yielding two other candidates at the same redshift (P96a). Fortuitously, the existence of a nearly identical medium-band filter on HST (F410M, centered at 4100 Å or Lyα at $z \simeq 2.4$) allowed the same observations to be conducted with WFPC2 in Cycle 5 at much higher sensitivity and spatial resolution than could be achieved from the ground. The HST is able to achieve nearly the full point-source sensitivity gain, which is lacking in ground-based data, because of the characteristically small sizes of the $z \simeq 2.4$ candidates (Figs. 1 & 5a). These $z \simeq 2.4$ candidates are labeled in the color Plate of P96b, which covers about 2.5 × 2.5 arcmin of the sky or ~ 0.7 × 0.7 Mpc at $z \simeq 2.39$.

To confirm these results, two long parallel observations were made with WFPC2 using the F410M, F450W, & F814W filters in Cycle 6 (out of three fields scheduled thus far). The first parallel field was at 21^h–$5°$, and the second at 16^h+$82°$, each consisting of seven orbits (about 4 × 2400s in F410M, 2 × 1500s in F450W, and ~ 800s in F814W for color information). Because of the more limited number of orbits, proper image averaging and CR-rejection is crucial, and was done with a custom-written IDL routine to deal specifically with a small number ($n \leq 4$) of low-signal images (Co97). Photometry was done as in W91, utilizing user-input apertures and a sky box that is interpolated underneath the object aperture by fitting a sloped plane to the pixels unaffected by faint neighbors (P96b). Approximately 40 objects were detected simultaneously in the three filters in each parallel field, and 115 in the deeper 53W002 field. Each photometric "curve of growth" was individually examined, and apertures were adjusted to give the best possible total magnitudes, while maintaining the *same* size aperture in the three filters for each object. Compared to the interactive package of O96, sky-estimates were produced consistent within 0.07% and aperture magnitudes grown-to-total consistent within 0.05 mag (with an rms $\simeq 0.22$ mag for each algorithm).

We believe that the discovery of this $z \simeq 2.39$ group is not likely strongly biased by targeting the weak radio galaxy 53W002, because the field galaxy counts in this region are consistent with those in several other randomly selected fields (Ca95, D95b, O96), and the surface density of radio sources at this flux level (~ 5 mJy at 8.4 GHz) is high enough that one such source would be found in every few WFPC2 fields (W93). Despite its steep spectral index, 53W002 is $\sim 30\times$ weaker than most of the luminous 3CR sources, which are known to cluster strongly, and therefore its presence is likely to have less of an effect on the surroundings.

3.2. *Details of the WFPC2 F410M search for Lyα emitting $z \simeq 2.4$ candidates*

The HST data are presented in the color-color diagram of Fig. 4, which shows photometry for 115 objects detected simultaneously in the Cycle 5 broad-band WFPC2 filters V_{606} and B_{450}, and in the medium-band filter F410M. Results from the two Cycle 6 HST F410M fields are also plotted in Fig. 4. (For consistency with the Cycle 5 F410M field, the $(B_{450}–V_{606})$ colors were interpolated from the $(B_{450}–I_{814})$ colors using the effective central wavelengths of the three broad-band filters involved, following Ho95b). As with the ground-based data, the F410M passband is contained entirely inside the B_{450} filter, so that proper continuum subtraction is possible. At $z \simeq 2.39$, these bands sample the emitted wavelengths of 1200–2100 Å, where the spectra of young star-forming galaxies are relatively flat (W91). The solid line in Fig. 4 shows the expected relation for objects with a flat power-law spectrum ($F_\nu \propto \nu^{-\alpha}$) across these three adjacent filters. Indeed most of the general field objects are distributed around it. There is a group of 18 objects which are located $\gtrsim 2\sigma$ below this line, showing that they have significant emission in F410M, which is likely Lyα emission at $z \simeq 2.39$, as argued by P96b and below. Eight of these $z \simeq 2.4$ candidates are spectroscopically confirmed with the MMT and the KPNO 4m (P96a, P96b, Co97, K97), out of ten objects for which spectra of high enough quality were obtained, indicating this method of finding compact $z \simeq 2.4$ candidates is at least $\gtrsim 70\%$ reliable. Apart from the three brighter $z \simeq 2.4$ objects, which were easily seen in ground-based photometry (P96a) and contained weak active galactic nuclei (AGN), most of the sample has Lyα emission with average restframe equivalent widths (estimated from the F410M photometry) of only 40–50 Å—more typical of ionization arising purely from star formation (cf. S96a, S96b).

Two open circles in the region below the lower 2σ error boundary populated by the $z \simeq 2.4$ candidates were spectroscopically confirmed *not* to be at $z \simeq 2.39$, but are stars (M stars can turn out to be apparent F410M-emitters given their complicated spectra around 4000 Å). Another ~250 objects were detected in V_{606} and B_{450} (with $B_{450} \lesssim 26.0$ mag), but not in the F410M image. They have 2σ lower limits in (F410M–B_{450}) $\gtrsim -0.2$ mag and are not plotted here. For $26.0 \lesssim B_{450} \lesssim 27.5$ mag, the B_{450}-band sample is $\gtrsim 90\%$ complete (O96), but the underexposed F410M image does not provide useful upper limits for such faint objects. The Cycle 6 fields are shallower, and have fewer objects, but have similar fractions of lower limits in F410M–B_{450} ($\gtrsim 0.0$ mag). These limits are indicated by the dotted arrows at the right axis of Fig. 4. Both the observed and the restframe ultra-violet reddening vector expected at $z \simeq 2.39$ are indicated by the arrows, and suggest that the few reddest candidates may have a visual absorption of $A_V \lesssim 2$–3 mag. Most of the compact $z \simeq 2.4$ Lyα emitting candidates, however, have rather blue colors and are likely not very reddened internally.

The differential volume element at $z \simeq 2.39$ is $\sim 20\times$ larger than at $z \simeq 0.097$, making it less likely that F410M traces [O II] at $z \simeq 0.097$ than Lyα at $z \simeq 2.4$. Any other emission lines between 1216 Å and 3727 Å strong enough to be detected (e.g., C-IV or Mg II) arise from objects known to have much lower surface densities (e.g., powerful luminous radio galaxies or QSOs). No other strong emission lines exist in this range for star-forming objects (K96), implying that most of the 18 significant candidates are likely at $z \simeq 2.39$. The reliability of the 18 $z \simeq 2.39$ candidates (Fig. 4) is further strengthened by the fact that three of the spectroscopically confirmed members are at the lower boundary in the (Lyα_{410}–B_{450}) color of the general field population.

3.3. *Faint blue sub-galactic clumps*

3.3.1. *Sizes*

Fig. 5a shows the size distribution for the $z \simeq 2.39$ candidates from both the Cycle 5 (shaded, P96b) and Cycle 6 (open, P97) fields in F410M. Most $z \simeq 2.4$ candidates are very compact, with half-light radii $r_{hl} \lesssim 0\farcs1$–$0\farcs3$ or $\lesssim 0.5$–1.5 kpc at $z \simeq 2.4$. Their typical $r_{hl} \lesssim 0\farcs2$ is smaller than the typical WFPC2 galaxy scale-lengths at these faint magnitudes ($\lesssim 0\farcs3$–$0\farcs4$ at $B \lesssim 27$; see Fig. 1 & O96). Several of the Cycle 6 $z \simeq 2.4$ candidates fall into the r_{hl} $=0\farcs3$–$0\farcs35$ bin, likely due to the lower signal-to-noise of these images and the corresponding tendency to overestimate their true sizes. The $z \simeq 2.4$ candidates, whose cores are in many cases barely resolved by the HST, are typically detected at least 3–4 scale lengths out in the deeper Cycle 5 images, suggesting that here the scale-length measurements are more robust (see Fig. 6a, which also shows a typical stellar profile for comparison). Due to the compactness of these objects, they do not suffer the full cosmological $(1+z)^4$ SB dimming as truly extended objects do. (Their wings of course do, which is why any disks may have disappeared into the background noise of the WFPC2 CCD; cf. Fruchter et al. 1997, this volume; see also P96b and §4.3 below). To derive a mean intensity profile with the greatest dynamic range, intensity-weighted composite images of 14 compact and isolated candidates were produced in each filter, *assuming* that the candidates are to first order all similar in shape and size at $z \simeq 2.39$ (cf. G96). The total effective exposure times of the image stacks are 14×5.7 hours in F606W and F814W and 14×16 hours in F450W, and have the best SB sensitivity. To measure the scale-length of the mean observed profile in Fig. 6a, model profiles were convolved with an empirical PSF taken from a star in the images. The B_{450} and F410M data were almost properly sampled, owing to 0.5 pixel substeps between individual exposure sequences, and the measurements were done on interleaved mosaics on 0.5-pixel centers. In each filter, the profiles are better fitted by an $r^{1/4}$-law than by a disk-like exponential over a range of almost five magnitudes in SB. The continuum scale-lengths of the candidates have a mean value of $0\farcs11$ ($\simeq 0.5$ kpc at $z \simeq 2.39$) and are very similar in BVI, showing no dependency on restframe wavelength below ~ 4000 Å.

The question arises as to whether or not one is seeing the full extent of these $z \simeq 2.4$ candidates. The K-correction for young spectral-energy distributions at $z \simeq 2.39$ could have compensated for at least some of the cosmological $(1+z)^4$ SB dimming (W91). Unlike the initial burst of star formation that may take place in ellipticals and the bulges of spirals, disk star formation rates can be almost constant with time (K89). Disks can, therefore, actually brighten toward decreasing redshift for a significant time, eventually reaching an approximately steady level until gas exhaustion. However, with the exception of 53W002 itself, which already has at $z \simeq 2.39$ a $r^{1/4}$-like profile with a relatively large scale-length (Figs. 5a & 6b), suggesting a massive early-type galaxy (W92, W97), and was by selection included in this WFPC2 field, none of the other candidates are yet fully assembled massive ellipticals or spirals. Their scale-lengths are quite comparable to those

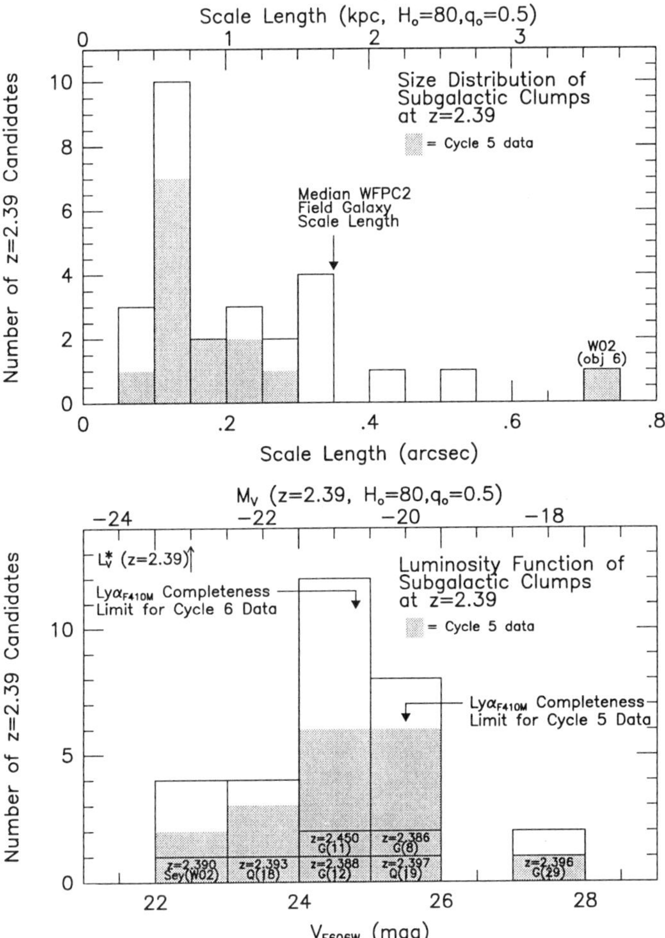

FIGURE 5. (a) [Top] Histogram of WFPC2 continuum scale-lengths for the significant Lyα emitting $z \simeq 2.4$ candidates in Cycle 5 (shaded, P96b) and Cycle 6 (white, P97). The scale-lengths were averaged over BVI and are typically $\lesssim 0''\!.1$–$0''\!.3$; (b) [Bottom] Histogram of WFPC2 V_{606} luminosities for the significant Lyα emitting $z \simeq 2.4$ candidates in both the Cycle 5 & 6 fields. The upper axis indicates their absolute magnitude distribution. The (evolving) L^* value is indicated by the upward arrow. AGN contributions were subtracted in a few cases as indicated (cf. W92, P96b). The labelled boxes represent the five spectroscopically confirmed $z \simeq 2.4$ candidates of P96b, with Seyfert, Quasar, Galaxy-like objects indicated.

of bulges in local spirals, which range from 0.2–4 kpc with a type-dependent median close to 1 kpc for S0–Sbc's. Given that the average value of the bulge-to-disk scale-length ratio of nearby late-type galaxies is $\sim 0.07 \pm 0.04$ (Co96), these $z \simeq 2.39$ candidates may be subgalactic-sized (compact) and young (blue) spheroids, possibly representing the bulges of young galaxies that have not (yet) developed significant disks around them, and/or disks that are reduced in brightness in the HST images by the severe SB-dimming (§4.3).

3.3.2. *Luminosities*

Fig. 5b shows the luminosity distribution of all significant F410M $z \simeq 2.4$ candidates from Cycles 5 (shaded, P96b) and 6 (open, P97). The implied luminosities at $z \simeq 2.39$ for all candidates range typically from $M_V \simeq -23$ to -18 mag, based on the stellar population models, age estimates, and K-corrections from W91. With BVI photometry,

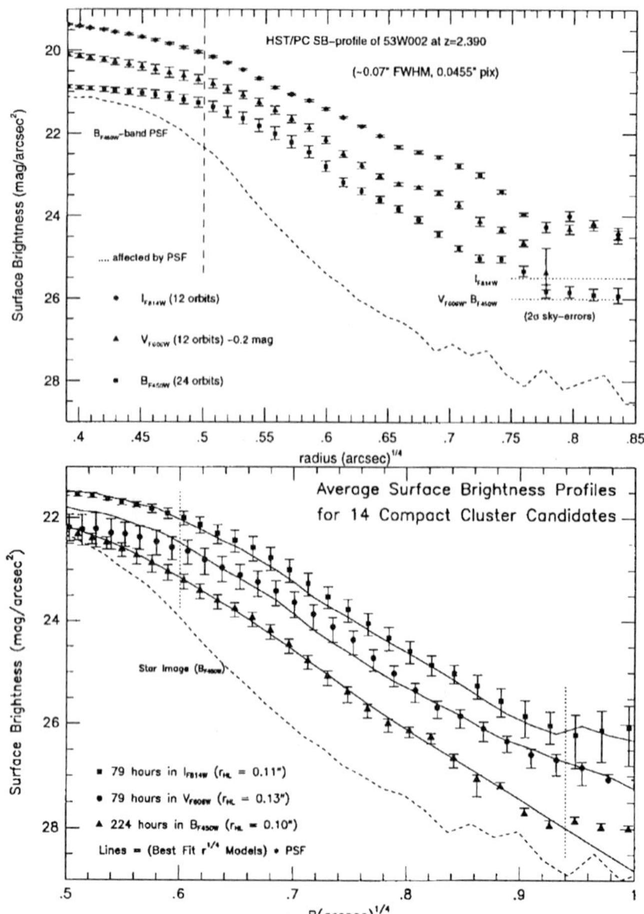

FIGURE 6. (a) [Top] Average light profiles vs. radius (in $r^{1/4}$ units) of the 14 $z \simeq 2.39$ candidates without neighbors in BVI (see P96b). The profiles are reliable for radii between the two vertical dotted lines, but are affected by the WFPC2 PSF etc. at smaller radii, and by sky-errors at larger radii. The stellar PSF is plotted in B_{450} (short dashes) to illustrate that these compact clumps are indeed barely resolved by HST. Solid lines indicate best-fit $r^{1/4}$ model profiles; (b) [Bottom] Light profiles in BVI for the extended component of the radio galaxy 53W002 at $z = 2.390$ in the quadrant that is relatively free of the blue clouds (§4.2 & W97). The central AGN and blue clouds were subtracted in each filter, using a faint star as PSF (W97). Short dashes represent the 25% AGN component in B, which affects the profile for $r^{1/4} \lesssim 0.5$ ($r \lesssim 0''.06$). The horizontal dotted lines indicate the 2σ sky-subtraction errors, which affect the profiles for $r^{1/4} \gtrsim 0.8$ ($r \gtrsim 0''.4$). For $0.5 \lesssim r^{1/4} \lesssim 0.8$ ($0''.06 \lesssim r \lesssim 0''.4$), 53W002's profile is consistent with an $r^{1/4}$-law, but at large radii it is affected by the blue clouds (§4.3.4).

the K-corrections are straightforward, provided that redshifts are known to be at $z \simeq 2.39$ either from spectroscopy or from the (F410M–B_{450}) colors. Subtraction of any contributions from an AGN was done following W92. The (evolving) absolute magnitude of an L^* galaxy at $z \simeq 2.4$ is $M_V \simeq -23$ mag, assuming that there would have been ~ 2 mag of luminosity evolution since $z \simeq 2.39$ for a typical starburst $\sim (0.3\text{–}0.5) \times 10^9$ years earlier, as the candidates' unreddened blue colors suggest (Fig. 4 & P96b).

The $z \simeq 2.4$ sample is not complete for $V_{606} \gtrsim 25.5$ mag ($M_V \gtrsim -20$ mag), as shown by the downwards arrows, which indicates the shallower completeness limit of the underexposed F410M images, compared to B_{450} or V_{606}. Given this completeness limit, the true luminosity distribution of the $z \simeq 2.4$ candidates thus likely continues for $M_V \gtrsim -20$ mag, and the initial (luminous) mass spectrum of this group of $z \simeq 2.4$ candidates could be quite steep. The Cycle 5 and Cycle 6 F410M data sets contain mostly $z \simeq 2.4$ candidates with $M_V \simeq -20.5$ to -21.5 mag. Assuming that most significant F410M emitters are at $z \simeq 2.4$ and given the K-corrections discussed above, these are sub-L* luminosities. Therefore, if indeed their stellar populations are young, most of the $z \simeq 2.4$ candidates have luminosities of $\lesssim 0.1$–$1\ L^*$, and so possibly had only $\sim 10^9$–$10^{10}\ M_\odot$ processed into stars at $z \simeq 2.39$. For these parameters, the free-fall time expected for these clumps is $\sim (2$–$4) \times 10^8$ years, long enough that the original population of short-lived O and early B stars are gone, but comparable to the age of the dominant stellar population (late B and early A stars). Hence, there was indeed enough time for their mass distributions to settle into regular $r^{1/4}$-like light profiles.

3.4. *Discussion of surveys for faint Lyα emitting candidates at $z \simeq 2.4$*

3.4.1. *Assembling giant galaxies from fragments?*

The redshift at which P96b find their $z \simeq 2.4$ candidates is close to the peak redshift of star formation in disk galaxies of $z \simeq 1$–2 (M96). These sub-galactic-sized objects likely existed throughout the entire redshift range $z \simeq 1$–4, and could have grown into the luminous giant galaxies (ellipticals and early-type spirals) that we see today through the process of repeated hierarchical merging (cf. NW94). An epoch-dependent merger rate proportional to $(1+z)^{2.5}$ would result in a time integral of ~ 10–20 mergers of compact objects from $z \sim 3.5$ to $z = 0$ (B94). This process of repeated merging would have to be largely complete by $z \sim 1$ if ellipticals and early-type spirals were in fact formed from such subgalactic clumps, as their HST morphological counts show little evolution since $z \lesssim 1$ (Fig. 3; D95, O96, Mu94). The 18 $z \simeq 2.4$ candidates could thus have merged to produce a few L^* galaxies today, given that objects separated by several hundred kiloparsecs will have many crossing times in the interval $z \simeq 2.39$ to $z = 0$. The total V_{606} luminosity for the 18 $z \simeq 2.4$ candidates is $M_V \simeq -24.7$ mag at $z \simeq 2.39$ (Fig. 5b), which agrees rather well with the combined luminosities of a few typical L^* galaxies today of $M_V \simeq -24.3$ mag (again including the expected ~ -2 mag from their K-corrections plus evolution, W91).

In models of hierarchical galaxy formation, the luminous galaxies we see today were built up through the repeated merging of smaller protogalactic pieces. Such subgalactic clumps appear to have been found at $z \simeq 2.4$ by P96b and P97. In addition, there have been recent reports of high redshift, "normal" star-forming galaxies with possibly larger scale-lengths and higher luminosities (e.g., H96, G94, S96a, F97, Low97, Tr97). About 40–50% of the S96a,b & Low97 samples are Lyα emitters, the remainder are Lyα absorbers or show no Lyα at all, and may be more extinguished by dust and possibly be a more evolved population of objects that has gone through more generations of star-formation. The relatively stronger Lyα emitters like the subgalactic clumps of P96b and the objects of H96, Y96, and F97 may have had fewer generations of O stars, and so fewer supernovae to produce significant dust to resonantly scatter the Lyα light. These findings suggest that galaxies likely formed over a large range of redshift rather than at one special time, and likely according to several different formation scenarios. If a large percentage of galaxies did in fact form hierarchically, one would expect to find their building blocks at high redshifts ($z > 1$–2), consistent with the large number of star-forming objects currently being found at such high redshifts. Many studies seem to point toward a merger rate that was significantly higher in the past, following $\propto (1+z)^m$ with $m \simeq 2$–3 (B94), and suggesting that there must have been many more subgalactic clumps around at earlier epochs in order to produce the observed local luminosity function. Taken together with the fact that most luminous early-type galaxies appear to have been "in place" by $z \sim 1$ (Mu94, D95a, O96, L95, H97), these two points imply that a large fraction of today's luminous ellipticals and early- to mid-type spirals formed at $z \gtrsim 1$, possibly from the merging of these sub-galactic sized objects. Indeed, it has been shown that disks can be regenerated (or generated) after such mergers (HM95).

3.4.2. *How much large scale structure could there be at $z \sim 2.4$?*

Currently, eight $z \simeq 2.4$ candidates have been spectroscopically confirmed with the MMT or the KPNO 4 meter out of the 17 candidates surrounding 53W002 (P96b). In total, there are two negative confirmations. Deeper Keck spectroscopy is presented by A97 on the F410M fields. If one assumes this same $\gtrsim 70\%$ success rate for the Cycle 6 parallel fields, for which spectroscopy is in progress, then the yield will be about two $z \simeq 2.4$ candidates (out of the three candidates found) from the 21-hour field and about eight $z \simeq 2.4$ candidates (out of the 12 candidates found) from the 16-hour field. The latter is consistent with the 53W002 field, while the former has a much lower density, given the relative exposure times. The difference in the numbers of candidates from the two (approximately) equally-deep Cycle 6 F410M fields is only significant at the $\sim 2\sigma$ level. There are two possible reasons for this discrepancy. First, it may be an indication that the 53W002 region is *not* representative for the general field, but that it could be a $\sim 2\sigma$ fluctuation compared to other random fields, based on the numbers estimated above (see P97). Given that the Cycle 6 parallel F410M fields are only $\sim 40\%$ as deep as the original 53W002 observations, one would expect to find 11–12 $z \simeq 2.4$ candidates in any random part of the sky if these objects are indeed a widespread population. If the existence of the weak radio source or the other two $z \simeq 2.4$ AGN surrounding 53W002 (see §4) make the W02 field not representative of the universe at $z \simeq 2.4$, then one should find very few (if any) significant $z \simeq 2.4$ candidates in any of the random F410M searches, not consistent with the number of $z \simeq 2.4$ candidates found in the F410M parallel fields. If there were large-scale structure at high-redshift as has been found at $z \lesssim 1$, one would expect to find ~ 10 $z \simeq 2.4$ candidates in a random field whose F410M line-of-sight is approximately tangential to a sheet or passing through the intersection of two sheets or a particularly rich sheet, as possibly in 53W002 field. Conversely, if one happens to choose a field whose F410M line-of-sight passes in between sheets, one should find very few (although not necessarily zero) candidates, consistent with the observed statistics. The F410M parallel fields of P97 allow both of the latter possibilities.

As noted in P96b, the confirmed 53W002 candidates show a remarkably small velocity dispersion (at $z \simeq 2.391 \pm 0.004$ with $\sigma_v \simeq 286$ km s^{-1}, corrected to $z = 0$), despite the much larger width of the F410M filter. Lyα emission could have been seen in the F410M filter from objects at $z \simeq 2.28$–2.45, although some dropoff in transmission would occur for $z \lesssim 2.30$ and $z \gtrsim 2.42$. This implies that the subgalactic clumps may have existed to some extent in groups or proto-clusters at high redshift, or else we should have seen a larger number of objects at other redshifts ($z \neq 2.391 \pm 0.004$) *inside* the F410M filter. Currently, only one object was found (out of eight) which has a discrepant redshift ($z = 2.451$), surprisingly far out into the red wing of the filter.

The 16-hour field could be a factor of 4–10× more dense than the 21-hour field, far in excess of any effect from its 1.2× greater F410M exposure time. The variation in number density among the Cycle 5 and 6 fields, if real, is indeed suggestive of some kind of structure (e.g., groups, clusters, or "sheets"). This structure may have been hit with the F410M filter "face-on" in the 53W002 and 16-hour fields, but we are perhaps looking "in between" any such major structures in the 21-hour field. A picture is beginning to develop which is quite consistent with that of RHS97, in which luminous galaxies at $z = 0$ are broken up into several individual Lyα emitting clumps at higher redshifts, and are embedded in sheet-like structures, often lying along filaments or ribbons where these sheets intersect (cf. RHS97; Ostriker, this volume). These sheets appear to be group or sub-cluster size, and may represent the preferred environment for high-redshift objects.

The extremely narrow redshift distribution for the spectroscopically confirmed objects thus far ($z \simeq 2.391 \pm 0.003$), as compared to the much broader width of the F410M filter ($\Delta z = 0.12$) leads one to question why objects at other redshifts in the range $2.30 \lesssim z \lesssim 2.42$ were not seen in the MMT or KPNO spectroscopy samples thus far. There are no major ground-based night-sky lines hampering the spectroscopic follow-up around 4100 Å. There is good agreement of the space density ($\simeq 0.027$ Mpc^{-3}) and velocity dispersion ($\simeq 286$ km s^{-1}) of the objects with measured MMT & KPNO redshifts with models based on the Press-Schechter theory (see Fig. 1 of WF91)—which predict the abundance of dark-matter halos as a function of redshift—suggesting that galaxies may have existed preferentially in groups or "proto"-clusters at $z \simeq 2.4$. In redshift surveys out to $z \lesssim 1$ (B90; Co94); LF94), and most recently to $z \gtrsim 1$ (CHS95a, C96, Cohen et al., this volume) with the Keck 10-meter telescope, often more than a few objects were found at very similar redshifts in these small survey fields, supporting the possible existence of clumpiness in the redshift distribution in small fields out to $z \gtrsim 1$. In addition, groups or other large structures have recently been found at redshifts comparable to the P96b group (F96). It is possible that the "spikes" or "frothiness" observed in the redshift distribution of small fields out to $z \sim 1$ may have also existed at some level out to $z \sim 1$–3, although the clustering amplitude at higher redshift would have to be lower according to most CDM models. Recent deep KPNO 4m imaging in the F410M filter (K97) and spectroscopy (Co97) shows that the that the 53W002 "cluster" stretches over $7'$($\sim$ 5 Mpc), with two more spectroscopic confirmations at $z \simeq 2.39$ (again out of an entire allowed redshift range of $2.33 \lesssim z \lesssim 2.46$!), and may be part of some larger-scale structure at $z \simeq 2.39$. Further investigations are needed to see if such large scale structure at these high redshifts are inconsistent with all CDM models.

3.4.3. *Relation to Lyα PG searches and chain galaxies*

It is possible that the P96b sub-galactic clumps, and possibly a large fraction of the elusive primeval galaxies, may be hiding as many of the compact FBGs (Fig. 1), and have escaped proper recognition from the ground until now because they are so small and faint. The P96b group of faint, compact star-forming subsystems at $z \simeq 2.39$ could thus be the galactic building blocks long sought after by proponents of the 'bottom-up' galaxy formation models. Recent ground-based narrow-band Lyα imaging surveys (TDT95) may have missed any such groups or structures because the filters used were generally too narrow ($\sim$ 30 Å) to detect one by chance, unless some previous knowledge of the redshift already existed from, for example, one or more known quasars or radio galaxies. The WFPC2 *medium*-band filter F410M would sample typically one or two such redshift structures (discussed in §3.4.2) at $z \simeq 2.39$ (and is not limited by sky-noise as most ground-based images are). Together with the extreme compactness of the P96b $z \simeq 2.4$ candidates, this explains the higher detection rate of faint Lyα emitting candidates with HST/WFPC2 in F410M.

Comparing the P96b findings to those of CHS95a, one finds that about 5% of the faint blue objects in the 53W002 WFPC2 field can be classified as 'chain galaxies,' which is lower than the 20%–50% estimated in their sample. We believe that such objects are likely the short-lived ($\lesssim 3 \times 10^8$ years out of a total of $\lesssim 5 \times 10^9$ years available for z

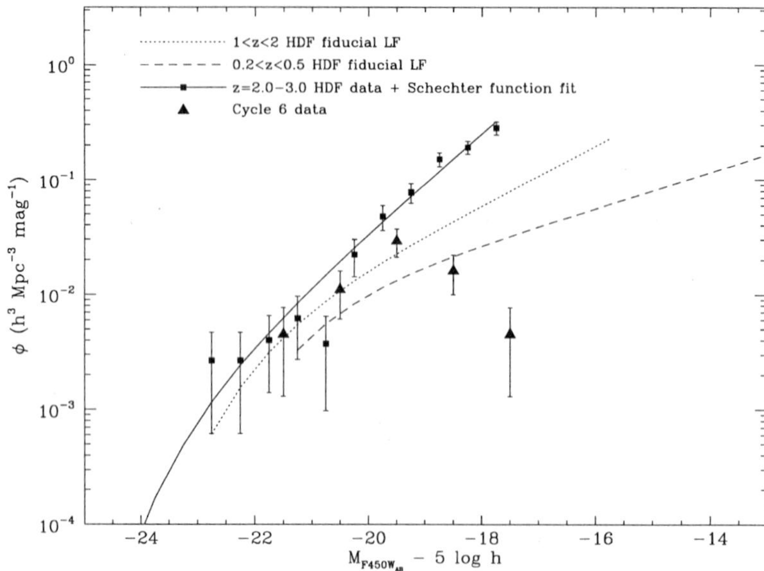

FIGURE 7. Luminosity function of the Cycle 5 & 6 Lyα emitting candidates (filled triangles) together with the LF at $z \sim$ 2–3 in the HDF from SLY97 (filled squares). The data were converted to AB magnitudes, and space densities were calculated using the comoving volume defined by the depth of the F410M passband and the size of the WFPC2 field at $z \simeq 2.4$. The solid line is the best fit Schechter HDF LF of SLY97. Dashed and dotted lines are fiducial photometric LFs of SLY97 at $z \sim$ 0.2–3. The Cycle 6 F410M Lyα LF becomes incomplete for $M_B \gtrsim -18.5$ mag.

$\gtrsim 1$) merger events among the many faint blue sub-galactic clumps, in which the gas is drawn out of the merging objects during the encounter (NW94; MH95).

3.4.4. *The luminosity function of Lyα $z \simeq 2.4$ emitting candidates*

The space density of the confirmed $z \simeq 2.4$ candidates from P96b is $\sim$ 0.19 Mpc^{-3} (assuming a gravitationally-bound group). If one assumes a similar velocity dispersion for the Cycle 6 fields, the space densities would be $\sim$ 0.064 Mpc^{-3} and $\sim$ 0.29 Mpc^{-3} for the 21-hour and 16-hour fields respectively. Making no assumptions about the existence of any structure or groupings at $z \simeq 2.4$, the entire width of the F410M filter is used to derive a space density of subgalactic clumps for the 53W002 field (again, assuming a $\gtrsim 70\%$ success rate) of $\sim$ 0.041 Mpc^{-3}. In the same way, space densities are derived for the Cycle 6 fields of $\sim$ 0.0058 Mpc^{-3} and $\sim$ 0.026 Mpc^{-3} for the 21-hour and 16-hour fields respectively.

The LF of the $z \simeq 2.4$ candidates was derived from the candidates in all three Cycles 5 and 6 fields together (Fig. 7), and binned according to their F450W luminosities following SLY97. The $z \simeq 2.4$ Lyα LF is quite steep ($\alpha \sim$ 1.8–2.0), despite the small-number statistics involved. SLY97 show a similar LF from *photometric* redshifts in the HDF that changes with redshift, in agreement galaxy redshift surveys at $z \lesssim 1$ (e.g., L95), showing both a brightening and steepening with increasing redshift up to $z \sim 3$, a result expected in hierarchical models of galaxy formation in which merging plays an important role in removing the subgalactic clumps at lower redshifts. The LF at $z \simeq 2.4$ of P97 is consistent with that of the HDF for $2 < z < 3$ (Fig. 8 of SLY97), given the following two caveats from the F410M selection: (a) the completeness limit of $M_{F450W_{AB}} \simeq -19.5$ at $z \simeq 2.4$ is imposed by the shallower F410M observations; and (b) given these completeness limits in F410M and B_{450}, the F410M selection allows to only detect significant Lyα emitters, *not* absorbers (the KPNO F410M images of K97 are deeper compared to B, and *do* allow to detect Lyα absorbers). According to the Keck spectroscopy of S96a,b and Low97, $\sim$ 40–50% of the higher luminosity U-band drop-out objects at z = 2–3 have (weak) Lyα in emission. If we assume the same fraction of Lyα emitters for the fainter $z \simeq 2.4$ candidates of P96a & P97, we may correct the F410M LF upwards by $\sim$ 0.3 dex for the fact that it was selected by Lyα in emission only, which would put the F410M points exactly on top of the SLY97 LF at $z \simeq$ 2–3.

3.5. *Future prospect for space-based Lyα surveys*

Because of the existence of only two medium-band filters in WFPC2 suitable for finding compact emission-line objects (F410M and F467M), we are limited in the number of redshifts we could search ($z \simeq 2.4$ and $z \simeq 2.85$). The Cycle 5 data offered a unique opportunity to examine the $z \simeq 2.4$ epoch, but similar observations with the F467M filter on HST can be carried out at $z \simeq 2.85$. The scientific return from a modest investment of exposure time in the HDF in the F410M and F467M filters would be considerable, since all the broadband *UBVI* data is in hand. This would be an excellent direct determination of the space density of $z \simeq 2.4$ and $z \simeq 2.85$ subgalactic clumps. A more extensive medium-band filter set on the Next Generation Space Telescope will be needed to find these objects at $z \simeq$ 2–7. Indeed, more random observations with WFPC2 in F410M are necessary to confirm these initial findings at $z \simeq 2.4$, and to better determine the luminosity function of the $z \simeq 2.4$ clumps. We also note that these studies can no longer be done with HST if the WFPC2 gets turned off after installation of the ACS in 1999, since the ACS does not have the required medium-band filter set, or ramp-filters with a large enough FOV to do the exercise.

4. Deep PC imaging of a compact radio galaxy at z = 2.390

4.1. *WFPC2 observations and astrometry*

The WFPC2 observations used to study 53W002 at the highest possible resolution were the same as those discussed in §2.1.1, except that 53W002 itself was put inside the PC. The PC noise is largely limited by read-noise and dark-current, and large-scale gradients in the flat-fielded images are $\lesssim$ 3% of sky (cf. W94a, Windhorst et al. 1997; W97). For the interpretation below, one needs to know the exact location in the PC-images of 53W002's highest-resolution VLA source, which has 0$''$.15 FWHM (W91). W97 show that, after correcting the WFPC2 positions for a small systematic VLA$_{FK5}$–HST$_{GSSS}$ offset, the high resolution VLA 8.44 GHz (W91) and 15 GHz (S97; S97) images coincide with 53W002's optical center—and likely its AGN (§4.2.1)—to within $\lesssim$ 3 PC pixels.

4.2. *Structure of the PC images of 53W002*

Here we discuss the components of this young galaxy revealed in the PC images:

4.2.1. *The maximum AGN contribution*

To constrain the maximum possible AGN contribution to 53W002's continuum, a scaled PSF was subtracted from its image core, using the faint red star "S", noted by W92. The maximum possible point source that can be subtracted from 53W002's core without making its central flux negative is 25 $\pm$ 2% of the total light in B, 21 $\pm$ 3% in V, and 20 $\pm$ 2% in I. The upper error boundaries of these fractions are firm, so that the AGN contribution to 53W002's total continuum is definitely $\leq$ 30% (W91, W92),

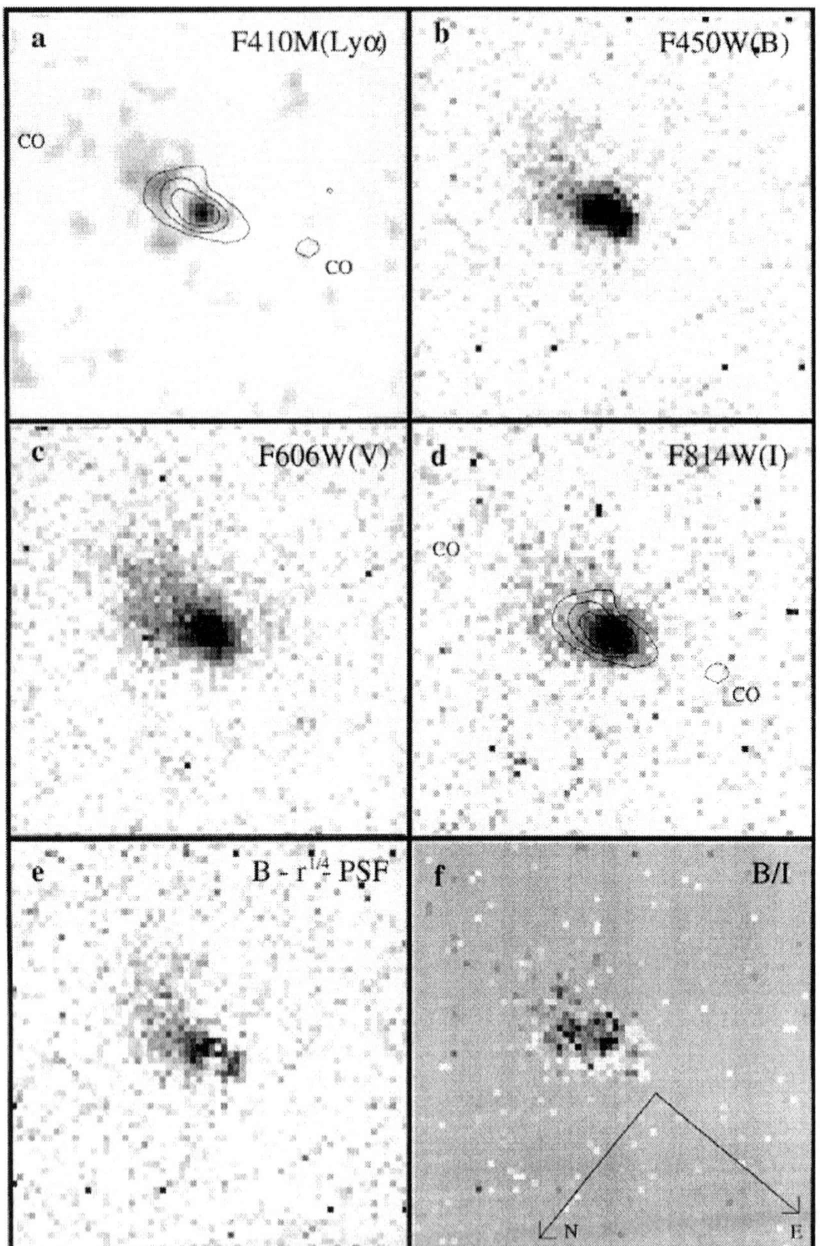

FIGURE 8. Grey scale images of 66 × 66 pixel sections (≃3.″0×3.″0) of the PC-exposures on 53W002 (see W97): (a) 15 × 2700s orbits in $Ly\alpha_{410}$; (b) 24 × 2400s orbits in B_{450}; (c) 12 × 1700s orbits in V_{606}; (d) 12 × 1700s orbits in I_{814}; (e) equal to (b) after subtraction of a central point source (with 25% of the total flux) and of the best fit $r^{1/4}$-profile (Fig. 6b); (f) the $(B-I)$ color image [=(b)/(d)]. The VLA 8.4 GHz contours of W91 are superimposed and the CO peaks of S97 are indicated in (a) & (d), as explained in W97. The AGN is located at each panel's central pixel. *Note:* the low SB and "arclike" feature of the Lyα cloud in (a); the dominant $r^{1/4}$-like profile and brighter one-sided blue cloud in (b)–(d); the brighter *and* fainter blue clouds in (e); and the blue (dark) central AGN and the reddish (white) linear feature (a "dust lane"?) separating both blue clouds in (f). The radio source covers most of the blue clouds plus the Lyα "arc", suggesting jet-induced starformation; the remainder of the Western continuum and Lyα cloud may be AGN light seen in reflection.

but the lower boundaries are soft. The AGN contribution could be $\lesssim 15\%$ of its total continuum, which would in Fig. 6b produce light-profiles that are straighter in $r^{1/4}$-space. This light is contained within 0.″06 FWHM or $\sim$ 500 pc and has blue colors $(B - I \simeq 0.06$ mag), as expected for an AGN at $z \simeq 2.4$. The moderately weak radio source 53W002 ($S_{1.4} = 50$ mJy, $\log P_{1.4} = 27.5$ $W\ Hz^{-1}$) thus has a similarly modest optical-UV AGN contribution: $V^{AGN} \simeq 24.3$, $M_V^{AGN} \simeq -21.8$ (power-law K-correction with $\alpha \simeq 1.0$; W91).

4.2.2. *The inner resolved continuum "core"*

To address the symmetric extended component, only the "clean" quadrant between the blue clouds discussed in §4.2.4 are fit (see Fig. 8a–8f). The central parts of the galaxy in F450W cannot be simultaneously fit by a single $r^{1/4}$ or exponential law. Comparing these profile-fits to the PSF shows that a small additional central light distribution (with $r_{hl} \sim$ 0.″05) is required, containing about half the flux of the nuclear point source, but with redder colors ($(B - I) \simeq 0.76 \pm 0.1$). The distinctions among this small extended component, the nuclear point-source (§4.2.1), and the more extended $r^{1/4}$-like distribution (§4.2.3) are artificial to some extent, reflecting a radial structure more complex than our simple library of pieces (see §4.2.4).

4.2.3. *The remaining $r^{1/4}$-like profile*

After subtraction of the central unresolved AGN component (§4.2.1), the underlying galaxy can be measured only in the quadrant *between* the two blue clouds (§4.2.4). W97 concentrate on a region south of the nucleus, where nearly a full 90° quadrant is clear of these contaminating sources. Details of the elliptical profile fitting technique and its errors are given by W92, W94b, Mu97, & Sc97. The BVI light-profiles are shown in Fig. 6b. Most of the deviation from an $r^{1/4}$-law at $r^{1/4} \gtrsim 0.75$ ($r \gtrsim$ 0.″32) occurs in B & V, and is due to the faint blue cloud leaking into the uncontaminated quadrant, and not only due to sky-subtraction errors, which affect the profile for $\mathrm{SB}_{BVI} \gtrsim 25.5$–$26.0$ mag arcsec^{-2} (see Fig. 6b). The BVI light profiles (Fig. 6b) follow an $r^{1/4}$-like profile closer than an exponential disk, although an early-type galaxy with a bulge-to-disk ratio $\gtrsim$ 3–5 cannot be ruled out from the PC data. The best $r^{1/4}$ fit has (a/b) $= 1.25 \pm 0.1$ and $r_{hl} \simeq$ 0.″20 $\pm$ 07 in B & V and $r_{hl} \simeq 0.27 \pm 0.05$ in I (or 1.8–2.5 kpc). The total flux of this symmetric component is well constrained at $27 \pm 3\%$ of the total. Its position angle ($PA \sim 110°$) is uncertain, but consistent with the orientation of the aligned clouds ($PA \sim 95°$), which appear to be separated by a redder feature (§4.2.4 & Fig. 8f). 53W002's average color is $(B - I) \simeq 1.3$. Its $(V - I)$ color ($\simeq 0.70$ mag) is less contaminated by the blue cloud. Within the errors, both colors are relatively constant with radius, so that any color gradient must be small for $r \lesssim$ 1.″0 ($\lesssim 0.3$ mag across the PC image). W91 & W94b present 12-band (Lyα $UBVRIgriJHK$) photometry for 53W002 and surrounding objects, and W94b, Mu97, & Sc97 discuss spectral evolution model fits to these and similar color-redshift data to constrain stellar population ages (defined as the onset of the major, currently visible starburst). Fig. 8f suggests that, with the exception of the region possibly affected by a "dust lane", 53W002 is not enormously reddened by dust (§4.2.4). Following these spectral evolution models, the colors of the symmetric component of 53W002 –if interpreted as coming from stars only—would suggest a stellar population with an average age of $\sim$ 0.4 Gyr (W91, W94b), which is of the same order as its dynamical time scale (§4.3). The lack of discernable color gradient does not allow us to distinguish whether 53W002 formed through a sudden, global halo collapse (cf. ELS62) or through rapid merging of many sub-galactic sized units (e.g., SZ78, P96b). Any color gradient between the current PC and NICMOS images would help decide between these scenarios. Below we show that there may be other processes triggering the (star)formation of 53W002.

4.2.4. *The nature of the blue clouds*

The residuals after removing the symmetric pieces described in §4.2.1–4.2.3 trace the aligned blue "clouds" nicely. The larger continuum cloud to the west—in the direction of the extended Lyα distribution (Fig. 8a & W91)—is quite extended and vaguely triangular (see Fig. 8e and the high contrast insert in the color Plate of W97). It peaks $0''.45$ west of the nucleus and extends $\gtrsim 1''$ from the core with an opening angle of about 45° (Fig. 8b), and has a brighter "arc" which is dominated by Lyα emission (Fig. 8a) about $0''.6$ from the core. On the opposite side is a very small blue object—possibly a "counter-cloud"—elongated perpendicular to the nucleus-cloud direction and confined within $0''.2$ from the core (Fig. 8e)—at the very limit of the HST/PC resolution. With the exception of the "arc" at the edge of the larger cloud, the aligned components and the core are essentially free of Lyα line-contamination, as seen in Fig. 8a–8f. The nucleus is a weak Lyα source, contributing only about 20% of the total Lyα flux (Fig. 8a). Smoothing of the medium-band Lyα image shows emission associated with the arc in the western aligned cloud, and otherwise very diffuse Lyα emission only (Fig. 8a). The arc at the outer edge of the western cloud contributes as much as 93% of the B-band light. This is the only feature seen in the Lyα image with significant contrast against the rest of the galaxy in terms of equivalent width. These two blue clouds could represent:

(1) Reflection of the AGN-light shining through a cone, including Lyα and C-IV emission lines from gas lit up by the cone. The asymmetry in size and flux between eastern and western clouds (Fig. 8b & 8e) may represent obscuration or genuine physical differences. The fact that we can see a much larger and somewhat symmetric ground-based Lyα cloud (compare our Fig. 8a to Fig. 3 of W91) argues against obscuration, although this extended Lyα gas could be mostly in front of—or away from—any dust, and in part unrelated to the AGN. Two other $z \simeq 2.40$ objects—Nos. 18 & 19 of P96b—also are AGN with continuum reflection cones (see the color Plate of W97), but with a relatively stronger AGN component compared to the surrounding material (at the 50–80% level of the total flux; P96b). The presence of these reflection cones implies the existence of a substantial amount of gas and/or dust well beyond the optical extent of these galaxies.

(2) A star-bursting region induced by the weak radio jet. Compared to the spectral evolution models described by W94b, the much bluer colors of the cloud—if caused by stars—would suggest a star-bursting region $\lesssim 10^8$ years old. This is similar to the typical radio source lifetime, but younger than the galaxy's dynamical time scale. A color map—produced by matching the registration, sampling and resolution of the B_{450} & I_{814} images—shows the color contrast between the inner and outer regions of 53W002 (Fig. 8f). A red, almost linear feature appears to separate the smaller cloud from the nucleus. In nearby galaxies, this would suggest an organized dust lane. If this feature is indeed a dust lane, it would have a differential optical depth between 1300 and 2400 Å ranging from $\tau \simeq 0.75$–1.5 averaged over the resolution limit. This is rather mild by standards of present-day galaxies. The visual or blue extinction expected for this amount of far-UV extinction would be easy to miss in nearby radio galaxies, so the total amount of dust required is not excessive for objects like 53W002, which might be chemically younger and correspondingly more metal-poor (KW91, K97).

4.3. *Discussion and conclusions*

The implication of the HST/PC components of 53W002 are discussed here in combination with clues from other recently published ground-based and HST data.

4.3.1. *Nature of the alignment effect in 53W002*

A recent interferometric OVRO image of 53W002 in redshifted CO (J 3–2)—which has 3″ FWHM—provides an important clue to the nature of the alignment effect (Scoville et al. 1997; S97). CO was detected $\sim$ 2–3″ away on both sides of 53W002's AGN, *and in the same direction as both blue HST clouds and the extended 8.4 GHz radio source, but not perpendicular to this direction.* Since the CO extends further in both directions than the two aligned blue clouds and the currently visible extended radio source (see Fig. 3 of S97), the CO was likely deposited there by physical processes related to the jet—possibly through a previous incarnation of the jet. Since carbon and oxygen had to be formed in massive stars, jet-induced star-formation thus likely played a role at some stage in the evolution of 53W002. Its overall $r^{1/4}$-like stellar population is extended in the same direction as the radio source, so that the jet possibly triggered a non-negligible fraction of 53W002's mass to form stars in these two directions. As long as this all happened within a few$\times 10^8$ years, there would have been just enough time for the stellar population to settle into a $r^{1/4}$-like profile (vA82). Together with the continuum and Lyα morphology of the blue clouds (§4.2.4 & Fig. 8), it thus appears that *both* reflection cones from an AGN and jet-induced star-formation are responsible for the alignment effect in 53W002.

The structure of 53W002 may be compared to the HST images of powerful radio galaxies of Lo95 and B97. From a set of eight 3CR radio galaxies at $z \approx 1$, they suggest that the morphologies change systematically with (projected) radio source size, and interpret this as evidence for jet-induced star formation in the aligned component. The weak radio galaxy 53W002 shares both properties with these powerful 3CR sources—a compact component identified with the AGN in the center of an extended starlight distribution, and a pair of clouds aligned with the projected radio axis. Contrary to many of these luminous 3CR sources, we can already see a fairly relaxed symmetric distribution of starlight centered around 53W002's nucleus. In hindsight, it is perhaps surprising to find the aligned component to be important even at these low radio powers, but the continuity of structure with the powerful radio galaxies is striking, and required the extra resolving power of the WFPC2/*PC* in the *B*-band to observe in detail.

4.3.2. *53W002's gas+dust content, its surrounding cluster, and possible evolution*

The measured Lyα fluxes and broad-band *UBVRIgriJHK* colors of 53W002 constrain its dust absorption ($A_V \lesssim 0.2$ mag) and star formation rate (SFR; W91, W94b). Similar arguments have been made by P96b for the other 17 surrounding blue $z \simeq 2.4$ candidates. The SFR of 53W002 is of order $\sim$ 100 M$_\odot$/year (W91), and $\gtrsim$ 5–10$\times$ less for the other $z \simeq 2.40$ candidates (P96b). The total stellar mass of 53W002—integrated over its assumed exponentially declining SFR—is $\sim 1.8 \times 10^{11}$ M$_\odot$ (W91). The $r^{1/4}$-like light-profile of 53W002 suggests that at $z = 2.39$ the object had already converted a non-negligible fraction of its gas mass into stars—rather efficiently on a $\sim$ 0.4 Gyr time-scale—suggesting a young early-type galaxy. The OVRO CO-flux of 1.5 Jy/km s^{-1}implies $\sim 2.1 \times 10^{11}$ M$_\odot$ in gas around 53W002 alone (S97). The velocity widths of the CO clouds are $\sim$ 250 km s^{-1}(HWHM), extending $\sim$ 1″.5 or $\sim$ 13 kpc on each side of 53W002 in the direction of the blue clouds *and* of the extended radio source, and possibly indicating a forming rotation curve (S97). These numbers imply an enclosed Keplerian mass of 1.5–3.8 $\times 10^{11}$ M$_\odot$, consistent with its total stellar mass above (W91). Taken together, this means that 53W002's H_2(+CO) gas-mass could be $\sim$ 30–60% of its total (luminous+gas+dark) matter.

What could this mean for the evolution of 53W002? If all this gas settled into disk stars within a few free-fall times ($\sim$ 1 Gyr), 53W002 could evolve into an mid-type spiral galaxy today (with B/D-ratio $\sim$ 0.5), or into an earlier-type galaxy ($B/D \gtrsim 1$) if most of the gas was used up during the initial starburst, and/or if a substantial fraction of the gas remained neutral (as seen in some nearby ellipticals and merger remnants, cf. H95a). The small velocity dispersion ($\lesssim$ 300 km s^{-1}) in P96b's group of $z \simeq 2.4$ candidates with measured redshifts, and the small area ($\lesssim$ 1 Mpc^2) over which the 17 sub-galactic $z \simeq 2.4$ candidates surrounding 53W002 are seen, suggests that many of these objects will likely merge into a few larger galaxies during the next half Hubble time after $z = 2.4$. Hence, while 53W002 may have formed as a $r^{1/4}$-dominated galaxy during a relatively quick and sudden collapse that started at $z \simeq 3$ (or $\sim$ 0.4 Gyr before $z = 2.4$)—possibly induced by star-formation along its radio jet—it appears to be also developing a massive disk at $z \simeq 2.4$ (S97). This disk may have completely settled $\sim$ 1–2 Gyrs later (or at $z \simeq 1.5$), but possibly be destroyed again during future mergers (at $z \lesssim 1.5$) with the surrounding sub-galactic sized $z \simeq 2.4$ candidates (P96b), so that 53W002 may end up as a massive early-type galaxy today. 53W002 may thus provide important clues as to how the luminous nearby early-type (radio) galaxies could have formed and evolved.

5. Summary

In this review, we presented results from parallel and other deep HST surveys. In §2, we summarized HST's morphological evidence for the evolution of different galaxy types for 19 $\lesssim B \lesssim$ 27.5 mag: early types (E/S0's and Sabc's) evolved the least since $z \lesssim$ 1, and late types (Sd/Irr's+Mergers) most rapidly. With better restframe ANN's, these results can be extended in the HDF and in future deep NICMOS and ACS images for $B \lesssim 29$ or H $\lesssim$ 25 mag. This will cover the bulk of galaxies in the range $z \sim$ 1–3. In §2 we summarized the WFPC2 F410M medium band searches for faint compact Lyα emitting objects at $z \simeq 2.4$. The unexpected success of finding these objects provides an important confirmation of CDM theories, leads us to the question whether or not all high z star-forming systems are obscured by dust, and provides clues as to where the long sought Lyα emitting primeval galaxies may be hiding—in many small pieces. The existence of any larger scale structure at $z \simeq$ 2–3 clearly begs further investigation. In §3, the radio galaxy 53W002 provided important clues as to how a luminous early-type galaxy seen today may have formed. For this object at least, this may have started with jet-induced star-formation (with an aligned cone associated with its AGN), soon followed by the formation of an $r^{1/4}$-like bulge and within a few Gyr by a massive disk, which may get destroyed later by subsequent mergers with the surrounding sub-galactic objects.

We thank Lee Armus, Nick Scoville, and Dave Burstein for helpful discussion, Doug VanOrsow, Harry Ferguson, Ray Lucas, and the ST ScI staff for their continuous help in these projects, and ST ScI for an inspiring HDF Symposium. Spectroscopic observations were obtained at the Multiple Mirror Telescope Observatory, a joint facility of the University of Arizona and the Smithsonian Institution. Kitt Peak National Observatory is operated by the Association of Universities for Research in Astronomy, Inc., under contract with the NSF. From 1994–1997, much of this work was supported by NASA grants GO-5308.0*-93A, GO-5985.0*-94A & GO-6610.0*-95A (to RAW and WCK), GO.2684.03.94A (to RAW), and GO-6609.0*-95A & AR.6385.01.95A (to RAW and SCO) from ST ScI, which is operated by AURA, Inc., under NASA contract NAS5-26555.

REFERENCES

Abraham, R., et al. 1996 *MNRAS* **279**, L47 (A96).

Armus, L., Scoville, N. Z., et al. 1997 *ApJ* in prep. (A97).

BABUL, A., & REES, M. J. 1992 *MNRAS* **255**, 346 (BR92).

BEST, P. N., LONGAIR, M. S., & RÖTTGERING, H. J. A. 1997 *MNRAS* **286**, 785 (B97).

BROADHURST, T., ELLIS, R., KOO, D., & SZALAY, A. 1990 *Nature* **343**, 726 (B90).

BROADHURST, T. J., ELLIS, R. S., & GLAZEBROOK, K. 1992 *Nature* **355**, 55 (B92).

BRUZUAL A., G., & CHARLOT, S. 1993 *ApJ* **405**, 538 (BC93).

BURKEY, J. M., ET AL. 1994 *ApJ* **429**, L13 (B94).

CALZETTI, D. & KINNEY, A. L. 1992 *ApJ* **399**, L39 (CK92).

CARLBERG, R. G. ET AL. 1994 *ApJ* **435**, 540 (C94a).

CARLBERG, R. G., & CHARLOT, S. 1992 *ApJ* **397**, 5 (CC92).

CASERTANO, S., ET AL. 1995 *ApJ* **453**, 599 (Ca95).

CHAMBERS, K. C., MILEY, G. K., & VAN BREUGEL, W. J. M. 1990 *ApJ* **363**, 21 (C90).

COHEN, J. G., ET AL. 1996 *ApJ* **471**, L5 (C96).

COHEN, S. H., WINDHORST, R. A., ODEWAHN, S. C., ET AL. 1997 *ApJ* in prep. (Co97).

COLLESS, M., ET AL. 1994 *MNRAS* **267**, 1108 (Co94).

COURTEAU, S., DEJONG, R. S., & BROEILS, A. H. 1996 *ApJ* **457**, L73 (Co96).

COWIE, L. L., HU, E. M., & SONGAILA, A. 1995a *Nature* **377**, 603 (CHS95a).

DEJONG, R. S., & VAN DER KRUIT, P. C. 1994 *A&ApS* **106**, 451 (dJ94).

DRIVER, S., ET AL. 1994 *MNRAS* **266**, 155 (D94).

DRIVER, S. P., WINDHORST, R. A., ET AL. 1995 *ApJ* **449**, L23 (D95a).

DRIVER, S. P., WINDHORST, R. A., & GRIFFITHS, R. E. 1995 *ApJ* **453**, 48 (D95b).

EGGEN, O. J., LYNDEN-BELL, D., & SANDAGE, A. 1962 *ApJ* **136**, 748 (ELS62).

ELLIS, R. S., ET AL. 1996 *MNRAS* **280**, 235 (E96).

FRANCIS, P. J. ET AL. 1996 *ApJ* **457**, 490 (F96).

FRANCIS, P. J., WOODGATE, B. E., & DANKS, A. C. 1997 *ApJ* **482**, L25 (F97).

GIAVALISCO, M., STEIDEL, C. C., & SZALAY, A. S. 1994 *ApJ* **425**, L5 (G94).

GLAZEBROOK, K., ET AL. 1995a *MNRAS* **275**, L19 (G95a).

GLAZEBROOK, K., ET AL. 1995b *MNRAS* 273, 157 (G95b).

GRIFFITHS, R. E., ET AL. 1994 *ApJ* 435, L019 (Gr94).

HEYL, J., ET AL. 1997 *MNRAS* **285**, 613 (H97).

HIBBARD, J. E. & MIHOS, J. C. 1995a *AJ* **110**, 140 (H95a).

HOLTZMAN, J. A., ET AL. 1995 *PASP* **107**, 1065 (Ho95b).

HU, E. M. & MCMAHON, R. G. 1996 *Nature* **382**, 281 (H96).

KEEL, W. C., PASCARELLE, S. M., WINDHORST, R. A. 1997 *ApJ* in prep. (K97).

KEEL, W. C., & WINDHORST, R. A. 1991 *ApJ* **383**, 135 (KW91).

KENNICUTT, R. C. 1989 *ApJ* **344**, 685 (K89).

KINNEY, A. L., ET AL. 1996 *ApJ* **467**, 38 (K96).

KOO, D. C., & KRON, R. G. 1992 *ARA&A* **30**, 613 (KK92).

KRON, R. G. 1982, *Vistas in Astronomy*, **26**, 37 (K82).

LACEY, C. G., ET AL. 1993 *ApJ* **402**, 15 (L93).

LE FÈVRE, O., ET AL. 1994 *ApJ* **423**, L89 (LF94).

LILLY, S. J., ET AL. 1995 *ApJ* **455**, 108 (L95).

LONGAIR, M. S., BEST, P. N., & RÖTTGERING, H. J. A. 1995 *MNRAS* **275**, P47 (Lo96).

LOVEDAY, J., ET AL. 1992 *ApJ* **390**, 338 (Lo92).

LOWENTHAL, J. D. ET AL. 1997, *ApJ* **481**, 673 (Low97)

MADAU, P., ET AL. 1996 *MNRAS* **283**, 1388 (M96).

MARZKE, R. O., ET AL. 1994 *AJ* **108**, 437 (Ma94).

McCarthy, P. J. 1993, *ARA&A* **31**, 639 (M93).

Metcalfe, N., Shanks, T., Fong, R., & Roche, N. 1995 *MNRAS* **273**, 257 (M95).

Mutz, S., Windhorst, R., et al. 1997 *AJ* **113**, 1537 (Mu97).

Mutz, S. B., et al. 1994 *ApJ* **434**, L055 (Mu94).

Naim, A., et al. 1995 *MNRAS* **275**, 567 (N97).

Navarro, J. F. & White, S. D. 1994 *MNRAS* **267**, 401 (NW94)

Neuschaefer, L. W., & Windhorst, R. A. 1995a *ApJ* **439**, 14 (NW95a).

Odewahn, S., Windhorst, R., Driver, S., & Keel, W. 1996 *ApJ* **472**, L013 (O96).

Odewahn, S. C. 1995 *PASP* **107**, 770 (O95).

Odewahn, S. C., et al. 1992 *AJ* **103**, 318 (O92).

Pascarelle, S., Windhorst, R., Keel, W., Odewahn, S. 1996 *Nature* **383**, 45 (P96b).

Pascarelle, S. M., Windhorst, R. A., et al. 1996 *ApJ* **456**, L21 (P96a).

Pascarelle, S. M., Windhorst, R. A., et al. 1997 *ApJ* in preparation (P97).

Pei, Y. C. & Fall, S. M. 1995 *ApJ* **454**, 69 (PF95).

Phillipps, S., & Driver, S. P. 1995 *MNRAS* **274**, 832 (PD95).

Rauch, M., Haehnelt, M. G., & Steinmetz, M. 1997 *ApJ* **481**, 601 (RHS97).

Roche, N., et al. 1997 *MNRAS* **28**, 200 (R97).

Sawicki, M. J., Lin, H., & Yee, H. K. C. 1997 *AJ* **113**, 1 (SLY97).

Schmidtke, P. C., et al. 1997 *AJ* **113**, 569 (Sc97).

Scoville, N. Z., et al. 1997 *ApJL* 485, in press (astro-ph/9706291) (S97).

Searle, L. & Zinn, R. 1978 *ApJ* **225**, 357 (SZ78).

Steidel, C. C., et al. 1996a *ApJ* **462**, L17 (S96a).

Steidel, C. C., et al. 1996b *AJ* **112**, 352 (S96b).

Thompson, D., Djorgovski, S. G., & Trauger, J. 1995 *AJ* **110**, 963 (TDT95).

Trager, S. C., et al. 1997 *ApJ* (in press) (Tr97).

Tyson, J. A. 1988 *AJ* **96**, 1 (T88).

van Albada, T. S. 1982 *MNRAS* **201**, 939 (vA82).

White, S. D. M. & Frenk, C. S. 1991 *ApJ* **379**, 52 (WF91).

Windhorst, R. A., et al. 1991 *ApJ* **380**, 362 (W91).

Windhorst, R. A., et al. 1993 *ApJ* **405**, 498 (W93).

Windhorst, R. A., et al. 1995 *Nature* **375**, 471 (W95).

Windhorst, R. A., et al. 1994a *PASP* **107**, 798 (W94a).

Windhorst, R. A., et al. 1994b *ApJ* **435**, 577 (W94b).

Windhorst, R. A., Keel, W. C. & Pascarelle, S. M. 1997 *ApJL* submitted (W97).

Windhorst, R. A., Mathis, D. F., & Keel, W. C. 1992 *ApJ* **400**, L1 (W92).

Yee, H. K. C. & Ellingson, E. 1995 *ApJ* **445**, 37 (YE95).

Large ground-based redshift surveys in the context of the HDF

By SIMON LILLY[1] and the CFRS-LDSS TEAM: ROBERTO ABRAHAM,[2] JARLE BRINCHMANN,[3] MATTHEW COLLESS,[4] DAVID CRAMPTON,[5] RICHARD ELLIS,[3] KARL GLAZEBROOK,[6] FRANCOIS HAMMER,[7] OLIVIER LE FEVRE,[7] GABRIELA MALLEN-ORNELAS,[1] DAVID SHADE,[5] AND LAURENCE TRESSE[3]

[1]University of Toronto, Canada

[2]Royal Greenwich Observatory, UK

[3]Institute of Astronomy, Cambridge, UK

[4]Australian National Observatory, Canberra, Australia

[5]Dominion Astrophysical Observatory, Canada

[6]Anglo-Australian Observatory, Australia

[7]Observatoire de Paris Meudon, France

Systematic redshift surveys of large numbers of galaxies are producing an increasingly good picture of galaxy evolution in the $0 < z < 1$ redshift interval, onto which the view of the deeper Universe gained from the Hubble Deep Field may be grafted. The available evidence is that the largest and most massive galaxies were largely in place by $z \sim 0.8$. Although these larger galaxies are evolving, most of the evolutionary changes seen in the galactic population to this redshift involves smaller galaxies.

---

1. The relevance of large ground-based surveys at $z < 1$ to the HDF

The remarkable Hubble Deep Field (HDF) image has been produced during a period of great progress towards the goal of understanding the evolution of galaxies over cosmic time. The progress over the last five years has been driven in large measure by the introduction in the early 1990's of highly multiplexed spectrographs on existing 4-m class telescopes, by the commissioning of the Keck 10m telescope and by the attainment of the full capabilities of the HST/WFPC. A primary motivation for the HDF was to build on these advances through the deepest possible look at galaxies in the distant Universe.

In recent years, there have been several large and deep redshift surveys carried out on ground-based 4-m class telescopes (Table 1). Broadly speaking these contain several hundred galaxies from magnitude limited samples selected in various wavebands. Different programs have been based in different wavebands: the LDSS (Glazebrook et al. 1995) in the B-band, the CFRS (Le Fèvre et al. 1995 and references therein) in the I-band, and the Hawaii Deep program (Cowie et al. 1996) in the K-band with extensions in the other bands also. Most of the galaxies in the 4-m samples lie above broadly equivalent flux density limits of $B_{AB} \sim 24$, $I_{AB} \sim 22.5$ and $K_{AB} \sim 21.5$. Cowie et al. (1996) have pushed another 0.5–1.0 magnitudes deeper with Keck on smaller samples. The surveys have generally achieved a reasonably high success rate in spectroscopic identification (around 80%) producing a high degree of statistical completeness in their sampling of the galaxy population.

Program	Selection	Galaxies	Redshift range and median z at limit
Broadhurst et al. (1988)	$B < 21.5$	~400	$0.03 < z < 0.9$, $< z_{lim} >= 0.4$
Colless et al. (1990)	$B < 22.5$		
Glazebrook et al. (1995)	$B < 24$		
CFRS (1995)	$I_{AB} < 22.5$	730	$0.1 < z < 1.3$, $< z_{lim} >= 0.6$
Hawaii Deep Survey	$K < 20$	390	$0.1 < z < 1.7$, $< z_{lim} >= 0.7$
Cowie et al. (1996)	$I < 23; B < 24.5$		

TABLE 1. Published large deep redshift surveys extending to $z \gg 0.5$

In the context of the HDF and the spectroscopic work that has been done in it, these existing ground-based survey are noteworthy for the much larger area of sky covered. For instance the effective area of the CFRS is 110 arcmin^2, almost 20 times larger than the HDF. Thus the spectroscopic surveys contain many more relatively bright galaxies (e.g., $I_{AB} < 22.5$ for the CFRS) than HDF-based studies.

Despite, by the standards of the HDF, being rather bright, the galaxies sampled by the redshift surveys are nevertheless an important population for evolution studies. First, and in quite general terms, the redshift surveys reach down to the brightness level where the integrated surface brightness of the galaxy population per magnitude interval is maximized (see Fig. 1) and the surveys are therefore sampling a good fraction of the extragalactic background light that is produced by discrete objects. It should be recalled that in the idealized case of galaxies with flat spectra ($f_\nu \propto \nu^0$), the production of heavy elements by a particular component of the galaxy population is proportional to the contribution of that population to the extragalactic background light (Lilly and Cowie 1987). Thus the population sampled by the redshift surveys is likely to be playing an important role in the overall history of star-formation in the Universe. Second, these surveys yield a median redshift $< z >\sim 0.6$ with the more luminous L* galaxies extending to redshifts $z > 1$. Thus these samples cover the evolution of luminous galaxies over the last 2/3 of the age of the Universe, and probe to a point not far from the peak in the global luminosity density that has been inferred from estimated redshifts in the HDF (Madau et al. 1996, Connolly et al. 1997). The median redshift is similar to that found in the HDF galaxies with measured redshifts.

2. The ground-based view of evolution at $z < 1$

2.1. *Existing deep surveys*

To a large degree, the different selection wavelengths employed in the surveys of Table 1 have been complementary: the B-band will preferentially select the bluer galaxies which are known to contribute most strongly to the evolution of the galaxy population, the I-band is best matched at $0.5 < z < 1.0$ to the rest-frame B and V wavebands, allowing the most precise delineation of evolutionary effects with respect to local galaxy samples. The K-band is anchored to that part of the spectrum which shows least evolution. In broad terms there is a large degree of concordance between the results of the above surveys. There are nevertheless some detailed differences which remain to be resolved. One important issue is the location of the bright end of the luminosity function—the CFRS-based LF is significantly brighter than that of the LDSS and allows a greater degree of conventional luminosity evolution in the most luminous galaxies seen today.

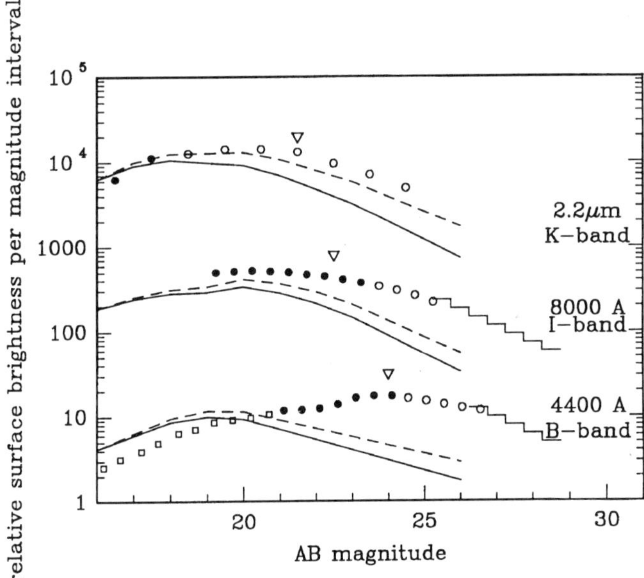

FIGURE 1. The relative surface brightness per magnitude interval of the galaxy population as a function of AB magnitude at three wavelengths (offset vertically for clarity). The limiting magnitudes of the large systematic redshift surveys are shown as triangles. These surveys sample galaxies that contribute most of the extragalactic background light in discrete galaxies.

2.2. *The luminosity function*

The luminosity functions derived from the above surveys (see e.g., Lilly et al. 1995, Ellis et al. 1996, Cowie et al. 1996) are broadly similar. There is general agreement that the principal evolutionary change in the galaxy population back to $z \sim 1$ is manifested as a substantial increase with redshift in the number density of blue galaxies with luminosities close to present day L* (i.e., $M_{AB}(B) \sim -21$ for $H_0 = 50$ km s^{-1} Mpc^{-1}). This phenomenon had been suggested by the first exploratory spectroscopy of faint galaxies almost a decade ago (see e.g., Colless et al. 1990, Lilly et al. 1991) but is now placed on a firm statistical footing.

The appearance of these galaxies at high redshift produces a substantial change in the luminosity function of galaxies selected to have blue colors, but little change in the luminosity function of redder (Lilly et al. 1995, Heyl et al. 1997). The same differential behavior is seen if selection is made by the equivalent width of [O II] 3727 (Ellis et al. 1996). Small et al. (1997) have shown that this differential behavior is evident even at low redshifts, $z < 0.5$. The division of the luminosity function color classes is somewhat arbitrary and is likely masking a more complex behavior in which individual objects cross the color-divides, in either direction (see Section 3.3 below).

In particular, it is important to remember that the luminosity function is just a statistical description of the galaxy population and inferences about the evolution of individual galaxies based on the form of the changes in the luminosity function may be misleading. As a ready example, the changes in the CFRS B-band luminosity function (i.e., not split by color) appear as almost pure density evolution, yet splitting by color reveals one (red) unchanging component and one (blue) component behaving as if undergoing pure luminosity evolution! Similarly, computation of an ultraviolet luminosity function (e.g., at 2800 Å) reveals changes that are best represented as pure luminosity evolution (see also Cowie et al. 1997).

Determining the physical evolution of individual objects (as opposed to the statistical evolution of the population) thus requires additional data over and above the simple color-magnitude-redshift output of the redshift surveys. Ideally, an observable property of the galaxies could be found that would remain constant throughout the evolution of individual galaxies. Unfortunately no such invariant property has been identified. The (halo) mass is certainly attractive, but even this may be expected to increase with time in hierarchical models (Kauffmann et al. 1993, Baugh et al. 1996). There are several programs producing kinematic estimates of masses (see e.g., Rix et al. 1997, Vogt et al. 1996, 1997, Guzman et al. 1997), although the observational difficulties are still formidable.

Attempts have also been made to use the near-infrared luminosity as a measure of stellar mass, although the correspondence must be much weaker for blue galaxies dominated by young rapidly evolving stellar populations than it is for more mature systems in which the infrared light is dominated by old red giants. If luminosity is closely associated with mass, then the luminous blue galaxies at higher redshifts must be massive forming galaxies and galaxy formation is following the down-sizing picture described by Cowie et al. (1996).

Observations of the morphologies and sizes of distant galaxies that are possible with HST provide a useful new perspective on this problem—though by no means do they provide a silver bullet! The "size" of a galaxy is a physical parameter that may or may not remain constant through its evolution and the "morphology" of the galaxy tells us about the nature of the star-formation activity in that galaxy as well as the occurrence of physically important processes such as merging. We describe below the results of our own program of HST observations of galaxies from the CFRS and LDSS redshift surveys.

Ultimately additional diagnostics such as metallicity measurements and estimates of the dust and gas content of the galaxies will also be brought to bear, and presumably a synthesis will emerge. The accounting methods for following the long-lived products of star-formation are sufficiently imprecise that it will be as important to focus as much on the presence or absence of passive galaxies as a function of redshift as on the properties of actively star-forming galaxies.

2.3. *The luminosity density*

Averaging over the luminosity function to compute a comoving emissivity of the galaxy population (i.e., the luminosity density of the Universe) as a function of redshift in a particular waveband has proved to be a powerful tool (Cowie et al. 1995, Pei and Fall 1995, Lilly et al. 1996,). This approach sidesteps many of the present uncertainties regarding the evolution of single objects and unresolved issues such as whether star-formation occurs in bursts or as a steady process, and the degree of merging between galaxies, have no effect on the derived luminosity density of the Universe. Even the effects of the cosmological model are weak, $(1+\Omega z)^{0.5}$. In effect, the Universe is considered as a single stellar system and the identity of individual galaxies is submerged. This is both the strength and weakness of this approach. The luminosity density approach is particularly powerful when combined with other global measures of the Universe, such as changes in the quantity and metallicity of neutral Hydrogen in the Universe (e.g., Pei and Fall 1995, Fall and Pei 1996).

Using the CFRS redshift sample and attempting to correct for sources below the survey limits, Lilly et al. (1996) found that the luminosity densities in the ultraviolet, optical and far-red wavebands increase strongly with redshift out to $z = 1$. We estimated $L_{2800} \propto (1+z)^{3.9\pm0.7}$, $L_B \propto (1+z)^{2.7\pm0.5}$ and $L_{10000} \propto (1+z)^{2.2\pm0.5}$ for $\Omega_0 = 1$ (the exponents are reduced by 0.5 for $\Omega_0 = 0$). The increase in [O II] 3727 luminosity density (Hammer et al. 1997, Cowie et al. 1995) is comparable to that seen in at 2800 Å.

For a constant initial mass function (i.m.f.) and constant extinction by dust, the ultraviolet luminosity density should reflect the instantaneous star-formation rate, suggesting that the Universe was forming stars of order 10 times faster, per comoving volume, than it is now. An interesting question was whether a self-consistent model for the stellar content of the Universe as a whole can be found. A self-consistent model with a Salpeter initial mass and a turn-down in the τ^{-2} increase in star-formation rate at around $z \sim 2$ was found to work (Lilly et al. 1996). Models with i.m.f.s that are more deficient in high mass stars or with ever increasing star-formation rates at early epochs overproduced the near-infrared light from long-lived stars (see also Piero Madau's contribution to these proceedings).

The luminosity density approach has been used and extended to higher redshifts by Madau et al. (1996) and Connolly et al. (1997) using photometrically estimated redshifts of galaxies in the HDF. Both studies suggest that the rise seen in the CFRS does indeed not continue indefinitely and that a peak in the ultraviolet luminosity density exists in the region $1.0 < z < 2.5$.

A major uncertainty in interpreting the changes in ultraviolet luminosity density is the role of dust. The νL_ν luminosity density at 60 μm at the present epoch (Saunders et al. 1990) is about 2/3 as large as that at 4400 Å and 2.5 times larger that at 2800 Å. It also probably increases as roughly $(1+z)^3$ towards $z \sim 1$ (Pearson & Rowan-Robinson 1996). Further information on the role of dust will come from ISO (see Rowan-Robinsons contribution to these proceedings) and particularly from sub-millimeter measurements at high redshifts that will come from a new generation of sensitive array bolometers such as SCUBA on JCMT (see Smail et al. 1997). Early indications from both of these are that dust continues to play a significant role in absorbing and re-radiating ultraviolet light from high redshift galaxies.

3. Recent results from HST observations of CFRS and LDSS survey fields

Clearly the spatial resolution of HST has opened up a new dimension to the study of galaxy evolution at high redshift. Hitherto, HST morphological data on large numbers of distant galaxies has generally been confined to samples of galaxies without redshift information, using the Medium Deep Survey (Glazebrook et al., 1995) or comparable material (Driver et al., 1995) or the HDF itself (Abraham et al. 1996). A general conclusion from these has been that the number counts N(m) of galaxies classified morphologically as ellipticals and spirals are roughly as predicted for models with a magnitude or less of luminosity evolution, whereas those of irregular/peculiar galaxies are much steeper, suggesting that it is galaxies with these late-type morphologies that are responsible for the evolution seen in the redshift surveys.

3.1. *HST observations of galaxies from the CFRS and LDSS redshift surveys*

A new collaborative project between the CFRS and LDSS redshift survey teams has brought these two approaches together, combining the exceptional morphological information from HST with the extensive redshift data from the ground-based CFRS and LDSS redshift surveys. We have obtained HST WFPC2 images of 25 fields from the CFRS and LDSS surveys which have yielded high quality images in F814W for a total of 341 objects (251 CFRS and 90 LDSS) that had previously been observed spectroscopically. These 341 objects have been morphologically classified, both by eye and by machine-based algorithms, and the 2-dimensional light distributions have been fit by multi-parameter bulge+disk models (as in Schade et al. 1995). Several new results have emerged from these analyses.

3.2. *Galaxies with late-type morphologies as the major evolving component*

Computation of the N(z) redshift distributions and luminosity functions for different morphological types has shown (Brinchmann et al. 1997, hereafter Paper 1) that the degree of evolutionary change increases from elliptical through spiral to irregular/peculiar. Furthermore, the steep number count N(m) of the irregular/peculiar galaxies in the earlier studies is now clearly seen to be due to the appearance of these galaxies at high redshifts rather than the alternative possibility of a steep local luminosity function. An important point is that the redshift information for individual galaxies now enables us to quantify the effects of the redshift on the morphologies (i.e., the morphological k-correction). The decreasing rest-wavelength generally makes objects appear of later type at higher redshifts, but in Paper 1, we show that this effect plays only a small part in the observed shift to irregular/peculiar morphologies at high redshift.

Of course, the morphology of a galaxy can change with time, and the irregular/peculiar galaxies seen at high luminosities at high redshift are not necessarily the antecedents of galaxies with similar appearance seen in the local Universe at the present epoch. The morphology of a galaxy really tells us only about its dynamical state and the nature of star-formation activity at the time that it is observed. The largest change in the galaxy population is associated with star-formation activity that is occurring in irregular morphological structures.

3.3. *The number density and evolution of large disk galaxies at high redshift*

The 2-dimensional fitting approach (Schade et al. 1995, 1996) introduces a quantitative parameter, size, to the analysis of the galaxy population. It also offers the possibility of isolating different components within galaxies (e.g., spheroid and disk) that are likely to have different evolutionary histories, and also of isolating particular subsets of the galaxy population (e.g., "large-disk" galaxies). Accordingly multi-parameter bulge-plus-disk models have been fit to all of the sources that have been observed with HST.

Lilly et al. (1997, hereafter Paper 2) have examined the properties of those galaxies in the sample that appear to be disk dominated (i.e., with bulge-to-total ratio less than 0.5), with particular emphasis on the large disks that have fitted exponential scale lengths greater than 3–$4h_{50}^{-1}$ kpc). The bulge and disk decomposition of these relatively large galaxies is more straightforward than for the smaller galaxies since their morphologies are better defined. Furthermore, a magnitude-limited sample will be most complete at a given redshift for these large and luminous galaxies. The distribution of disk ellipticities (i.e., of implied inclination angles) suggest that we are dealing with true two-dimensional disks.

Fig. 2 shows the size function for these large disks in the interval $0.5 < z < 1.0$ computed for two different cosmologies and compared with the local estimate of de Jong (1996b). There is little evidence in these data for a change in the size function to redshifts approaching unity, suggesting that disks have roughly constant size and number density to these redshifts. Internal to the CFRS sample, the average scale length at constant number density has likely increased by no more than 30% since $z \sim 0.8$, and a tighter constraint is obtained by including the de Jong (1996b) local size function. If disks have grown on average by more than this, then their number density must have fallen with time, presumably through the effect of merging.

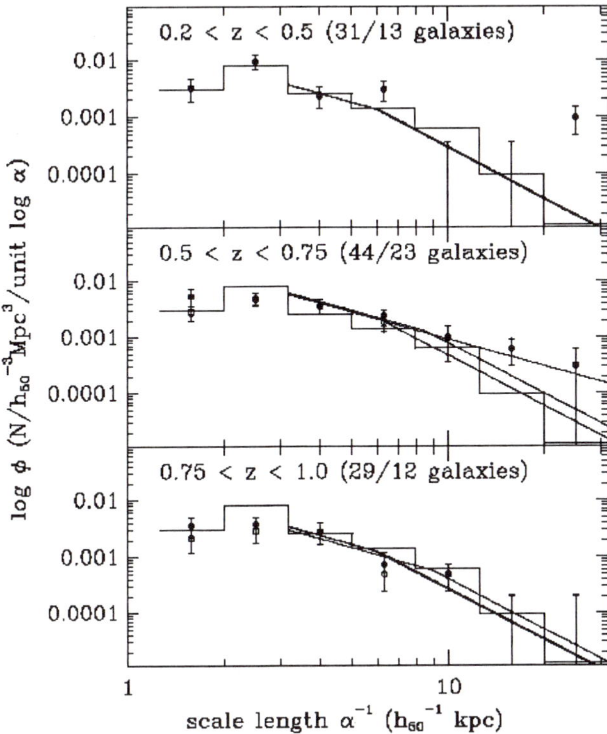

FIGURE 2. The size function of galactic disks in the CFRS sample in three redshift bins (data points), compared with the local size function derived by de Jong et al. (1996b) (histogram representation).

In addition to a roughly constant number density, other properties of the large-disk galaxies such as the distribution of bulge-to-total B/T ratios are consistent with their being a single "identifiable" population seen at a range of epochs. It is thus reasonable to suppose that changes in the average properties of these galaxies with redshift will reflect evolutionary changes in individual members of the population, although care must be taken to address biases introduced by the selection criteria of the original redshift surveys.

We find that the average properties of the large disk galaxies (with $\alpha^{-1} > 4h_{50}^{-1}$ kpc) have indeed changed with time (Figs. 4*a*–*d*). In comparison with the local disks studied by de Jong (1996a), the high redshift disks show an elevated rest-B surface brightness (by 0.5–0.8 mag at $z = 0.7$), bluer rest-frame (U-V) colors, later type morphologies and somewhat elevated rest-frame equivalent widths of [O II] 3727. The change in B-band surface brightness is consistent with other recent estimates (Forbes et al. 1996, Vogt et al. 1996), and also with the original estimate of Schade et al. (1995) when it is realized that the latter was an average based on a luminosity limited sample rather than a size limited one as here.

In Paper 2, it is shown that these changes are all consistent with a model in which the star-formation rate in the disks has declined with an e-folding time of around 5 Gyr, equivalent to a factor of 4 since $z \sim 1$, and are inconsistent with a scenario in which the star formation rate in the disks has been constant with time. The Milky Way Galaxy would be a member of the high redshift sample (unless $H_0 < 50$ km s^{-1} Mpc^{-1} and the modest change in scale length and the declining star-formation history that is implied

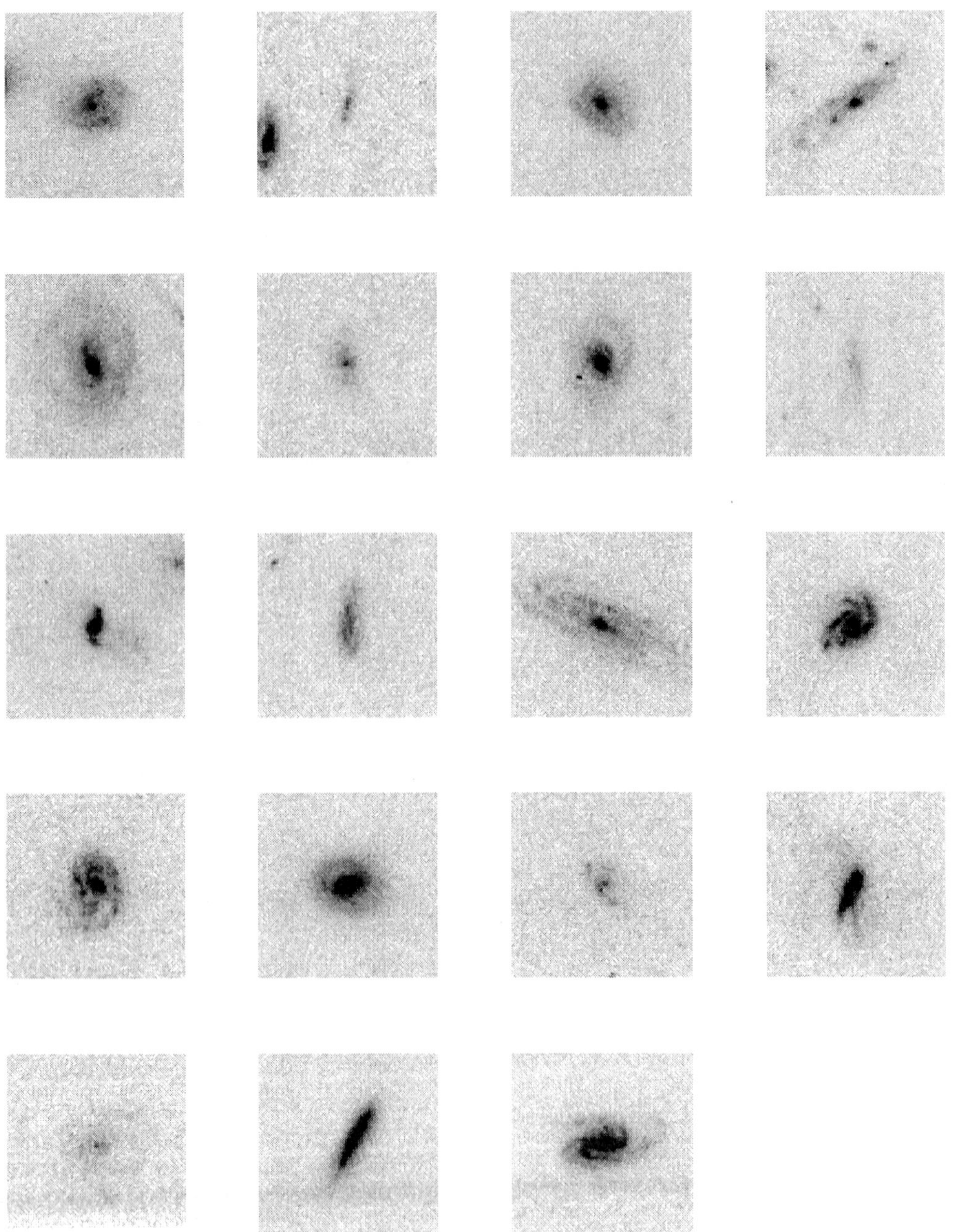

FIGURE 3. Montage of disk galaxies with $0.5 < z < 0.75$ in the CFRS and LDSS samples. The galaxies have $B/T < 0.5$ and disk scale lengths $\alpha^{-1} > 4h_{50}^{-1}$ kpc.

by our observations of high redshift galaxies are entirely consistent with models for the evolution of the Milky Way disk (see e.g., Twarog 1980, Frantzos & Aubert 1995).

The B-band luminosity of the large disks is evidently increasing roughly as $(1+z)^{1.4}$. This is smaller than the change observed in the galaxy population as a whole, $(1+z)^{2.7\pm0.5}$ for $q_0 \sim 0.5$. This suggests that other, presumably smaller, galaxies are making a larger

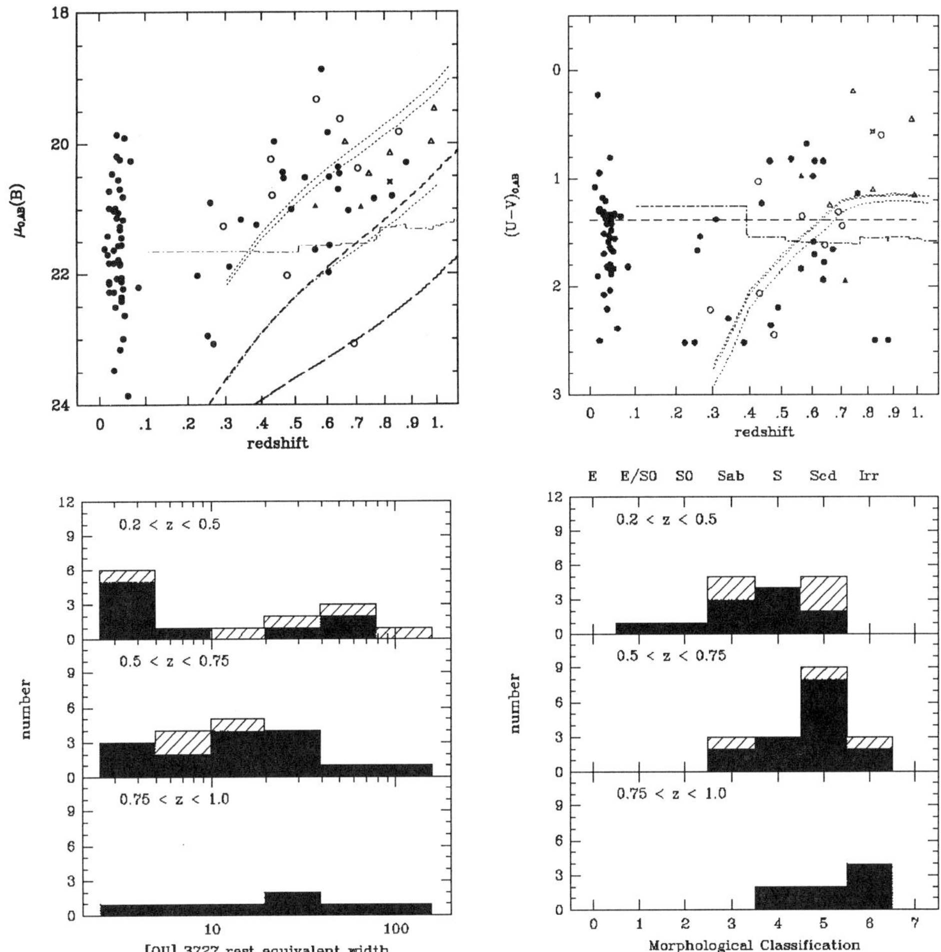

FIGURE 4. *(a–top left)* The disk central surface brightness of disks with $\alpha^{-1} > 4h_{50}^{-1}$ kpc as a function of redshift. Data is CFRS+LDSS sample ($z > 0.2$) and the local sample of de Jong et al. (with a stretched redshift scale for the latter). The dashed lines represent various aspects of selection effects and are discussed in detail in Paper 2. *(b–top right)* As for *a*, the overall rest-frame $(U-V)_{0,AB}$ color of large disk galaxies is plotted as a function of redshift. *(c–bottom left)* The distribution of [O II] 3727 rest-frame equivalent widths in the large disk galaxies in the CFRS (solid) and LDSS (hatched) samples, in three redshift bins. *(d–bottom right)* As for *c*, the visual morphological classification of large disk galaxies in the CFRS and LDSS samples, in three redshift bins.

contribution to the overall change in the galaxy population. However, it should be noted that this shortfall is narrowed significantly if $q_0 \sim 0$, since the disk brightening is essentially independent of q_0 but the luminosity density decreases as $(1+z)^{0.5}$.

3.4. *The size-luminosity distribution function*

The suggestion that it is smaller galaxies that are driving the increase in star-formation rate to $z \sim 1$ is more directly illustrated by the bivariate size-luminosity function. The Fig. 5 (from Paper 2) shows the bivariate size-luminosity function for all galaxies in the CFRS sample that has been observed with HST, computed at $0.2 < z < 0.5$ and $0.5 < z < 1$. For generality, the size parameter used here is a half-light radius, generated

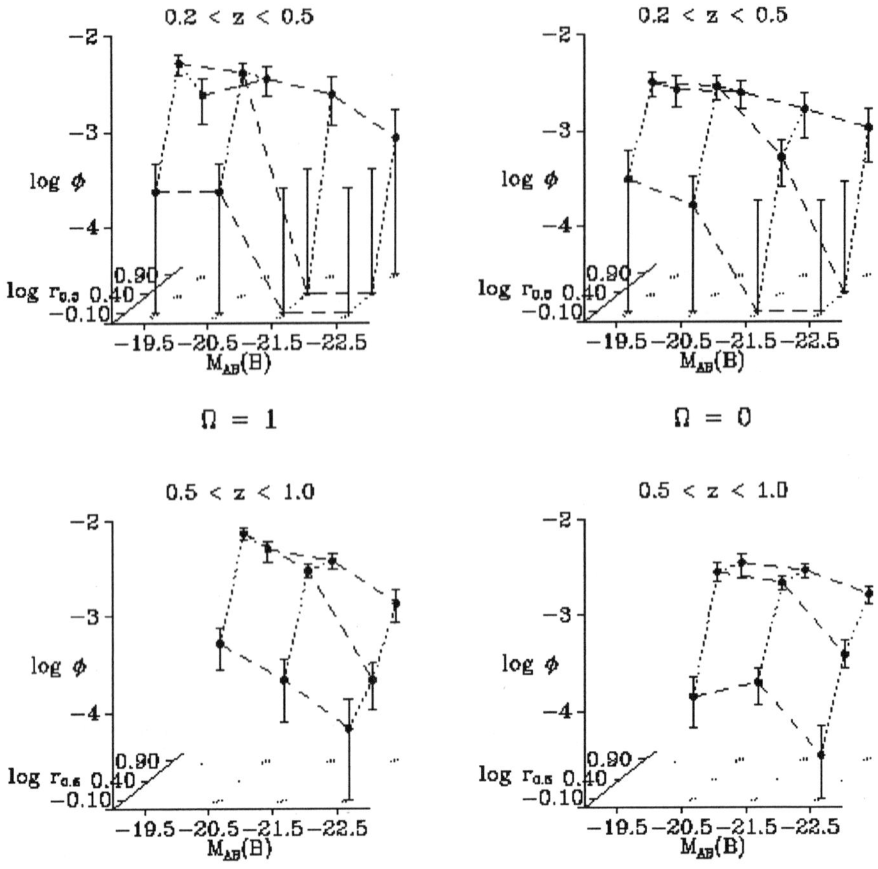

FIGURE 5. The bivariate size-luminosity function for all CFRS galaxies, computed in two redshift intervals and for two cosmologies. The size is a half-light radius. The largest evolutionary changes occur for galaxies with moderate sizes and luminosities.

from the 2-dimensional fits. No particular distinction has been made between spheroidal and disk components.

This diagram shows how the largest evolutionary change is the filling in of the central area of the diagram by the appearance of numerous luminous but moderately sized galaxies. These galaxies have $M_{AB}(B) \sim -21.5$ and $r_{0.5} < 5h_{50}^{-1}$ kpc. The "ridge-line" at large sizes at the rear of the figure stays more or less unchanged aside from the effects of the modest brightening described above. Again, however, it should be noted that the impression of differential behavior is weakened for low q_0: This is because a given galaxy appears to be larger and more luminous, but rarer, as q_0 is decreased.

4. The evolution of galaxies

The morphological analysis reported above reinforces the view that emerged from the original luminosity function analyses—that field galaxy evolution is differential. Galaxies with different sizes and present-day luminosities evolve differently. Kinematic studies (Vogt et al. 1997, Guzman et al. 1997, Illingworth in these proceedings) also point in this direction. The largest changes in a multi-dimensional ϕ(L, color, size, morphology,

σ_v) distribution function are occurring, out to $z = 0.8$, for moderately sized, moderate luminosity galaxies rather than for the largest most luminous galaxies. These latter appear to be in place by $z \sim 0.8$ and to then evolve only moderately through to the present.

That evolution should be differential is not at all surprising since even a cursory examination of the galaxy population reveals that the average properties of present-day galaxies vary with luminosity and/or size and mass. As has been noted for several years, the sense of the differential behavior is interesting: at least superficially it is the opposite to what might have been expected in hierarchical models for galaxy formation, in that it is the larger galaxies that appear to be in place first.

4.1. *How important is merging in the evolution of the galaxy population?*

An important unanswered and highly relevant question is to what extent the merging of galaxies drives the evolution of the galaxy population. One direct approach is to study the fraction of galaxies observed at any epoch to be undergoing a merger, and we look at this below.

A complementary approach is to study whether the number density of massive elliptical galaxies has increased with epoch (cf. Kauffmann et al. 1996). The present samples are too small and too limited to answer this latter question definitively. Two analyses of the V/V_{max} statistic based on color-selection in the CFRS sample (Lilly et al. 1995, Kaufmann et al. 1996) give different values which appear to be largely due to different philosophies concerning the redshift range that should be used. In our original analysis (Lilly et al. 1995), with an unevolving cut in rest-frame color, $(U - V)_{0,AB} > 2.05$, with photometrically estimated redshifts for the spectroscopic failures and limiting the redshift range to $0.3 < z < 1.0$, we found $V/V_{max} = 0.50 * 0.04$. In contrast, Kauffmann et al., with no redshift cutoff, with evolving color selection criteria (from Bruzual and Charlot models) and including the effects of passive luminosity evolution, found $V/V_{max} = 0.41 \pm 0.03$. In the new sample with morphological information from HST we find, for $0.3 < z < 1.0$, $V/V_{max} = 0.53 \pm 0.04$ for galaxies with $r^{0.25}$ profiles, decreasing to 0.49 ± 0.05 when an additional color and morphological cut is applied. The effects of adding luminosity evolution decreases this to 0.46 ± 0.05, which is still insignificantly different from a value of 0.5. Im et al. (1995) have sought to use a larger sample using color estimated redshifts for a morphologically selected sample of ellipticals found evidence for luminosity evolution at constant comoving density.

Returning to the direct approach, the primary difficulty is in translating any observationally defined "merger fraction" (e.g., the fraction of galaxies in close pairs) to a merger rate. This requires an indication of the length of time for which the particular phenomenon will be observed during a particular merger event. Our own analysis of the HST images of the CFRS-LDSS sample is not yet completed (Le Fevre et al., in preparation, Paper 4). However, there are already many indications in the literature that the fraction of galaxies undergoing mergers or major interactions increases with redshift (Zepf & Koo 1989, Carlberg et al. 1994, Patton et al. 1997). The recent study of the CNOC-1 sample by Patton et al. (1997) yielded a merging pair fraction as a function of redshift as $f_m \sim 0.02(1+z)^3$.

Is this merger activity consistent with the above observations of a roughly constant comoving number densities of large disk galaxies (cf. Toth & Ostriker 1992)? The probability that a given galaxy undergoes a merger in the redshift interval dz is given by:

$$\frac{dP}{dz} = f_m \times \frac{d\tau}{dz} \times \frac{1}{t_m}$$

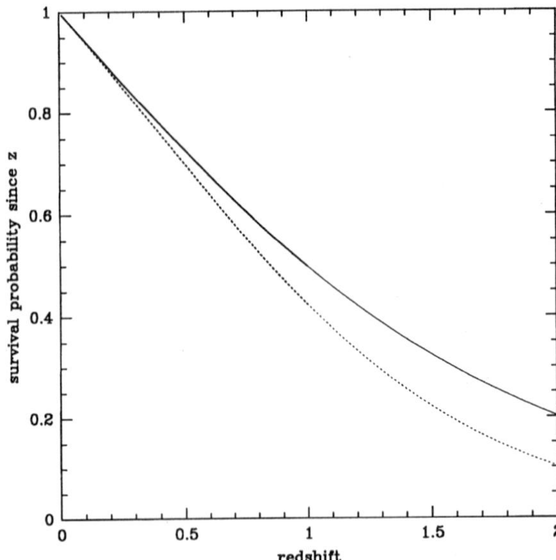

FIGURE 6. The probability for galaxies to survive to the present epoch without having undergone a major merger since a redshift z, based on the parameters describe in the text.

where f_m is the fraction of galaxies exhibiting some merger signature and tm is the (uncertain) time for which that phenomenon will be observable. The resulting survival probability $P_s(z)$ for $f_m = 0.02(1+z)^3$, $H_0 = 70$ km s^{-1} Mpc^{-1} and $t_M = 5 \times 10^8$ years is shown in Fig. 6.

This suggests that it is plausible that most disks have survived since an epoch corresponding to $z \sim 0.8$, consistent with the roughly constant size function discussed above, and also with the survival of the Milky Way disk. However, if the trend continues with redshift to $z \gg 1$ (a regime of particular relevance to the HDF) we would expect merging to be increasingly important. Relatively few galaxies will have survived from this epoch without having undergone a major merger.

Indeed, it is tempting to associate the peak in the $L(2800$ Å$)$ luminosity density at $1 < z < 2$ with the point at which the merging of galaxies becomes a dominant evolutionary driver. This is would be consistent with the often highly disturbed morphologies of HDF galaxies. In this picture, the rise in the global star-formation rate with epoch at $z > 2$ could be associated with the production of stars which ended up in spheroidal components through extensive merging whereas the subsequent decline at $z < 1$ would involve primarily the consumption of gas in disks more or less isolated systems.

5. Summary

The observational view of galaxy evolution since $z \sim 1$ has improved considerably over the last few years as a result of several independent redshift surveys containing several hundred galaxies in the $0 < z < 1.5$ range. The galaxies in these samples are representative of the brighter galaxies in the HDF (e.g., $I_{AB} < 22.5$) but are nevertheless playing a major role in the history of star-formation in the Universe.

There is general agreement that the ultraviolet luminosity density of the Universe increases as roughly $(1+z)^{3-4}$ due to the appearance at high redshifts of luminous blue galaxies with colors similar to those of present-day Irr galaxies and luminosities around present-day L*.

The population of large disk galaxies (i.e., those with $\alpha^{-1} > 3.2h_{50}^{-1}$ kpc) has changed relatively little to $z \sim 0.8$. The size function of galactic disks (the number per unit comoving volume per interval in disk scale length) seen at $0.5 < z < 1.0$ is very similar to that measured locally by de Jong et al. (1996b). On the other hand, the average properties of these large disk galaxies show several indications of modestly increased star-formation activity: elevated surface brightness, bluer overall colors, increased [O II] 3727 equivalent widths and later type morphologies. These are consistent with an increase in the average star formation rate by a factor of 3–4, in accord with the history of the Milky Way disk that has been inferred from analysis of the solar neighborhood.

To redshifts $z \sim 0.8$, the largest changes to the bivariate size-luminosity function are due to fairly small galaxies with half-light radii $r_{0.5} < 5h_{50}^{-1}$ kpc and $M_{AB}(B) \sim -20.5$. These generally have late type morphologies (i.e., irregular/peculiar). The nature of these galaxies is not well understood at this point, but they may simply be the small disk galaxies seen today at $M_{AB}(B) < -19$.

REFERENCES

Abraham, R. J., Tanvir, N. R., Santiago, B. X., Ellis, R. S., Glazebrook, K., van den Bergh, S. 1996 *MNRAS* **279**, L47.

Baugh, C. M., Cole, S., Frenk, C. S. 1996 *MNRAS* **283**, 1361.

Brinchmann, J., Abraham, R., Schade, D., Tresse, L., Ellis, R. S., Lilly, S. J., Le Fevre, O., Glazebrook, K., Hammer, F., Colless, M., Crampton, D. 1997 *ApJ*, submitted.

Bruzual A., G., & Charlot, S. 1993 *ApJ*, **405**, 538.

Carlberg, R. G., Pritchet, C. J., Infante, L. 1994 *ApJ* **435**, 540.

Connolly, A., Szalay, A., Dickinson, M., SubbaRao, M., Brunner, R. 1997, preprint.

Cowie, L. L., Hu, E. M., Songaila, A. 1995 *Nature* **377**, 603.

Cowie, L. L., Songaila, A., Hu, E. M., Cohen, J. 1996 *AJ* **112**, 839.

de Jong, R. S. 1996a *A&AS* **118**, 557.

de Jong, R. S. 1996b *A&A* **313**, 45.

Driver, S. P., Windhorst, R. A., Ostrander, E. J., Keel, W. C., Griffiths, R.E., Ratnatunga, K. U. 1995 *ApJ* **449**, L23.

Ellis, R. S., Colless, M., Broadhurst, T. J., Heyl, J., Glazebrook, K. 1996 *MNRAS* **280**, 235.

Forbes, D., Phillips, A. C., Koo, D. C., Illingworth, G. D. 1996 *ApJ* **462**, 89.

Frantzos, N., Aubert, O. 1995 *A&A* **302**, 69.

Glazebrook, K., Ellis, R. S., Colless, M., Braodhurst, T. J., Allington-Smith, J. R., Tanvir, N. 1995 *MNRAS* **273**, 157.

Glazebrook, K., Ellis, R. S., Santiago, B., Griffiths, R. E. 1995 *MNRAS* **275**, L19.

Guzman, R., Gallego, J., Koo, D., Philips, A., Lowenthal, J. D., Faber, S. M., Illingworth, G., Vgt, N. 1997, preprint, astro-ph 97-04001.

Hammer, F., Crampton, D. C., Le Fevre, O., Lilly, S. J. 1995 *ApJ* **455**, 88.

Hammer, F., Flores, H., Lilly, S. J., Crampton, D., Le Fevre, O., Rola, C., Mallen-Ornelas, G., Schade, D., Tresse, L. 1997 *ApJ* **481**, 49.

Heyl, J., Colless, M., Ellis, R. S., Broadhurst, T. 1997 *MNRAS* **285**, 613.

Im, M., Ratnatunga, K., Griffiths, R. E., Casertano, S. 1995 *ApJ* **445**, L15.

Kauffmann, G., White, S. D. M., Guiderdoni, B. 1993 *MNRAS* **264**, 201.

Le Fevre, O., Crampton, D., Lilly, S. J., Hammer, F., Tresse, L. 1995 *ApJ* **455**, 60.

Lilly, S. J., Le Fevre, O., Crampton, D., Hammer, F., Tresse, L. 1995a *ApJ* **455**, 50.

LILLY, S. J., HAMMER, F., LE FEVRE, O., CRAMPTON, D. 1995b *ApJ* **455**, 75.

LILLY, S. J., TRESSE, L., HAMMER, F., CRAMPTON, D., LE FÈVRE, O. 1995c *ApJ* **455**, 108.

LILLY, S. J., LE FEVRE, O., HAMMER, F., CRAMPTON, D. 1996 *ApJ* **460**, L1.

MADAU, P., FERGUSON, H. C., DICKINSON, M. E., GIAVALISCO, M., STEIDEL, C., FRUCHTER, A. 1996 *MNRAS* **283**, 1388.

PATTON, D. R., PRITCHET, C. J., YEE, H. K. C., ELLINGSON, E., CARLBERG, R. G. 1997 *ApJ* **475**, 29.

PEARSON, C., & ROWAN-ROBINSON, M. 1996 *MNRAS* **283**, 174.

RIX, H-W., GUHATHAKURTA, P., COLLESS, M., ING, KRISTINE 1997 *MNRAS* **285**, 779.

SAUNDERS, W., ROWAN-ROBINSON, M., LAWRENCE, A., EFSTATHIOU, G., KAISER, N., ELLIS, R. S., FRENK, C. S. 1990 *MNRAS* **242**, 318.

SCHADE, D., LILLY, S. J., CRAMPTON, D., HAMMER, F., LE FEVRE, O., TRESSE, L. 1995 *ApJ* **451**, L1.

SCHADE, D., LILY, S. J., LE FEVRE, O., HAMMER, F., CRAMMPTON, D. 1996a *ApJ* **464**, 79.

SCHADE, D., CARLBERG, R., YEE, H. K. C., LOPEZ-CRUZ, O., ELLINGSON, E. 1996b *ApJ* **464**, L63.

SMAIL, I., IVISON, R. J., BLAIN, A. 1997 *ApJL*, in press.

TOTH, G., OSTRIKER, J. P. *ApJ* **389**, 5.

TWAROG, B. 1980 *ApJ* **242**, 242.

VOGT, N., FORBES, D., PHILLIPS, A. C., GRONWALL, C., FABER, S. M., ILLINGWORTH, G. D., KOO, D. C. 1996 *ApJ* **465**, L15.

VOGT, N., FORBES, D., PHILLIPS, A. C., FABER, S. M., GALLEGO, J., GRONWALL, C., GUZMAN, R., ILLINGWORTH, G. D., KOO, D. C., LOWENTHAL, J. D. 1997 *ApJ* **479**, 121.

ZEPF, S. E., KOO, D. C. 1989 *ApJ* **337**, 34.

The properties of Lyman-break galaxies at redshift $z \sim 3$

By MAURO GIAVALISCO†

The Observatories of the Carnegie Institution of Washington,
813 Santa Barbara Street, Pasadena, CA 91101, USA

The Lyman-break technique has opened the very high-redshift universe (e.g., $z > 2$) to the empirical investigation. The efficiency of the technique in detecting very distant star-forming galaxies has allowed the compilation of large samples at redshifts as high as $z \sim 4.5$ with well controlled selection criteria and homogeneous characteristics. Such data sets are essential to explore the spectrum of properties of early galaxies, including spectra, luminosity function, morphology, star-formation history, clustering and mass, that are necessary to constrain the theories of galaxy formation. Developed for a ground-based survey, the Lyman-break technique has been successfully transported to *HST* observations during the HDF. Although the HDF sample of Lyman-break galaxies is very small compared to the ground-based survey, it is considerably deeper and covers a larger redshift range. This has allowed us to study the luminosity distribution of these galaxies on a large dynamic range and probe their evolution with redshift. Here we discuss the salient features of this combined exploration, from the ground and from space, of star-forming galaxies during the first $\approx 15\%$ of the age of the universe, emphasizing how the two surveys have complemented each other to help us unveil the fundamental traits of this population. More empirical information will be necessary (at high as well as intermediate redshifts) before we can map the Lyman-break galaxies onto systems at lower redshifts along an evolutionary sequence with some confidence. However, there is a good deal of fruitful work that can be started now by comparing the predictions of cosmological theories to the properties of the galaxies that the Lyman-break technique, both in its ground-based and space-borne incarnation, has allowed us to explore.

1. The Lyman-break technique and the HDF: How did they come to be?

Exploiting the Lyman limit as a clear spectral feature to identify distant star-forming galaxies is hardly a new idea (e.g., Partridge & Peebles 1967). However, its implementation as an observational technique had to wait some 25 years or so, until detectors became sensitive enough (particularly in the U band) to make it practical with the available telescopes. Early observations that made use of deep CCD multi-band imaging to detect very high-redshift galaxies by their faintness in the U-band (see Figure 1) and constrain the abundance of galaxies at $z > 3$ were obtained by Guhatakurta et al. (1990) and Songaila et al. (1990). However, these works only set approximate upper limits to the surface density of these systems that turned out to be of rather limited use.

The first to exploit the Lyman-break technique to actually single out very high-redshift galaxies from the deep counts with relatively high efficiency were Steidel & Hamilton (1993). Their first observations were obtained in rather poor seeing conditions, were only of modest depth (by modern standards), and covered a limited sky area to produce a large sample of robust candidates. Significantly deeper and better quality data were obtained in 1994 by Giavalisco et al. (unpublished), exploiting the image quality available at the ESO New Technology Telescope; and by Steidel et al. (1995) at the William Herschel Telescope. Later in 1995, additional deep images were obtained using the 200-inch Hale

† Hubble Fellow

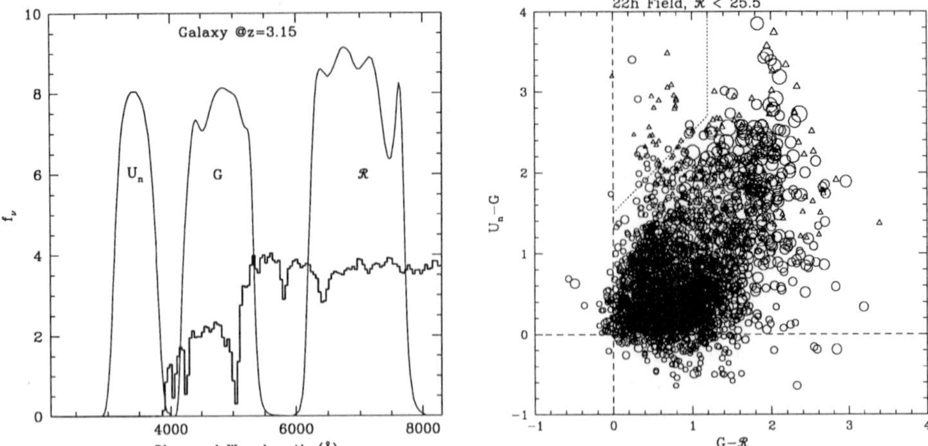

FIGURE 1. **a)** Color-selection of star-forming galaxies at $z \sim 3$. The transmittance of the $U_nG\mathcal{R}$ filters are plotted. The histogram shows the UV SED of an unreddened continously star-forming galaxy at $z = 3.15$, with the flux in an arbitrary scale. **b)** Color-color plot from deep $UnG\mathcal{R}$ imaging in the SSA22 field obtained at the 200-inch Palomar telescope with the COSMIC camera. Magnitudes are in the AB scale. The selection window for Lyman-break galaxies is marked by the dashed line.

telescope and the large field-of-view COSMIC camera at its prime focus. Although these data resulted in relatively large samples of robust $z \sim 3$ Lyman-break galaxy candidates (or U-band "dropouts" in the current parlance), they had to remain such for one more year. Due to the faintness of these candidates (typically $\mathcal{R} \gtrsim 24$), 4-m class telescopes are impractical to secure follow-up spectroscopy, and only redshifts of unusual candidates were known by the end of 1995 (Giavalisco et al. 1994a). No definitive proof could be provided for the existence of a population of galaxies at significant redshifts without the power of the 10-m class telescopes. Meanwhile, the HDF survey was being conceived and designed.

As Richard Ellis has recounted in this volume, there was initially little consensus on how to design what (as it was already clear) was to become a totally unprecedented survey of faint galaxies. The HDF would have detected galaxies too faint for spectroscopy to be practical in the foreseeable future. Deep imaging per se can only be of limited use if some information on the properties and the redshifts of the galaxies is not available. Plus, very deep *HST* images were already available in the archive (e.g., Dickinson et al. 1995) and simply improving their depth by a magnitude or so seemed only a relatively modest achievement, given the investment in telescope time. A multi-band survey that provided color information seemed the way to proceed if progress had to be made from deep imaging. But which passbands to use and how many? How to divide the available exposure time among the various, differently sensitive passbands to maximize color information and depth?

As we have seen, U-band dropouts were already relatively well known in 1995, with a sample that counted ~ 100 galaxy candidates at $z \sim 3$, some with *HST* morphologies as well. The consistency of such properties as luminosity distribution, surface density (Steidel et al. 1995), *HST* morphology (Giavalisco et al. 1994b; 1996a) and also limited spectroscopic information (Giavalisco et al. 1994a) seemed powerful arguments that suggested the existence of a new and totally unexplored population of galaxies whose redshifts were likely to be well in excess of $z \sim 2$. Thus, transporting to the HDF the same observational techniques adopted for the ground-based survey appeared a rather sensible choice. This would provide the HDF survey with color information and at the same time opened the possibility to directly study unprecedentedly distant galaxies at very faint fluxes. However, for this to be achieved, a significant amount of the available observing time had to be spent with the poorly efficient U passband (i.e., the F300W filter of the WFPC-2 collection). Although it was initially received as bizarre spending as much as half of the available exposure time imaging at wavelengths where the throughput of *HST*+WFPC-2 is very low, in the end considerations of practical nature (see Ferguson and Dickinson, this volume) made the decision of including the F300W filter in the HDF survey a rather easy one to take. Paradoxically, thanks to the low orbit of *HST*, the observing time that the Director of ST ScI allocated to the project could be effectively used imaging at wavelengths where the throughput is relatively high (i.e., F450W, F606W and F814W) and still get the precious F300W data essentially for free during the bright part of the orbit of the telescope. The HDF survey was made sensitive to the U-band dropouts.

In September 1995, we were able to follow up the NTT and Palomar samples of U-band dropouts with the Keck 10-meter telescope and LRIS spectrograph (Oke et al. 1995), confirming the existence of a massive population of star-forming galaxies at $z \sim 3$ (Steidel et al. 1996a). In December 1995, the HDF was successfully carried out, and by January 15, 1995 the first versions of the reduced images and of the source catalog were released to the community. During the same month, we obtained the first spectra of U-band dropouts selected from the HDF and confirmed their high redshifts (Steidel et al. 1996b). Other spectroscopic follow-ups of high-redshift candidates came shortly afterwards (Lowenthal et al. 1997).

2. Lyman-break galaxies

The Lyman-break technique consists of a set of color criteria to identify star-forming galaxies at high-redshift through multi-band imaging across the 912 Å Lyman-continuum discontinuity. At high redshifts ($z \gtrsim 2.5$ for ground-based telescopes and $z \gtrsim 2$ for the HDF) the Lyman limit is shifted far enough into the optical window—into the U passband in particular—to be detectable with broad-band photometry. By placing filters on either side of the redshifted break (see Figure 1) one can identify high-redshift objects by their faintness in the U band and by an otherwise blue spectral energy distribution. Using redder passbands, the technique can be used to look for galaxies at even higher redshifts. For example, one can define similar criteria for B-band dropouts which would be galaxies candidates at redshifts $3.5 \lesssim z \lesssim 4.5$ and so on.

More details can be found in Mark Dickinson's contribution in this volume, who describes the Lyman-break technique applied to the HDF. Here we will simply review the salient features of the ground-based survey. For this, we have adopted a custom photometric system, named $U_n\,G\,\mathcal{R}$, optimized for selecting LBGs with redshifts around $z \sim 3$ (Steidel & Hamilton 1993). An initial selection region in the $[G-\mathcal{R},\, U_n-G]$ plane where to expect high-redshift candidates was identified based on the expected colors of moderately unreddened star-forming galaxies computed using stellar population synthesis codes (Bruzual & Charlot 1998, in prep.), and including the effects of the opacity of interstellar gas and intervening absorption by HI (Steidel et al. 1995; see also Madau 1995). Our selection criteria were subsequently verified and refined after extensive spectroscopy with the Keck telescopes.

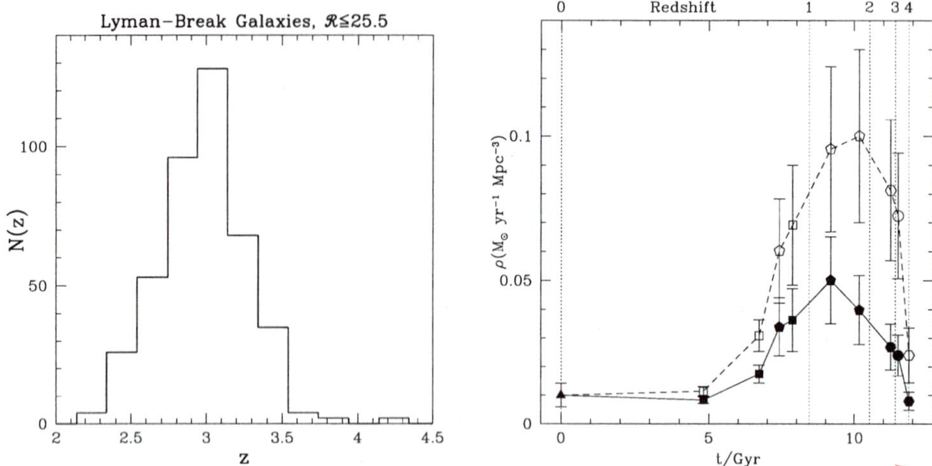

FIGURE 2. **a)** The redshift distribution function $N(z)$ of 418 LBGs obtained with the Keck telescope and the LRIS spectrograph. The bin size is $\Delta z = 0.2$, and the median redshift is $z = 3.01$. The interval $2.4 \lesssim z \lesssim 3.6$ contains $\approx 90\%$ of the galaxies. **b)** The cosmic star-formation activity as a function of time in linear scale. Filled symbols are uncorrected data. Open symbols are data corrected for dust obscuration (see text). The triangle is the $H\alpha$ point from Gallego et al. (1995). Squares are the CFRS points (Lilly et al. 1995). Pentagons are the photometric redshift from Connolly et al. (1997). Pentagons and circles are the HDF and $U_nG\mathcal{R}$ LBGs points, respectively. $H_0 = 50$ km^{-1} Mpc^{-1} and $q_0 = 0.5$.

An object is considered a Lyman-break galaxy if its colors satisfy

$$(U_n - G) \geq 1.0 + (G - \mathcal{R}); \quad (G - \mathcal{R}) \leq 1.2, \tag{1}$$

with an additional requirement $\mathcal{R} < 25.5$ imposed to produce a reasonably complete sample that is suitable for spectroscopic follow-up (magnitudes are in the AB scale of Oke & Gunn, 1983). Figure 1b shows an example of the color diagram and selection window that we have used to identify the high-redshift candidates. At the time of this writing, the ground-based survey consists of more than 1,200 U-dropouts brighter than $\mathcal{R} \sim 25.5$, of which 418 have been spectroscopically confirmed in the range $2 < z < 4$. Figure 2a shows the redshift distribution $N(z)$ of the current spectroscopic sample. In comparison, the HDF sample is significantly smaller, including 30 candidates with $V_{606} < 25.5$, 22 of which have secured redshifts (see Dickinson's paper). However, it is considerably deeper, with 187 candidates at $V_{606} < 27$, a flux level where its fractional completeness is approximately equivalent to the ground-based sample. Unfortunately, the redshift distribution of this faint sample is empirically poorly constrained.

As it turns out, the Lyman-break selection is particularly efficient to single out high-redshift galaxies from the deep counts, and we have found that at least 75% of the objects meeting the criteria of Eqn. (1) are indeed high-redshift galaxies. The median value of the redshift distribution in Fig. 2a is $\bar{z} = 3.01$ and the standard deviation is $\sigma_z = 0.29$, with $\approx 90\%$ of the objects having redshifts in the range $2.4 \lesssim z \lesssim 3.6$. About 5% of the objects meeting these criteria are stars, essentially all brighter than $\mathcal{R} \sim 24$. The remaining 20% of objects remain unidentified because of the low S/N of their spectra. However, it is important to note that these are still consistent with being at high-redshifts, either the same as the confirmed galaxies or slightly smaller (e.g., $1.8 \lesssim z \lesssim 2$). Also, note that our success in obtaining a redshift has no obvious dependence on luminosity or color.

3. The spectra of the Lyman-break galaxies

With 418 galaxies, the spectroscopic sample of Lyman-break galaxies is a rich source of information on the physical properties of star-forming galaxies at high redshifts. In addition to the optical data, the current sample also includes 5 near-IR (K-band) spectra obtained with the UKIRT telescope and the CGS4 spectrograph by Pettini et al. (1998, in preparation). In most cases, the S/N of the "discovery spectra," i.e., of the optical spectra obtained to secure the redshift of the galaxies, is relatively modest. Furthermore, the near-IR spectra, being obtained with a 4-m class telescope, have limited depth. Nevertheless, the amount of information contained in the data is sufficient to discuss important properties of the galaxies, as we shall now review.

3.1. *The rest-frame UV spectra*

At $z \sim 3$ the optical spectra sample the rest-frame far-UV radiation, approximately between 1000 and 2000 Å, carrying information on massive stars and on the interstellar gas. Overall, the spectra are all qualitatively similar and also bear a rather close resemblance with those of local star-forming galaxies, consistent with a relatively small dispersion of properties of the powering source and of the physical state of the interstellar absorbing medium. However, a closer inspection reveals differences, both between the Lyman-break galaxies themselves and with the local systems, particularly in the kinematics of the interstellar absorption lines and in the Lyman-alpha properties.

Figure 3 shows two examples of spectra that illustrate the variety observed among the Lyman-break galaxies along with the UV spectrum of NGC 4214, a Wolfe-Rayet galaxy observed with *HST* and the GHRS by Leitherer et al. (1996). As mentioned above, in both cases the similarity between the spectra of the Lyman-break galaxies and the local galaxies is striking. In each case, the dominant characteristics of the far-UV spectra are (1) the flat continuum, whose spectral index is approximately $f_\nu \sim \nu^0$ or slightly redder; (2) the weak or absent Lyα emission, whose equivalent width is, in general, substantially smaller than the predictions of the radiation-bound case-B recombination theory; (3) the strong interstellar absorption lines due to low-ionization species of C, O, Si and Al; (4) the prominent high-ionization stellar lines of HeII, CIV, SiIV, NV.

The two spectra, do show differences, however. The first has no detected Lyα emission and has a broad absorption feature at its wavelength together with deep and relatively narrow interstellar absorption lines. The second, on the contrary, is characterized by broader and shallower interstellar lines and by the Lyα emission line, whose equivalent width, however, is modest, ≈ 10 Å in the rest frame.

The S/N of these "discovery spectra" is insufficient to detect other stellar features. However, it is very interesting that whenever we could follow-up at higher S/N and dispersion some bright Lyman-break galaxy, we have always detected weak stellar features such as the photospheric lines of SVλ1502 and OIVλ1343, and SiIIIλ1417, characteristic of O stars.

The observed UV continua result from the integrated light of massive O and early-B stars. A galaxy at $z = 3$ with $\mathcal{R} \sim 24.5$ has a far-UV luminosity density $\mathcal{L} \sim 1 \times 10^{41}$ (2.5×10^{41}) h_{50}^{-2} erg s^{-1} Å^{-1} at $\lambda \sim 1500$ Å, with $q_0 = 0.5$ (0.1), respectively (we will follow this convention throughout this paper, unless explicitely indicated). This is about three order of magnitudes the luminosity density of the brightest local star clusters studied with *HST* (see, e.g., Leitherer et al. 1996). Given the short life-time of O and B stars, it is relatively straightforward to convert the observed absolute UV luminosities into instantaneous rates of star formation once an IMF has been assumed. Adopting a continuous star-formation activity with a Salpeter IMF from 0.1 to 100 M$_\odot$, a galaxy at $z = 3$ with $\mathcal{R} \sim 24.5$ forms stars at the rate of 12 (29) h_{50}^{-2} M$_\odot$ yr^{-1} if $q_0 = 0.5$ (0.1).

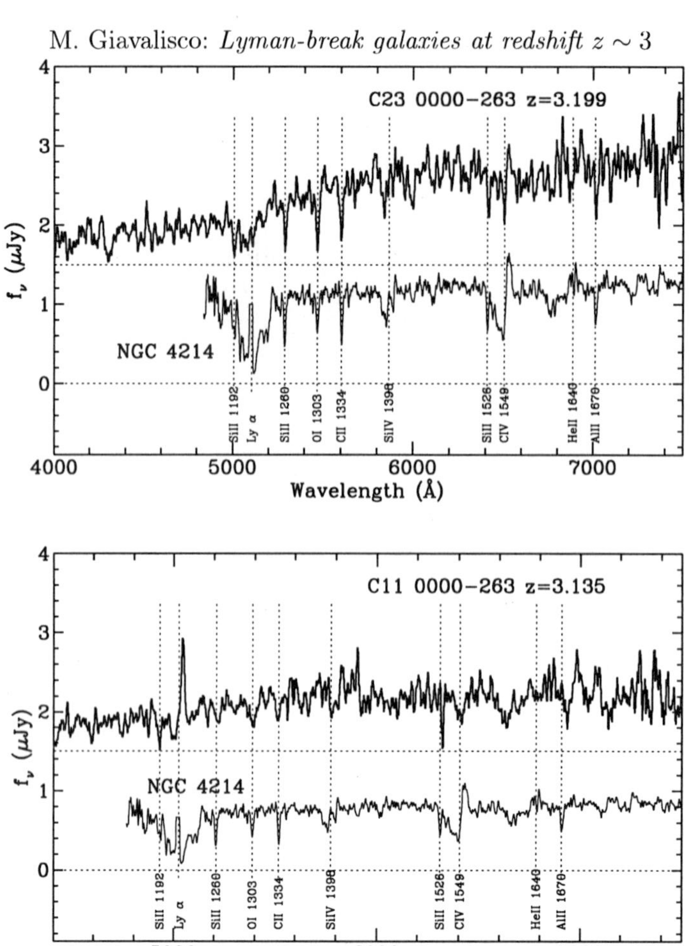

FIGURE 3. Two examples of Keck spectra of Lyman-break galaxies plotted together with the spectrum of a star-forming knot in the Wolfe-Rayet galaxy NGC 4214 (Leitherer et al. 1996) placed at the same redshift. The flux zero point of the high-z galaxies is offset relative to the zero point for the scaled spectrum of NGC 4214 (indicated with the lower dotted line). Some of the identified spectral features (both stellar and interstellar) have been labeled for comparison. These two cases give an idea of the variety of spectra observed among the LLG population. In the top panel Lyα is in absorption together with strong interstellar absorption lines. The CIV P-Cygni feature is also visible. The galaxy in bottom panel has Lyα in emission, albeit with rest-frame equivalent width $E_w = 8$ Å and less strong but broader interstellar features. There is a strong absorption feature at the wavelength of the CIV, but no emission.

These are almost certainly lower limits, since dust extinction and a shorter burst age result in larger rates (a burst with an age of $\approx 10^7$ yr implies a SFR 1.7× larger than the values above).

Calzetti et al. (1994) have shown that the far-UV spectral index is a robust indicator of dust reddening because it depends weekly on the exact shape of the IMF and the age of the burst. Both models and empirical templates of starburst galaxies shows that the spectral index expected in absence of dust obscuration is significantly bluer than what is typically observed in the Lyman-break galaxies, as is schematically illustrated in Figure 4. Although such a relatively red spectrum can be produced by stellar populations with a

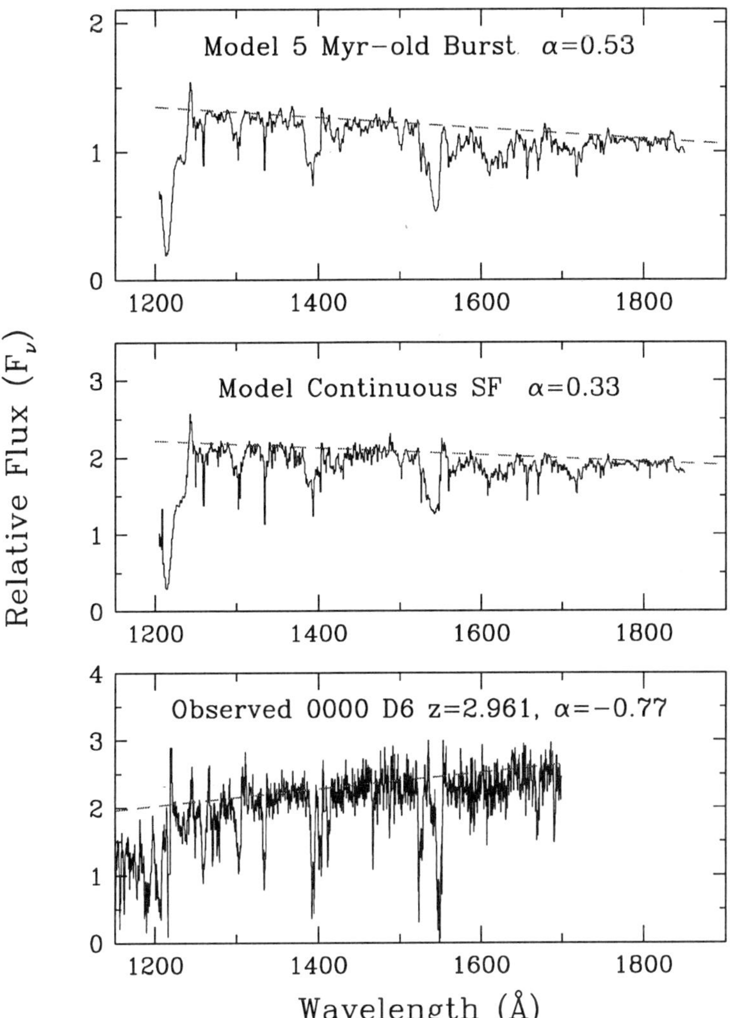

FIGURE 4. Spectral indexes of models of burst and continuous star-formation rates and of one observed Lyman-break galaxy. This last is characterized by the presence of photospheric lines of O stars, suggesting that dust obscuration, as opposed to lack of massive stars, is responsible for its reddening. The flux is in an arbitrary scale. The models are from Leitherer et al. (1995)

significant deficiency of massive stars (due either to a shallower IMF or an evolved burst), this possibility seems unlikely since even in the relatively red spectrum of Figure 4 itself we were able to directly see the photospheric lines of O stars. Thus, the presence of dust obscuration (as opposed to a deficiency of very massive stars) seems the explanation for the observed UV continua.

A more detailed description of the amount of dust obscuration and the correcting factors to recover the intrinsic star-formation rates are discussed in Dickinson's paper. Such corrections are in the range $\approx 3\times$–$7\times$, depending on the exact shape of the extinction curve, and although of modest amount, they are nevertheless of sufficient amplitude to affect our conclusions about the nature and possible evolutionary history of both the single Lyman-break galaxies and of the overall population in relation to the general problem of galaxy formation, as we will discuss later.

The Lyα emission of the LBGs is very often weak or absent despite the rather high star-formation rates implied by the UV continua, in agreement with the generally null results of search for "primeval galaxies" based on the presence of this emission line (e.g., Thompson et al. 1995). Such a relatively weak Lyα is very likely the result of resonant scattering in presence of dust in an outflowing interstellar medium (Charlot & Fall 1993). The Lyα, when detected, is generally redshifted respect to the interstellar absorption lines by several hundred km s^{-1} and its profile is often asymmetric. The redshifts of both the Lyα and of the interstellar lines are not likely to reflect the systemic redshift of the galaxies. In the few cases of high S/N spectra where photospheric lines from O stars have been observed (SVλ1502 and OIVλ1343) and also when the optical nebular lines ([OII], Hβ and [OIII]) have been observed with near-IR spectra, the redshifts of these features have been found to exceed that of the interstellar lines by up to ~ 500 km s^{-1}. Or, in other words, if the bulk of the stars define the systemic redshift of the galaxies, the interstellar lines are blueshifted respect to them. Taken together, the phenomenology of the Lyα the interstellar, stellar and nebular lines is consistent with the presence of large-scale outflows in the interstellar medium of the Lyman-break galaxies, very likely the consequence of the injection of kinetic energy by stellar winds and supernova events (Pettini et al. 1997). Large-scale motions of expanding shells are very likely the main reason for the large strengths of the interstellar lines, which typically have equivalent widths of a few Angstroms. Since these lines are saturated, these equivalent widths are much more sensitive to the velocity fields in the gas than to its metallicity.

3.2. *The rest-frame optical spectroscopy*

We have also obtained near-IR spectroscopy for some Lyman-break galaxies in our sample, which allowed us to study the optical rest-frame spectrum of these systems (Pettini et al. 1998, in prep.). These observations have been carried out at the UKIRT 3.8-m telescope with the CGS4 spectrograph and had, therefore, to be limited to the brightest galaxies of our sample. Even in this case, we could essentially only study the nebular emission lines of [OII], Hβ and [OIII] of the galaxies and only barely detect their continua (see Figure 5). Nevertheless, these data have provided very useful complementary information on the amount of extinction and the kinematics of Lyman-break galaxies.

There is a broad agreement between the star-formation rates derived from the Hβ fluxes and the ones obtained from the far-UV continuum luminosity density, once moderate corrections for dust obscuration are included. A typical net correction to the observed star-formation rates would be again in the range of 3$\times$ to 7$\times$, depending on the adopted obscuration law. This seems to exclude, at least in the galaxies for which near-IR spectroscopy is available, large amounts of dust obscuration, as have been suggested by some (e.g., Meurer et al. 1997). Interestingly, using the Calzetti's law (1997) to compute Hβ fluxes from the UV continuum results in values that are in line with the observed ones only if continuous star formation is assumed, while significantly higher fluxes are found in the case of a burst (Pettini et al. 1998, in preparation).

As discussed above, the near-IR spectra have also shown differences of up to ~ 400 km s^{-1} between the redshifts of the UV absorption lines and those of the optical nebular emission lines, with the latter having higher redshift than the former and being more likely to be closer to the systemic velocities of the galaxies than the UV absorption lines. The lines are all resolved and their widths, if interpreted in terms of velocity dispersion, yield values in the range $50 \lesssim \sigma \lesssim 160$ km s^{-1}. If the kinematics of these lines is a good tracer of the gravitational motions within the galaxies, then combining the velocity dispersion with the sizes derived from the *HST* images gives dynamical masses in the range 10^{10}–10^{11} M$_\odot$, comparable to the mass of the Milky Way bulge (Dwek et al. 1995)

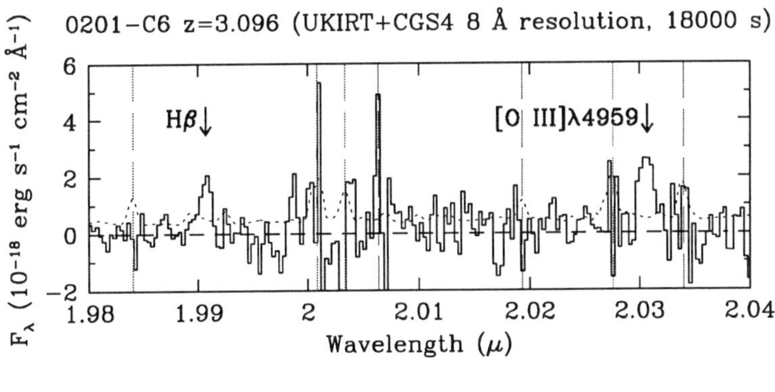

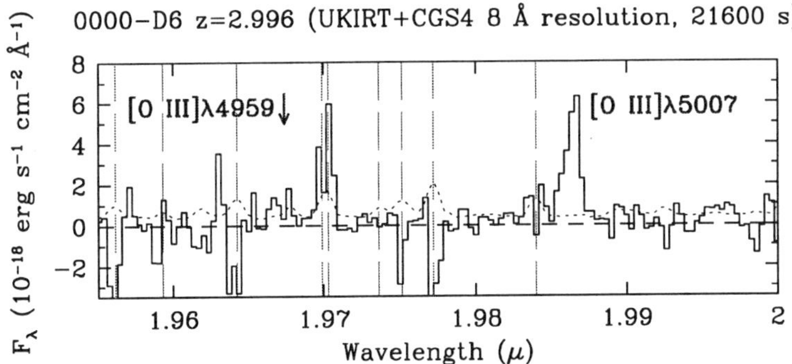

FIGURE 5. Two examples of near-IR spectra (K band) of Lyman-break galaxies obtained with the UKIRT telescope and the CSG4 spectrograph (Pettini et al. 1998, in preparation). The observed nebular emission lines are labeled. The vertical dashed lines mark the position of major night sky lines. In each case, the continuum emission is just barely detected. The emission lines are resolved and their velocity widths correspond to velocity dispersions in the range of $50 \lesssim \sigma \lesssim 160$ km s^{-1}.

and to the mass of the innermost few kpc of an L^* elliptical galaxy. While the extent to which the optical nebular lines can be used as dynamical indicators is not known, we note that the total masses involved are very likely to be substantially greater, given that the present IR observations sample only the innermost cores of the galaxies, where the star formation rates are highest.

4. The Lyman-break galaxy population

The Lyman-break galaxies represent a substantial population of star makers at a time when the age of the universe was about 15% of its current value. The ground-based survey is sensitive to systems at redshifts ~ 3 (see Figure 2a), whose surface density down to $\mathcal{R} < 25.5$ is ~ 1.2 arcmin^{-2}, or $\approx 5\%$ of the faint counts at the same magnitude. This corresponds to a comoving volume density of $8.5 \times 10^{-4}\ h_{50}^3$ Mpc^{-3} $(1.6 \times 10^{-4} h_{50}^3)$ if $q_0 = 0.5$ (0.05), or about the same density (1/5) of the present-day galaxies with $L \sim L^*$.

The HDF Lyman-break galaxies are more numerous, with a surface density at the same magnitude limit that is larger than that of their ground-based counterparts by a factor $\approx 7\times$ (see Mark Dickinson's contribution in this volume for a more complete discussion of this point). In part, this can be explained by the fact that the redshift distribution of the HDF galaxies extends to smaller values than the ground-based one, because the F300W filter is ~ 600 Å bluer than the U_n. Therefore, for a given magnitude limit, the

HDF probes a fainter portion of the absolute luminosity distribution and also a larger comoving volume per unit area. It is also likely that observational effects, such as the higher angular resolution and the deeper flux limit in the U band of the HDF respect to the ground-based survey, result in a higher number of candidates. However, we cannot exclude that the higher abundances of HDF galaxies can also be due to evolutionary effects. The HDF redshift distribution includes a larger fraction of galaxies around $z \sim 2$ and if at this epoch star-forming galaxies were more numerous than at $z \sim 3$ (particularly those with fainter UV absolute luminosities), this results in higher surface density of candidates.

With observed star-formation rates that reach ~ 50–$70\, h_{50}^{-2}$ M$_\odot$ yr^{-1} (with $q_0 = 0.5$) only for few very bright cases, at first sight it might seem that the Lyman-break population lacks the powerful star makers expected in the models of "monolithic collapse" for the formation of elliptical galaxies and spiral bulges. In these models, the galaxies assembled most of their stellar mass in powerful burst at high redshifts and subsequently evolved passively (e.g., Eggen, Lynden-Bell & Sandage 1962; Partridge & Peebles 1967). But as we have seen, dust obscuration is very likely directly responsible for a factor ~ 3–7 of attenuation, and the dereddened star-formation rates are actually in line with the expectations of models where the spheroids assembled most of their stars at high redshifts. For example, a galaxy with $\mathcal{R} = 24.5$ would be forming stars at a rate of ~ 60 (145) h_{50}^{-2} M$_\odot$ yr^{-1} after dereddening by a factor of 5, while the brightest galaxies ($\mathcal{R} \lesssim 23$) would have sfr ~ 400 (1000) h_{50}^{-2} M$_\odot$ yr^{-1} or more.

The age of the universe at redshift $z \sim 3$ is about 15% of its present value, so $\sim 10^9$ yr of sustained star formation with the above rates that started at $z \sim 4$ or so can easily accommodate an M^* (i.e., 10^{11} M$_\odot$) worth of stars by $z \sim 2$ which will, from that epoch on, evolve mostly passively. In view of the fact that approximately half of the local stellar mass density is segregated in spheroidal systems (Schechter & Dressler 1987) and that these stars are among the oldest known, it is very interesting to ask which fraction of the cosmic stellar mass-density was produced in the Lyman-break galaxies at $z \gtrsim 2$. Figure 2b shows a revision to the popular "Madau's plot" (Madau et al. 1996), namely the star-formation density as a function of redshift/cosmic time, once some dust corrections are taken into account. We have chosen to plot the data in linear scale and using the cosmic time in place of redshift to provide a more intuitive idea of the amount of evolution and the time scale of the cosmic star formation activity. The data relative to Lyman-break galaxies have been corrected by a factor of 5, while those from the CFRS a factor of 2, in general agreement with the recent comparison between Hα and UV luminosities discussed by Tresse & Maddox (1998).

The conclusion that the cosmic star formations peaks somewhere between $z = 1$ and $z = 3$ is probably still valid. However, the relative proportions of stars that have been produced at different epochs change considerably. By integrating the curve relative to the unreddened data it is found that about 15% of the stars was formed at $z > 2$, while using the dereddened curve yields $\approx 35\%$ of the stars. The exact fraction depends on the amount of correction of both the high- and moderate-redshift data points. The former is affected by the shape of the extinction law, the latter is uncertain because it depends on the star-formation history of the galaxies. However, despite these uncertainties its seems very likely that a significant fraction of the stellar content of the universe has been assembled at $z > 2$ in the Lyman-break population.

Even if the Lyman-break galaxies are responsible for the oldest stellar populations of the present-day universe (e.g., those segregated in the spheroids) we still do not know [illegible] these stars were collected together to form the stellar bodies of elliptical galaxies [illegible]l bulges. Are the LBGs sub-galactic fragments observed during an intense phase

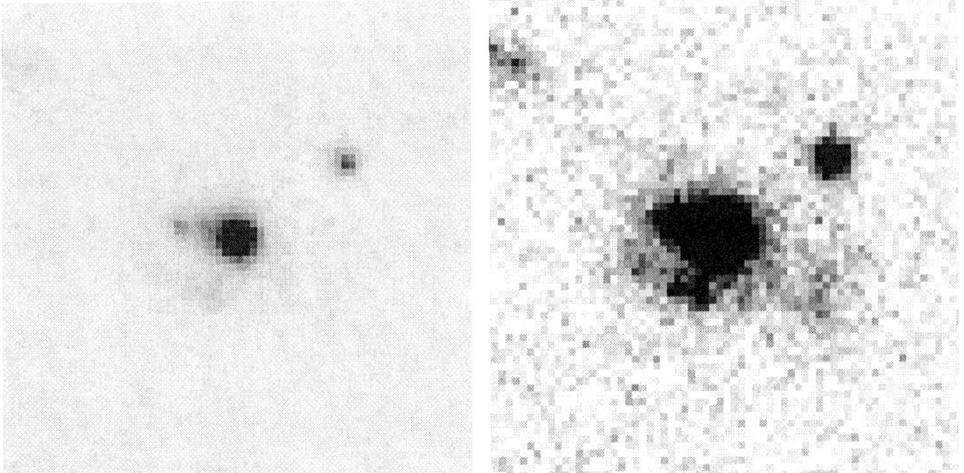

FIGURE 6. WFPC2/PC image through the F702W passband of galaxy 0000-263-D6 at redshift $z = 2.961$. **a)** The contrast is adjusted to show the compact and regular "core" of the galaxy, which is very well modeled by a $r^{1/4}$ radial profile, with $r_e = 0.25$ arcsec. **b)** A different contrast shows the diffuse and structured "halo", where disturbances are observed as a blob to the NE (actually merged with the core isophotes, but clearly visible in the left panel), a plume to the S and an extended nebulosity to the W. The compact source to the NW of the galaxy is at a significantly lower redshift. Both panels are 3.13 arcsec in size.

of star-formation that are destined to subsequently merge and form massive galaxies, as predicted by the hierarchical cosmological theory (e.g., Baugh et al. 1998)? Or are they already massive galaxies during an early evolutionary phase? Which fraction of them has passively evolved (i.e., did not undergo major merging events) until the present days? It is intriguing that these galaxies have overall morphologies and sizes that (as we are about to discuss) are similar to those of the present-day spheroids, and a volume density that can account for a significant fraction (depending on the choice of the cosmological parameters) of present-day galaxies with $L \sim L^*$. However, their evolutionary history is presently unconstrained on an empirical basis. Important pieces of information will come from reliable estimates of the duration of the star-formation activity of the Lyman-break galaxies and their mass spectrum. These issues can be tackled observationally through deep multi-band spectro-photometry and high-resolution spectroscopy at optical and near-IR wavelengths with 10-m class telescopes.

5. Morphology

At the time of this writing, only the morphology of the far-UV light of Lyman-break galaxies has been studied with *HST* in some detail in both the ground-based and HDF sample (Giavalisco et al. 1996a; Steidel et al. 1996b; Lowenthal et al. 1997). These studies have revealed a dispersion of properties. Galaxies from the ground-based sample are characterized by rather compact and relatively regular morphologies, while those from the HDF, which are on average considerably fainter, include a larger number of more diffuse, irregular and apparently "disturbed" cases.

Figure 6 shows the *HST* + WFPC2 image of 0000-263-D6 at $z = 2.961$, an example of Lyman-break galaxy with compact morphology from the ground-based sample (Giavalisco et al. 1998 in prep.). The image has been taken through the F702W passband and samples the rest-frame spectrum at $\lambda \sim 1700$ Å. We have displayed it with two different levels of contrast to show the compact "core" of the galaxy and the surrounding diffuse nebulosity or "halo" that extends up to a few arcsec. This is representative of the morphology of the compact Lyman-break galaxies. The cores segregate $\sim 90\%$ of the total light and tend to have regular morphologies, characterized by a relatively steep light profile that is often well approximated by a $r^{1/4}$ law, but exponential profiles have also been observed (Giavalisco et al. 1996b). Surrounding the core there is a diffuse "halo" with an irregular light distribution, sometimes harboring faint, compact sub-structure. The half-light radii of the cores are typically ~ 0.2–0.3 arcsec or ~ 2.2–3.4 (~ 1.5–2.2) h_{50}^{-1} kpc at $z = 3$, while the observed isophotal diameters (lower limits to the intrinsic sizes due to the extreme $(1+z)^4$ surface brightness dimming), are typically 1.5–2 arcsec or ~ 17–22 (~ 11–15) h_{50}^{-1} kpc.

The Lyman-break galaxies from the HDF have UV morphologies that tend to be less regular and more fragmented than their ground-based counterparts. In part, this can be due to the increased depth of the HDF over the WFPC-2 images of the ground-based sample. This has certainly made fainter features being detected with a relatively high S/N ratio and these can now contribute to the overall appearance of the structures. However, we must remember that most of the HDF galaxies belong to a fainter portion of the luminosity function and have, on average, a smaller redshift. Only a few of them have luminosity that can be directly compared with the ground-based ones (Steidel et al. 1996b). More importantly, as mentioned above and discussed by Dickinson, the HDF data seem to suggest that some amount of evolution in the number density of Lyman-break galaxies has taken place in the redshift range $2 \lesssim z \lesssim 3$. Because we do not know the redshift distribution of the fainter HDF U-band dropouts—this could plausibly be dominated by galaxies at $z \lesssim 2$—it is not entirely clear that the HDF and ground-based Lyman-break galaxies belong to the same "population". It would be quite plausible that a insurgent population at $z \sim 2$ be characterized by a more irregular UV morphologies than the one at $z \sim 3$.

Overall, the morphology of the UV light of the Lyman-break galaxies is suggestive of relaxed structures. In particular, the fact that these morphologies are relatively regular for galaxies with high absolute luminosity (large star-formation rates) and become more irregular and fragmented at fainter absolute luminosities reinforces the idea that star formation takes places in virialized dark halos with an efficiency that is directly related to their mass and/or the degree of relaxation. The presence of relaxed structures at these redshift is certainly not surprising and it is actually consistent with the expectations of most theories of galaxies formation, which predict that the most actively star-forming regions are located in correspondence of the bottom of the potential well of virialized dark halos (e.g., Baugh et al. 1997). What it is interesting here is that the observed sizes are relatively large. As discussed by Giavalisco, Steidel & Macchetto 1996) the half-light radii and isophotal diameters of LBGs are similar to the effective radii r_e and sizes of present-day elliptical galaxies of intermediate luminosities and bulges of spirals. At $z \sim 3$ these are about one order of magnitudes larger than the expectations from the semi-analytical models based on the CDM theory (e.g., Baugh et al. 1998). Furthermore, the theory predicts systems dominated by a disc morphology, while the observations show the predominance of spherically symmetric structures, with very few cases of highly elongated systems (an example of these rather rare cases is 0000-263-C10 whose optical and near-IR images are shown in Figure 8, top row).

In this respect it is interesting to compare the UV morphologies of high- and low-redshift galaxies (Giavalisco et al. 1996). Figure 7 shows local galaxies, observed with UIT and the SW camera at ~ 1550 Å, as *HST* would have observed them in the F606W passband if they were placed at redshift $z = 3$ (first and third column of the figure except

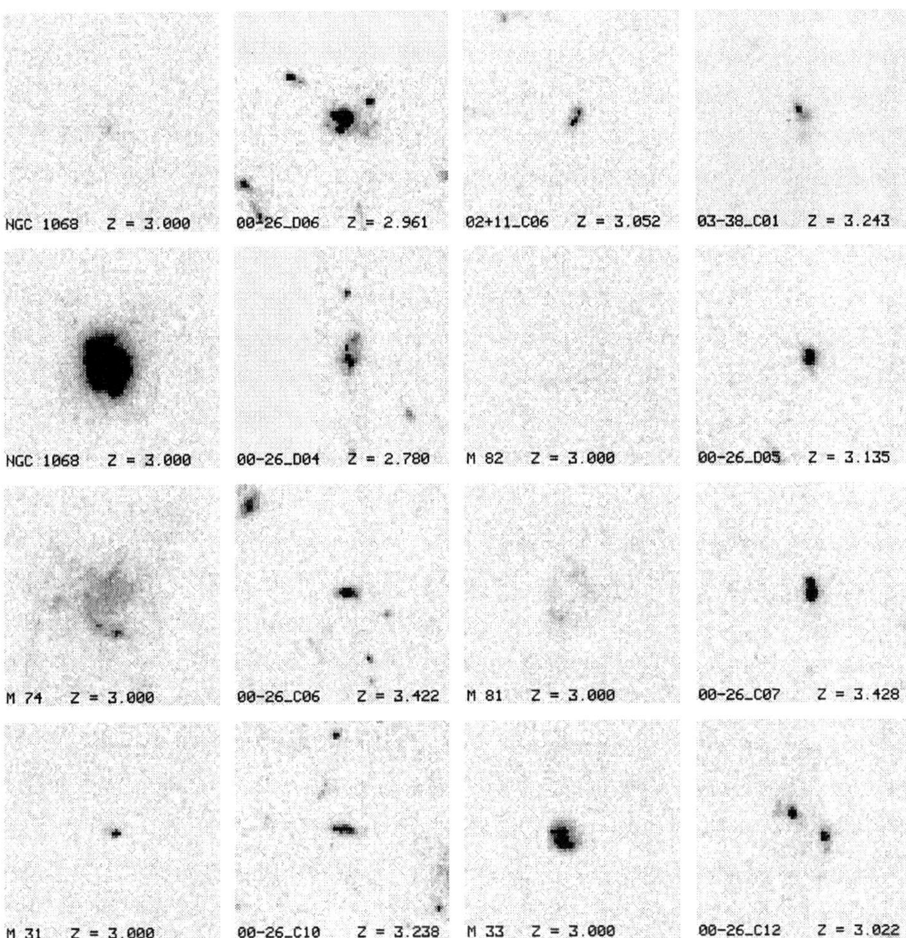

FIGURE 7. WFPC2+F606W images of LLGs and local galaxies as *HST* would observe them if placed at $z = 3$. We had to boost their surface brightness $100\times$ to detect them in 5 hr except NGC 1068, shown before and after the boost. Panels are 7 arcsec in size and $q_0 = 0.1$.

the top panel of 3rd column. See labels), compared against *HST* LBGs. The images are all photometrically consistent, however we had to boost the surface brightness of the local galaxies by a factor of 100 to detect them with the same exposure time of the *HST* observations (5 hours). This exercise shows that 1) the sizes of the inner regions (i.e., where the oldest stellar population are generally thought to be located) are similar to those of the LBGs; 2) the local galaxies have significantly less star formation per unit volume (proportional to the UV surface brightness) than their distant counterparts and, as the images show, this is distributed in the disc, while the bulges or central regions contribute very little.

In local galaxies star formation is generally localized in small regions that involve a small fraction of the stellar mass. Thus, the UV morphology can be substantially different form the optical one, and the effects of the *morphological k-correction* can be relatively large, shifting the position of the galaxy in the Hubble sequence towards later types at bluer wavelengths. The LBGs, on the other hand, are systems that are assembling the bulk of their stars, and star formation is propagated to a much larger fraction of their

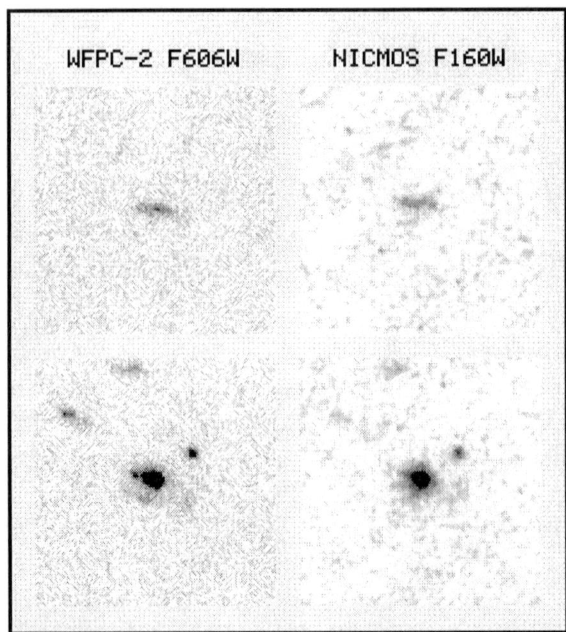

FIGURE 8. Comparison between the rest-frame UV and optical morphologies of 2 Lyman-break galaxies, 0000-263-C10 at $z = 3.238$ (top) and 0000-263-D6 at $z = 2.961$ (bottom). The UV images have been obtained with WFPC-2/PC and the F606W passband (rest-frame ≈ 1500 Å). The optical images are from NICMOS-2 and the F160W passband (rest-frame ≈ 4000 Å). The NICMOS images have been resampled to the PC resolution of 0.046 arcsec/pix. Panels are 4.6 arcsec in size. In both cases, sizes and overall morphology at UV and optical wavelengths are quite similar. Galaxy 0000-263-C10 is an example of the relatively rare cases with elongated morphologies, suggestive of a disc system. The UV light profile of this galaxy is well approximated by an exponential law (Giavalisco et al. 1996b).

structure. Therefore, in this case it is plausible that their UV morphology more closely traces that of the final structure. This seems directly supported by the similarity between the rest-frame UV and optical morphologies of the Lyman-break galaxies revealed by recent NICMOS images (Giavalisco et al. 1998, in prep.), as shown in Figure 8.

6. Clustering

Clustering is a fundamental property of galaxies in that it establishes the link with the background cosmology and with the properties and distribution of the dark matter. The extent to which the perturbations detected in the cosmic microwave background radiation have grown after the first $\sim 15\%$ of the age of the universe (i.e., at the epoch of the U-band dropouts) depends sensitively on the cosmological parameters. In general, interpreting clustering observations of galaxies is complicated by the lack of knowledge on how these trace the mass-density field. However, if one understands the evolution of some well-defined category of galaxies (of which the Lyman-break galaxies are an example), one can hope to constrain the background cosmology by following the clustering evolution of these systems over a substantial range of cosmic time.

The high efficiency of the Lyman-break technique and its well-controlled selection criteria are key features to study galaxy clustering at redshifts as high as $z \sim 4$. The large samples that can be put together with small effort probe the cosmic density-perturbation field much more densely and to smaller scales than other high-redshift galactic systems,

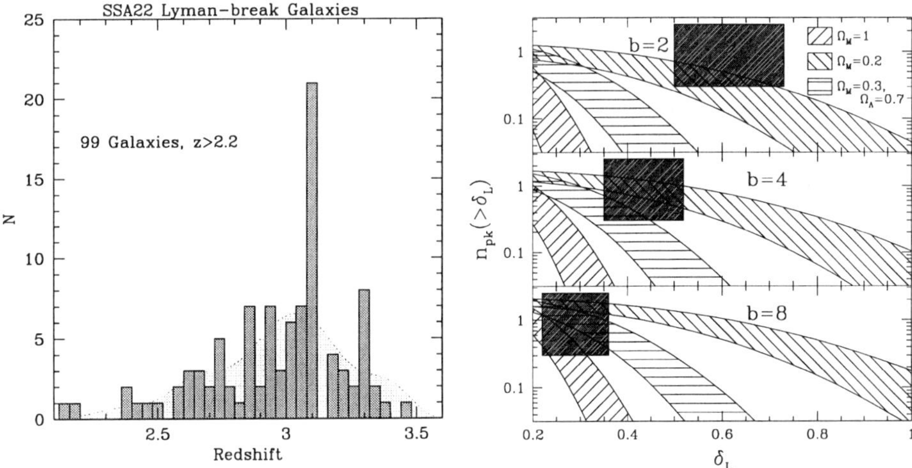

FIGURE 9. **a)** Distribution of LBG redshifts towards the SSA22 field plotted together with the redshift selection function of the whole survey (dashed line). The prominent spike at $z \sim 3.1$ is clearly visible. **b)** Volume density of redshift concentrations with density $\delta \geq \delta_L$ as a function of δ_L for three different CDM cosmologies and for a choice of the light-to-mass biasing parameter together with the values derived from the redshift distribution in the SSA22 field. The size of the boxes represent the error bars. As the figure shows, such redshift concentrations do not exist in the CDM cosmogonies, unless the LBGs are rather biased tracers of the mass.

such as quasars and radio-galaxies. With the LBGs one can study high-redshift clustering and morphology of large-scale structure on spatial scales that are directly relevant to the physics of galaxy formation. Furthermore, this allows for direct comparisons with studies of galaxy clustering in the local and intermediate redshift universe.

6.1. *Large structures at high redshifts*

The spectroscopic survey of Lyman-break galaxies has revealed the presence of large spatial concentrations of these systems already at redshifts $z \sim 3$ (Steidel et al. 1998). Along the line of sight (i.e., along the redshift axis) these concentrations overall resemble the redshift spikes observed at more modest redshifts (e.g., Broadhurst et al. 1990; Cohen et al. 1996). However, the comoving volumes that are probed in each pointing in the case of the Lyman-break galaxies are much larger. As a result, the observed spikes are directly associated to physical grouping of galaxies in narrow volumes and are not the result of the intersection of narrow pencil-beams with the complex topology of the large-scale structure (e.g., Kaiser & Peacock 1991). We have shown that the redshift spikes of Lyman-break galaxies are associated to structures with mass similar to that of rich clusters, or $M \sim 10^{15}$ $M_\odot$. At redshifts $z \sim 3$ this has interesting cosmological consequences.

Figure 9a shows an example of spikes in the SSA22 field, one of the best studied regions of our survey. The shaded regions represent the overall redshift distribution of the whole survey, which is, in practice, the redshift selection function. A large overdensity of galaxies at $z \sim 3.1$ is clearly visible, together with another smaller concentration at redshift $z \sim 3.4$. We find that similar structures are actually ubiquitous and we detect similar overdensities in essentially every field that we have surveyed. As discussed in Steidel et al. (1998), the statistics of these spikes and their size allow us to place constraints on the mass-to-light biasing parameter of the Lyman-break galaxies.

In Figure 9b we plot the function $n(\delta_l)$, namely the volume density of spikes with linear density contrast δ_l or larger, from the CDM power spectrum for three different choices of the cosmological parameters Ω and Λ along with the data from the SSA22 field. As the diagram shows, the concentrations that give origin to the observed redshift spikes would not exist in standard CDM cosmogonies unless the Lyman-break galaxies were very biased tracers of the mass-density field. The amount of bias required depends on the choice of the cosmological parameters and it is rather large—$b \sim 8$—in the case of the standard CDM model, while it is more moderate in an open universe, with $b \sim 2$ if $\Omega = 0.2$ and $\Lambda = 0$.

Thus, if a CDM power spectrum is a good approximation to the true one, the implications are that the Lyman-break galaxies are rather biased tracers of the mass and an immediate conclusion would be that they are more strongly clustered than the mass. As we will show in the next section, the correlation function of the LBGs provides direct empirical support for this. In turn, since the most massive dark halos are the most strongly clustered and heavily biased (Mo & Fukugita 1997), the derived bias implies that the characteristic mass associated with the Lyman-break galaxies is large. For the choices of cosmological parameter shown in Figure 9b, the derived values of the bias point to halos with mass $M \sim 10^{12}$ $M_\odot$. This is a strong evidence in support of the interpretation that the Lyman-break galaxies are associated with massive structures, comparable to present-day bright (L^*) galaxies.

6.2. *The clustering of Lyman-break galaxies*

The high efficiency of the Lyman-break technique and the relatively narrow range of redshifts/cosmic time that it probes make angular clustering a particularly economic means to study large-scale structure at high redshifts, once the redshift distribution $N(z)$ of the galaxy candidates has been measured. Not only is one free from securing a complete spectroscopic follow-up of all the candidates, but the systematics due to selection effects are easier to handle than those that affect studies of spatial clustering using the full redshift information. Knowing $N(z)$, the real-space correlation function can be accurately derived from the angular one through the Limber transform (Peebles 1980; Efstathiou et al. 1991). It turns out that the inversion of the $w(\theta)$ is rather insensitive to the still relatively large uncertainties on the angular function, and the spatial correlation length is much more tightly constrained than this (Giavalisco et al. 1998).

The angular correlation function $w(\theta)$ is defined in terms of the excess probability over the random (poisson) distribution of finding a companion in an angular shell of size $d\Omega$ placed at an angular separation θ from a selected galaxy, given the surface density of sources $\mathcal{N}$ (Peebles 1980):

$$dP = \mathcal{N}\left[1 + w(\theta)\right] d\Omega. \tag{2}$$

We used the 5 largest and deepest fields of our survey to produce a weighted average angular correlation function $w(\theta)$ of LBGs. Figure 10a shows the data points, their error bars and the fitted power-law model.

We subsequently inverted the angular correlation function using the Limber transform and the observed redshift distribution $N(z)$ to derive the parameters of the spatial correlation function at the epoch of observations, namely the correlation length r_0 and the slope $\gamma = 1 + \beta$, under the assumption of the power-law model $\xi(r) = (r/r_0)^{-\gamma}$.

To estimate confidence intervals on the parameters γ and r_0, we used montecarlo simulations. Figure 10b shows the distribution of values of r_0 and γ obtained from the simulations. As mentioned above, the correlation length turns out to be much more tightly constrained than the individual parameters of the angular correlation function,

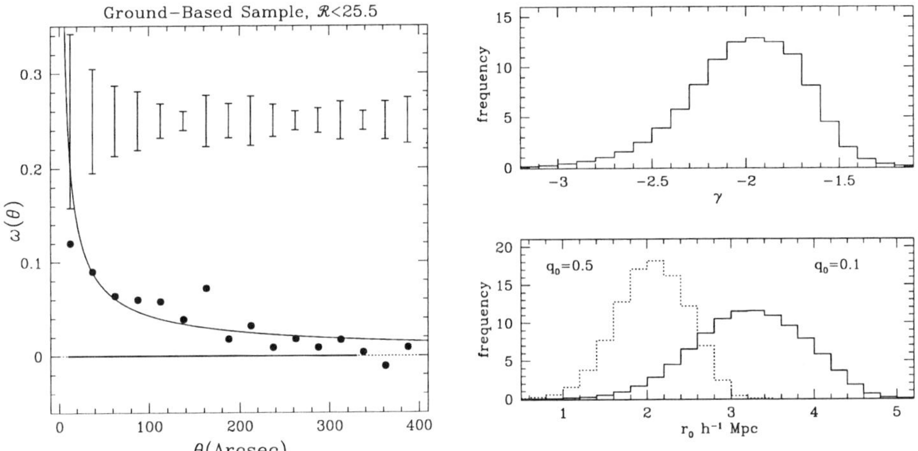

FIGURE 10. **a)** Weighted average angular correlation function of LBGs. The filled points are from the PB estimator, the open points from the LS one. The error bars are shown on the top of the figure. The continuous line is the best-fit power law to the PB data points, the dotted line is the fit to the LS. The thick horizontal continuous segment on the x axis marks the angular range over which the we computed the fits. **b)** The histogram of the correlation length r_0 (lower panel) and of the slope γ (upper panel) from the montecarlo simulations.

with a typical fractional error of $\approx 30\%$ at the 1-σ level. As our fiducial measure and error bar we adopt the median of the distributions of the simulations and its corresponding 68% confidence intervals. These are $r_0 = 2.1^{+0.4}_{-0.5}$ and $r_0 = 3.3^{+0.7}_{-0.6}$ h^{-1} Mpc (comoving coordinates) for $q_0 = 0.5$ and $q_0 = 0.1$, respectively, for the correlation length, and $\gamma = 1.98^{+0.32}_{-0.28}$ for the slope.

We can estimate the bias of these galaxies by comparing their correlation function ξ_g to the correlation function of the mass ξ_m:

$$b(r) = \sqrt{\frac{\xi_g(r)}{\xi_m(r)}}. \tag{3}$$

Although (as eq. 3 shows) the bias is in principle a function of scale, our constraint on the power-law exponent γ is relatively weak and we can only estimate a "typical" value of the bias over the scales of a few Mpc which are probed here. In practice, we use the ratio of the correlation length of the LBGs to that of $\xi_m(r)$ predicted by the CDM theory to compute the bias, which is therefore relative to $r = 1$ h^{-1} Mpc. Using a CDM power-spectrum with shape parameter $\Gamma^* = 0.25$, claimed to fit the shape of the local large-scale structure very well (Peacock 1997), and normalization of Eke, Cole, & Frenk (1996), we estimate $b \sim 4.5$ (1.5) for $q_0 = 0.5$ (0.1). Choosing $\Gamma^* = 0.20$ results in $b \sim 5$ (1.5), while adopting the normalization of White, Efstathiou & Frenk (1993) results in $b \sim 4$ (1).

Very interestingly, the correlation function of the Lyman-break galaxies has a slope that is comparable to or steeper than that measured at intermediate and low redshifts. The evolution of the slope of the correlation function of the mass (or, equivalently, that of the power spectrum at small scales) has a pronounced dependence on Ω. For a CDM-like power spectrum, it depends very weakly on the shape parameter Γ^* and, for flat models, on the normalization. As eqn. (3) shows, the slope of $\xi_g(r)$ differs from that of $\xi_m(r)$ because of the dependence of the bias parameter $b(r)$ with the spatial scale. The form of $b(r)$, its dependence on galaxy properties and how it evolves with redshift are still subjects of discussion (e.g., Mann, Peacock & Heavens 1997; Bagla 1997). If the scale dependence of $b(r)$ for the LBGs over the spatial scales probed by our correlation analysis, namely $1 \lesssim r \lesssim 10\ h^{-1}$ Mpc, is similar to that of the local galaxies, then our measures of γ are inconsistent with $\xi_m(r)$ from the CDM theory if $\Omega = 1$. With our choice of $\Gamma^* = 0.25$ we found $\gamma_m = 1.25$ (over the range $1 < r < 10\ h^{-1}$ Mpc), independently of the normalization. As mentioned above, the dependence on Γ^* is very weak. If $\Gamma^* = 0.1$ then $\gamma_m = 0.98$, while if $\Gamma^* = 0.6$, then $\gamma_m = 1.14$. Thus, the observed slope rules out the steepest CDM slope ($\gamma_m = 1.25$) is ruled at the 99.95% confidence level. Open CDM models with the same parameters as above produce slopes in the range $1.6 < \gamma_m < 2.1$ (in open models the slope of $\xi_m(r)$ has a more pronounced dependence on the normalization), which are all consistent with our data.

The above computations assume a bias constant with spatial scale. We stress, however, that the evolution of the slope of the correlation function is useful for constraining cosmological models only if the dependence of the bias with the spatial scale and its evolution with redshift are known. The function $b(r)$ also depends on the properties of the halos, which further complicates the interpretation of the data because of the difficulty of establishing an evolutionary sequence between the systems observed at high redshifts and the local galaxies. Bagla's (1997) N-body simulations seem to suggest that the bias will not be strongly scale-dependent—his $b(r)$ for $M \geq 2 \times 10^{12}\ M_\odot$ halos at $z = 0$ in standard CDM has a power-law slope of only ~ -0.18—and if $b(r)$ for Lyman-break galaxies is similarly flat, our conclusions about the slope would not be importantly changed. But until more is known about the scale-dependence of the bias they will remain speculative.

6.3. *Spatial clustering and its "evolution"*

The LBGs at $z \sim 3$ are characterized by strong spatial clustering, with a co-moving correlation length of $\sim 3 - 4h^{-1}$ Mpc for a low matter density ($\Omega_0 = 0.2$) Universe. This is comparable to the clustering of present-day spiral and IRAS galaxies, and a factor of ≈ 2 smaller than that of present-day ellipticals.

A simple comparison of the observed clustering properties with the expected clustering of a suitably normalized CDM density field, as shown in Figure 11, shows that, in the context of such models, the LBGs must be substantially biased with respect to the dark matter, with higher bias required in models with higher matter density. As discussed above, the strong clustering and the large bias of the Lyman-break galaxies are consistent with biased galaxy formation theories and provide additional evidence that these systems are associated with massive dark matter halos.

One might be tempted to try to fit the Lyman break galaxy clustering properties onto a general evolutionary sequence for galaxy clustering versus cosmic epoch. Figure 11 shows the galaxy clustering strength measured from various redshift surveys, including the LBGs, as a function of redshift. To quantify the clustering strength we have used the function $r_0^\gamma \times (1+z)^3$, where now r_0 is in *proper* coordinates. The figure also shows the corresponding plot of the bias derived using the CDM mass correlation function, which we have computed using the non-linear code for the evolution of the power spectrum by Peacock (1997). We used the shape parameter $\Gamma^* = 0.25$ and normalized the power spectrum to $\sigma_8 = 1.0$ for the open model and $\sigma_8 = 0.5$ in the Einstein-de Sitter case. The error bars have been computed by propagating the errors in the measure of the correlation length, but it is clear that the dominant uncertainty is in the choice of normalization of the theoretical curve.

As Figure 10 suggests, the variations in the value of the "effective" bias from sample to sample complicate the interpretation of the apparent evolution of galaxy clustering as due to the gravitational growth of structures, preventing us from deriving information on the

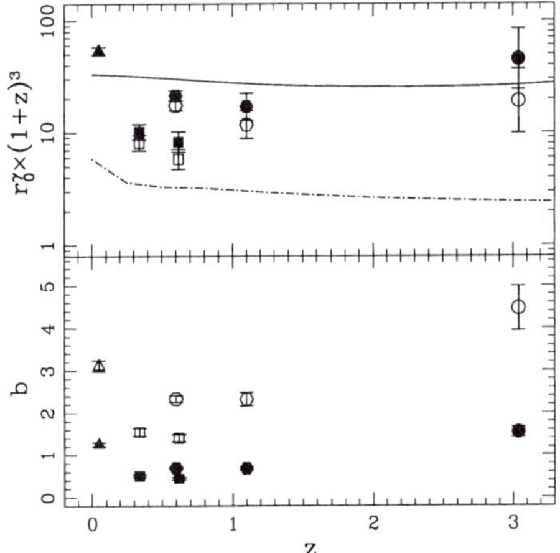

FIGURE 11. **Top.** The strength of galaxy clustering as a function of redshift. The continuous solid and dashed lines are the expectations from the CDM theory with $\Gamma^* = 0.25$, $\sigma_8 = 1.0$ and 0.5 for the two case of $q_0 = 0.1$ and 0.5, respectively. **Bottom.** Linear bias as a function of redshift, assuming the CDM correlation function. Filled symbols are for $q_0 = 0.1$, open symbols for $q_0 = 0.5$. Triangles are APM data (Loveday et al. 1995), squares are CFRS data (Le Fèvre et al. 1996), hexagons Keck K-band data (Carlberg et al. 1997), and circles are the LBG data.

clustering evolution of the mass. For example, the data show that the traditional power-law model $\xi(r, z) = \xi_0(r) \times (1 + z)^{-(3+\epsilon)}$ used to describe the gravitational evolution of clustering (Peebles 1980) is not a good representation for any value of ϵ over the redshift interval $0 \lesssim z \lesssim 3$. In fact, it is clear from N-body simulations (e.g., Brainerd & Villumsen 1994; Bagla 1997) that the behavior of the clustering of *halos*, as opposed to that of the overall mass, will have a non-trivial dependence on redshift that is not accounted for in the "ϵ" models. When one also notes the uncertainties in the way in which dark halos are effectively sampled in any given redshift survey, we question the ultimate usefulness of fitting values of this parameter to the results of redshift surveys over substantial redshift baselines.

Our results emphasize that the interpretation of apparent "evolution" of the clustering properties of galaxies in deep surveys, through either angular clustering analyses (e.g., Efstathiou 1995, Brainerd et al. 1995, etc.) or with comprehensive redshift surveys (e.g., Carlberg et al. 1997, Le Fèvre et al. 1996), cannot be directly interpreted as a reliable measure of the overall growth of structure. In a typical survey limited by apparent magnitude, the mix of galaxy types and the relative sensitivity of the survey to stellar mass, star formation rate, and bolometric luminosity (and the complicated manner in which these quantities are related to overall galaxy mass) will be strongly redshift-dependent. It is apparent now, if it has not always been, that the clustering properties of galaxies at moderate to high redshift cannot be used as a cosmological tool unless one is prepared to simultaneously understand where galaxies form and how they evolve relative to the underlying distribution of dark matter—cosmology and galaxy formation cannot be understood independently in this context.

7. Summary

Lyman-break galaxies at $z \sim 3$ represent both a large jump in redshift and a substantially different detection/selection technique than has been used previously in galaxy studies. While we do not yet have enough empirical evidence to allow us delineate an evolutionary scenario to link this population to galaxies observed at lower redshifts or in the present-day universe, several important traits of this population can already be traced:

- The Lyman-break galaxies form a substantial population of star formers already at $\approx 15\%$ of the cosmic age. They have assembled at least $\sim$ 10–15% of all the stars observed in the local universe, but the true fraction is very likely higher by a factor of several due to the presence of dust obscuration in the observed rest-frame far-UV light of these galaxies. These stars seem unavoidably linked to the oldest stellar systems observed in the present-day universe, i.e., the elliptical galaxies and the bulges of spiral galaxies.
- The rest-frame UV spectra of these galaxies bear close resemblance to those of local starburst galaxies. While the ISM of these galaxies has certainly undergone some degree of metal enrichment, as indicated by the numerous metal interstellar absorption lines, it is hard to make quantitative statements at this time, because the features are heavily saturated. Despite the high star-formation rates and the abundance of ionizing photons, the Lyα line is often weak or absent, very likely due to the presence of dust in the ISM coupled with resonant scattering. The kinematics of the absorption and emission lines shows the existence of large outflows of gas, with velocity in excess of $\sim$ 400–500 km s^{-1}, very likely due to stellar winds and supernova explosions. This stresses the importance of understanding the effects of feedback in the process of star (and galaxy) formation.
- Both the UV colors and optical nebular emission lines directly point to the presence of dust obscuration. Correcting factors very likely range between $3\times$ and $7\times$, but unfortunately the exact numbers depend quite strongly on the shape of the extinction law, which is poorly constrained in these galaxies. The unobscured star formation rates of the individual galaxies, if sustained for $\sim 10^9$ yr, have produced stellar masses that span the range observed in the local universe, with a typical value around M^*, or 10^{11} M$_\odot$.
- *HST* high-resolution imaging of the UV morphology of the Lyman-break galaxies shows a dispersion of properties. The brighter systems have compact morphologies with light profiles characteristic of collapsed systems. They appear characterized by a relatively high degree of spherical symmetry, and have half-light radii that compare with those of present-day bulges and elliptical galaxies of intermediate luminosity. At the fainter end of the luminosity distribution the morphology seems to become more irregular and fragmented. A preliminary inspection of rest-frame optical images obtained with *HST* + NICMOS seems to show morphologies that are rather similar to the UV ones.
- The Lyman-break galaxies have clustering properties expected from massive dark halos. As a consequence, they are also heavily biased tracers of the mass. Values of the mass of the dark halos that host Lyman-break galaxies at the bright end of the luminosity distribution are $M \sim 10^{12}$ M$_\odot$. Large concentrations of these systems with total mass typical of rich clusters (i.e., $\sim M \sim 10^{15}$ M$_\odot$) are also observed already at redshift $z \sim 3$. The Lyman-break galaxies are characterized by strong spatial clustering, with comoving correlation length $r_0 = 2.1$ (3.3) h^{-1} Mpc, only a factor of ≈ 3 (2) of the present-day correlation length. The slope of the correlation function is $\gamma = -1.9$, very close to the one observed in the local universe. If the biasing parameter is only mildly varying as a function of the spatial scale, this disfavors standard CDM models and points towards an open universe. We are currently studying the presence of clustering segregation with the UV luminosity (i.e., the dependence of the star-formation activity on the mass of the hosting halos) using the HDF sample of Lyman-break galaxies (Giavalisco et al. 1988, in preparation).

We would like to thank Charles Steidel, Mark Dickinson, Max Pettini, Kurt Adelberger and Mindy Kellogg for many useful discussions and for allowing us to use unpublished material in this paper. We would also like to thank the organizers of this ST ScI Symposium for the excellent meeting. We gratefully acknowledge support from the Hubble Fellowship program through grant HF-01071.01-94A awarded by the Space Telescope Science Institution, which is operated by the Association of Universities for Research in Astronomy, Inc. under NASA contract NAS 5-26555.

REFERENCES

Bagla, J. S. 1997 *MNRAS*, submitted, astro- ph/9711081.

Baugh, C. M., Cole, S., Frenk, C. S., & Lacey, C. G. 1998 *ApJ*, in press, astro-ph/9703111.

Brainerd, T. G. & Villumsen, J. V. 1994 *ApJ* **431**, 477.

Brainerd, T. G., Smail, I., & Mould, J. 1995 *MNRAS* **275**, 781.

Broadhurst, T., Ellis, R., Koo, D. & Szalay, A. 1990 *Nature* **343**, 726.

Calzetti, D., Kinney, A. L., & Storchi-Bergmann 1994 *ApJ* **429**, 582.

Calzetti, D. 1997 *AJ* **113**, 162.

Carlberg, R. G., Cowie, L. L., Songaila, A., Hu, E. M. 1997 *ApJ* **484**, 538.

Charlot, S., & Fall, M. 1993 *ApJ* **415**, 580.

Cohen, J. G., Cowie, L. L., Hogg, D. W., Songaila, A., Blanford, R., Hu, E. M., Shopbell, P. 1996 *ApJ* **471** L5.

Connolly, A. J., Szalay, A. S., Dickinson, M. E., Subbarao, M. U., & Brunner, R. J. 1997 *ApJ* **486**, L11.

Dwek, E. et al. 1995 *ApJ* **445**, 716.

Efstathiou, G., Bernstein, G., Katz, N., Tyson, A. J., & Guhatakurta, P. 1991 *ApJ* **380**, L47.

Efstathiou, G. 1995 *MNRAS* **272**, L25.

Eggen, O. J., Lynden-Bell, D., & Sandage, A. R. 1962 *ApJ* **136**, 748.

Eke, V. R., Cole, S., & Frenk, C. S. 1996 *MNRAS* **282**, 263.

Gallego, J., Zamorano, J., Aragon-Salamanca, A., & Rego, M. 1995 *ApJ* **455**, L1.

Giavalisco, M., Macchetto, D. F., & Sparks, W. 1994 *A&A* **288**, 113.

Giavalisco, M., Macchetto, D. F., Sparks, W., & Madau, P. 1994 *ApJ* **441**, L13.

Giavalisco, M., Livio, M., Bohlin, R., Macchetto D. F., & Stecher, T. P. 1996a *AJ* **112**, 369.

Giavalisco, M., Steidel, C. C., & Macchetto, D. F. 1996b, *ApJ* **470**, 189.

Giavalisco, M., Steidel, C. C., Adelberger, K. L., Dickinson, M. E., Pettini, M., & Kellogg, M. 1998 *ApJ*, submitted.

Guhatakurta, P., Tyson, J. A., & Majewski, S. R. 1990 *ApJ* **357**, L9.

Kaiser, N., Peacock, J. 1991 *ApJ* **379**, 482.

Le Fèvre, O., Hudon, D., Lilly, S. J., Crampton, D., Hammer, F., & Tresse, L. 1996 *ApJ* **461**, 534.

Leitherer, C., Robert, C., & Heckman, T. M. 1995 *ApJS* **99**, 173.

Leitherer, C., Vacca, W., Conti, P., Filippenko, A., Robert, C., & Sargent, W. 1996 *ApJ* **465**, 717.

Lilly, S., Tresse, L., Hammer, F., Crampton, D., & Le Fèvre, O. 1995 *ApJ* **455**, 108.

Loveday, J., Maddox, S. J., Efstathiou, G., & Peterson, B. A. 1995 *ApJ* **442**, 457.

Lowenthal, J. D., Koo, D. C., Guzman, R., Gallego, J., Phillips, A. C., Faber, S. M., Vogt, N. P., Illingworth, G. D., Gronwall, C. 1997 *ApJ* **481**, 673.

Madau, P. 1995 *ApJ* **441**, 18.

MADAU, P. FERGUSON, H. C., DICKINSON, M. E., GIAVALISCO, M., STEIDEL, C. C., & FRUCHTER, A. 1996 *MNRAS* **283**, 1388.

MANN, R. G., PEACOCK, J. A., & HEAVENS, A. F. 1997 *MNRAS*, in press, astro-ph/9708031.

MEURER, G., HECKMAN, T., LEHNERT, M., LEITHERER, C., & LOWENTHAL, J. 1997 *AJ*, in press.

MO, H. J., & FUKUGITA, M. 1996 *ApJ* **467** L9.

OKE, J. B., & GUNN, J. E. 1983 *ApJ* **266**, 713.

OKE, J. B., COHEN, J. G., CARR, M., CROMER, J., DINGIZIAN, A., HARRIS, F. H., LABRECQUE, S., LUCINIO, R., SCHAAL, W., EPPS, H., MILLER, J. 1995 *PASP* **107**, 375.

PARTRIDGE, R., & PEEBLES, P. J. 1967 *ApJ* **147**, 868.

PEACOCK, J. A. 1997 *MNRAS* **284**, 885.

PEEBLES, P. J. E. 1980 *The Large-Scale Structure of the Universe.* Princeton University Press.

PETTINI, M., STEIDEL, C. C., ADELBERGER, K., KELLOGG, M., DICKINSON, M. E., GIAVALISCO, M. 1997, in *Origins* (eds. M. J. Shull, C. E. Woodward, & H. Thronson). ASP Conference series. astro-ph/9708117

SCHECHTER, P., & DRESSLER, A. 1987 *AJ* **94**, 563.

SONGAILA, A., COWIE, L. L., & LILLY, S. J. 1990 *ApJ* **348**, 371.

STEIDEL, C. C., & HAMILTON, D. 1993 *AJ* **105**, 2017 (Paper II).

STEIDEL, C. C., PETTINI, M., & HAMILTON, D. 1995 *AJ* **110**, 2519 (Paper III).

STEIDEL, C. C., GIAVALISCO, M., PETTINI, M., DICKINSON, M. E., & ADELBERGER, K. L. 1996a, *ApJ* **???**, ???.

STEIDEL, C. C., GIAVALISCO, M., DICKINSON, M. E., & ADELBERGER, K. L. 1996b, *AJ* **112**, 352.

STEIDEL, C. C., ADELBERGER, K. L., DICKINSON, M. E., GIAVALISCO, M., PETTINI, M., & KELLOGG, M. 1998 *ApJ*, in press, astro-ph/9708125.

Thompson, D., Djorgovski, G., & Trauger, J. 1995 *AJ* **110**, 963.

TRESSE, L., & MADDOX, S. J. 1998 *ApJ*, in press, astro-ph/9709240.

WHITE, S. M., EFSTATHIOU, G., FRENK, C. S. 1993 *MNRAS* **262**, 1023.

Photometric redshifts of galaxies in the Hubble Deep Field

By
KENNETH M. LANZETTA, ALBERTO FERNÁNDEZ-SOTO† AND AMOS YAHIL

Department of Physics and Astronomy, State University of New York at Stony Brook, Stony Brook, NY 11794–3800, USA

We describe our application of broad-band photometric redshift techniques to faint galaxies in the Hubble Deep Field. To magnitudes $AB(8140) < 26$, the accuracy of the photometric redshifts is a few tenths and the reliability of the photometric redshifts approaches 100%. At fainter magnitudes the effects of photometric error on the photometric redshifts can be rigorously quantified and accounted for. We argue that broad-band photometric redshift techniques can be applied to accurately and reliably estimate redshifts of galaxies that are up to many magnitudes fainter than the spectroscopic limit.

1. INTRODUCTION

Between 2500 and 3500 objects (depending on exactly how they are counted) are visible in the Hubble Deep Field (HDF) images obtained by the Hubble Space Telescope (HST), to a limiting magnitude threshold that approaches $AB(8140) \approx 30$. Spectroscopic redshift identifications of just over 100 of these objects—or roughly 3% of the total—have been obtained with the Keck telescope, following nearly two years of intensive effort. The faintest objects that have been (or presumably can be) spectroscopically identified with the Keck telescope are of magnitude $AB(8140) \approx 26$, of which there are another 200 or so that remain unidentified in the HDF. In time, spectroscopic redshift identifications of perhaps 300 objects—or roughly 10% of the total—might be obtained with the Keck telescope. What of the other 90%—the objects fainter than the spectroscopic limit?

Fortunately, it appears to be the case that broad-band photometric redshift techniques can be applied to accurately and reliably estimate redshifts of galaxies that are up to many magnitudes fainter than the spectroscopic limit. At low redshifts, the most prominent broad-band spectral feature is the 4000 Å spectral discontinuity, while at high redshifts the most prominent broad-band spectral feature is the Lyman spectral discontinuity, which is produced by intrinsic and intervening neutral hydrogen absorption. But the broad-band photometric redshift techniques make use not only of discrete spectral discontinuities but also of overall galaxy spectral energy distributions and so are applicable to galaxies that do not exhibit prominent broad-band spectral features (e.g., low-redshift, late-type galaxies) as well as to galaxies that do exhibit prominent broad-band spectral features (e.g., low-redshift, early-type galaxies or high-redshift galaxies). To magnitudes $AB(8140) \lesssim 26$, at which photometric redshifts can be compared with spectroscopic redshifts, the accuracy of the photometric redshifts is a few tenths and the reliability of the photometric redshifts approaches 100%. To even fainter magnitudes, at which numerical simulations can quantify the effects of photometric error, the accuracy and reliability of the photometric redshifts are plausibly sufficient to establish statisti-

† Current address: School of Physics, University of New South Wales, P.O. Box 1, Kensington, NSW 2033, Australia

cal moments of the galaxy distribution versus redshift, even to magnitudes close to the limiting magnitude thresholds of the images.

Interest in broad-band photometric redshift techniques has in the past been driven by the prospect of determining redshifts of galaxies at relatively low cost. Our interest is instead driven by the prospect of determining redshifts of galaxies that are inaccessible to ground-based spectroscopy at any cost, i.e., galaxies of very low luminosity or very high redshift. To bring this about, it is necessary first to make the case for the photometric redshifts. This is, of course, a difficult agenda, because the aim is to establish the validity of the photometric redshifts in a regime in which they by definition cannot be independently verified. Yet we believe this is also a crucially important agenda. The truly phenomenal progress toward observing normal galaxies at high redshifts made over the past several years has simultaneously revealed the possibilities and the limitations of ground-based spectroscopy with the Keck telescope. It is now clear that the very most luminous galaxies at redshifts $z \lesssim 4$ are accessible to Keck spectroscopy, but it is also now clear that lower-luminosity galaxies at redshifts much above $z \lesssim 4$ are probably not accessible to Keck spectroscopy. To push toward lower luminosities and higher redshifts it will almost certainly be necessary to rely on broad-band photometric redshift techniques, perhaps supplemented by occasional spectra of very faint objects obtained with HST (which for faint, spatially unresolved objects at near-infrared and infrared wavelengths is a far more sensitive spectroscopic instrument than the Keck telescope).

The goal of this review is to make the case for the photometric redshifts. Happily, our task is made more manageable by recent progress on two fronts: First, at least six independent groups (besides ours) are estimating photometric redshifts of galaxies in the HDF—and are obtaining results that are in remarkable agreement with each other. (Comparisons of results of these independent analyses are presented by Ellis 1997 and Hogg et al. 1997.) Second, at least six independent groups are obtaining spectroscopic redshifts of galaxies in the HDF for comparison with the photometric redshifts. Although we concentrate on our own photometric analysis, much of what we report should apply to the various other photometric analyses, which are similar to our analysis in spirit if not in detail. We begin by briefly describing the technique in § 2 and the accuracy and reliability of the photometric redshifts in § 3. We continue by applying the broad-band photometric redshift technique to galaxies of redshift $z = 5 - 7$ in § 4 galaxies of redshift $z = 7 - 17$ in § 5 and to the redshift distribution of faint galaxies in § 6. We conclude by discussing possible future directions in § 7.

2. Technique

2.1. *Overview*

Two conceptually different approaches have been applied to estimate photometric redshifts of galaxies in the HDF. The first technique, which we designate the "spectral template technique," is to seek optimal agreement between the observed photometry and redshifted spectral templates. The spectral templates are constructed from observations of nearby galaxies or stellar population synthesis models and from observations of intervening neutral hydrogen absorption (measured by means of QSO absorption lines). The second technique, which we designate the "empirical technique," is to seek optimal agreement between the observed photometry and polynomial fits to empirical correlations between fluxes, colors, and redshifts. The empirical correlations are established and calibrated with respect to known redshifts of spectroscopically identified galaxies. Practitioners of the two techniques are listed in Table 1.

Group	Technique	Reference
Hawaii	spectral template	1
Imperial College	spectral template	2
Johns Hopkins	empirical	3
Stony Brook	spectral template	4
Toronto	spectral template	5
Victoria	spectral template	6

REFERENCES—(1) Cowie 1997; (2) Mobasher et al. 1996; (3) Connolly et al. 1997; (4) Lanzetta, Yahil, & Fernández-Soto 1996; (5) Sawicki et al. 1997; (6) Gwyn & Hartwick 1996.

TABLE 1. Practitioners of broad-band photometric redshift techniques

Group	Reference
Berkeley	1
Caltech	2
Elston	3
Hawaii	4
Lick	5
Steidel et al.	6,7

REFERENCES—(1) Zepf, Moustakas, & Davis 1997; (2) Cohen et al. 1997; (3) Elston 1997, private communication; (4) Cowie 1997; (5) Lowenthal et al. 1997; (6) Steidel et al. 1996; (7) Dickinson et al. 1997, this volume.

TABLE 2. Sources of spectroscopic redshifts

The two techniques are subject to certain strengths and weaknesses as follows:

(1) A strength of the spectral template technique is that it does not depend on the spectroscopic redshifts, whereas a weakness of the empirical technique is that it does depend on the spectroscopic redshifts. This is an issue because some fraction of the published spectroscopic redshifts are in error, and while the spectral template technique provides an independent check of the spectroscopic redshifts, the empirical technique can propagate errors in the spectroscopic redshifts into errors in the photometric redshifts. item A strength of the spectral template technique is that it can be applied to arbitrarily large redshifts, whereas a weakness of the empirical technique is that it cannot be applied to arbitrarily large redshifts.

(2) A strength of the empirical technique is that it accounts for evolution of galaxy spectra with redshift, whereas a weakness of the spectral template technique is that it does not account for evolution of galaxy spectra with redshift (although the analysis of Gwyn & Hartwick 1996 does attempt to model time-evolving spectral templates).

Despite the difference inherent in the techniques, there seems to be general consensus that both techniques, properly applied, yield similar results over the common redshift interval at which they can be compared (cf. Ellis 1997; Hogg et al. 1997).

In the following sections, we describe our implementation of the spectral template technique, which is described in more detail by Lanzetta, Yahil, & Fernández-Soto (1996), Lanzetta, Fernández-Soto, & Yahil (1997), and Fernández-Soto, Lanzetta, & Yahil (1997, in preparation). Specifically, we describe our methods of (1) photometry and (2) redshift estimation, which are in principle the only aspects of the analysis upon which results of the broad-band photometric redshift techniques can depend.

2.2. *Photometry*

Our current analysis is based on the Version 2 optical images obtained by HST in December, 1995 using the Wide Field Planetary Camera 2 (WFPC2) and the F300W, F450W, F606W, and F814W filters (Williams et al. 1996) and on the Version 1 infrared images obtained by the Kitt Peak National Observatory (KPNO) 4 m telescope in April, 1996 using the IRIM camera and standard J, H, and K filters (Dickinson et al. 1997, in preparation). First, we identify objects in the F814W image using the SExtract source extraction program (Bertin & Arnouts 1996). Next, we measure the noise characteristics of the images by calculating empirical covariances, excluding pixels associated with identified objects. Next, we use non-overlapping isophotal aperture masks determined by the SExtract program to measure fluxes and flux uncertainties, applying different methods in the optical and infrared images.

To measure fluxes and flux uncertainties in the optical images, which are characterized by point spread functions of FWHM ≈ 0.1 arcsec, we directly integrate within the aperture mask of every object detected in the F814W image. To measure fluxes and flux uncertainties in the infrared images, which are characterized by point spread functions of FWHM ≈ 1 arcsec, we (1) model the spatial profile of every object detected in the F814W image as a convolution of the portion of the F814W image containing the object with the appropriate point spread function of the infrared image and (2) determine a least-squares fit of a linear sum of the model spatial profiles to the infrared image. The advantages of this method over simple aperture photometry are that (1) the flux measurements correctly weight signal-to-noise ratio variations within the spatial profiles and (2) the flux uncertainty measurements correctly include the contributions of nearby, overlapping neighbors. The observed and modeled K-band images are shown in Figure 1.

Our analysis of the infrared images yields essentially optimal photometry under the assumption that the spatial profiles of the objects are independent of wavelength at observed-frame wavelengths $\lambda > 8140$ Å. This assumption may be justified as follows: At low redshifts, the infrared images measure rest-frame optical and infrared wavelengths, at which galaxy spatial profiles are observed to be more or less independent of wavelength. At high redshifts, galaxies are (as it turns out) physically small and unresolved by the infrared images at any wavelength. A similar analysis of deep ground-based optical images of the HDF (e.g., through medium-band filters) could certainly be applied to great advantage, making use of the unprecedented depth and spatial resolution of the F814W image to provide nearly ideal photometric spatial templates.

2.3. *Redshift estimation*

First, we construct spectral templates of E/S0, Sbc, Scd, and Irr galaxies, including the effects of intrinsic and intervening neutral hydrogen absorption. Next, we integrate the redshifted spectral templates with the throughputs of the F300W, F450W, F606W, F814W, J, H, and K filters, at redshifts spanning $z = 0$–7. Next, we construct the "redshift likelihood function" of every object detected in the F814W image by calculating the relative likelihood of obtaining the measured fluxes and uncertainties given the modeled fluxes at an assumed redshift, maximizing with respect to galaxy spectral type and arbitrary flux normalization. Finally, we determine the maximum-likelihood redshift estimate of every object detected in the F814W image by maximizing the redshift likelihood function with respect to redshift.

The spectral templates are constructed from different sources at far-ultraviolet, near-ultraviolet and optical, and infrared wavelengths. First, we adopt empirical spectra of Coleman, Wu, & Weedman (1980) at wavelengths $\lambda = 1400$–$10{,}000$ Å. Next, we add

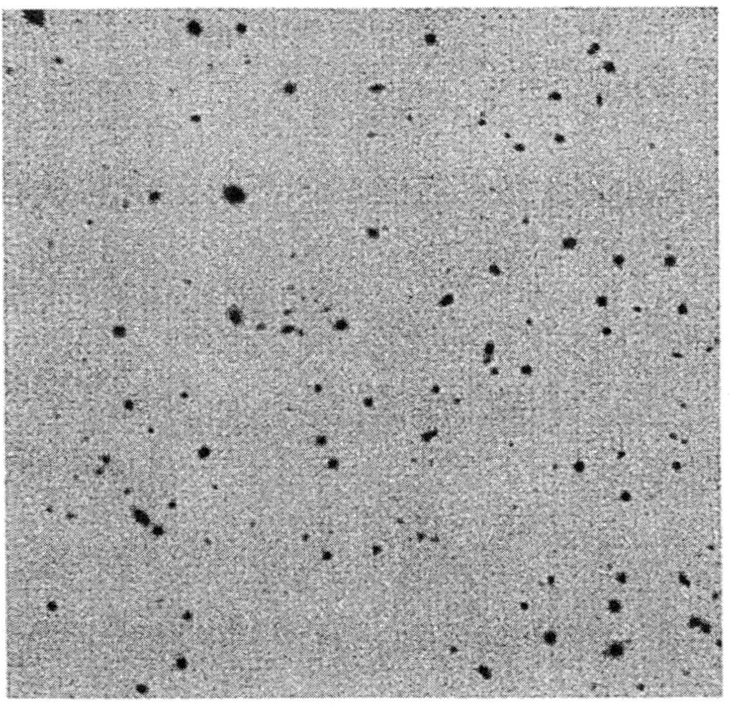

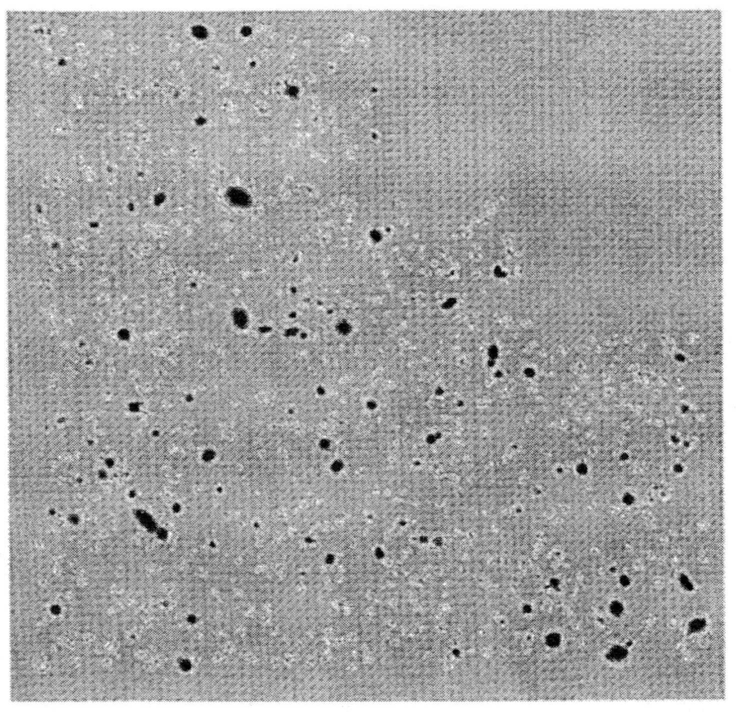

FIGURE 1. Observed (left panel) and modeled (right panel) K-band images. Angular extent is $x.xx \times x.xx$ arcmin^2, and spatial resolution is FWHM ≈ 1 arcsec.

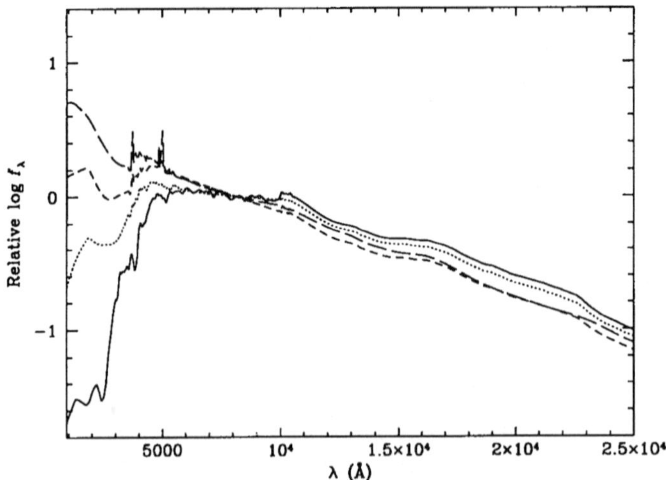

FIGURE 2. Spectral templates of E/S0 (solid curve), Sbc (dotted curve), Scd (short dashed curve), and Irr (long dashed curve) galaxies.

spectra derived from simple power-law relationships motivated by results of Kinney et al. (1993) at wavelengths $\lambda = 912$–1400 Å. Next, we add spectra derived from stellar population synthesis models of Bruzual & Charlot (1993) at wavelengths $\lambda = 10,000$–25,000 Å. Finally, we include the effects intrinsic and intervening neutral hydrogen absorption (as a function of redshift) by (1) assuming that galaxies are optically thick at wavelengths shortward of the Lyman limit and (2) adding absorption from intervening Lyα-forest Lyα and Lyβ absorption lines according to the "flux decrement" parameters D_A and D_B, measured by means of QSO absorption lines by Madau (1995) and Webb (1996, unpublished). The spectral templates are shown in Figure 2.

We experimented with spectral templates based at near-ultraviolet and optical wavelengths on (1) empirical spectra of Kinney et al. (1996) and (2) spectra derived from stellar population synthesis models of Bruzual & Charlot (1993). We abandoned these alternate spectral templates in favor of the adopted spectral templates because photometric redshifts determined with the alternate spectral templates did not compare as favorably with the spectroscopic redshifts as photometric redshifts determined with the adopted spectral templates.

3. Accuracy and reliability of photometric redshifts

In this section we assess the accuracy and reliability of the photometric redshifts by (1) direct comparison of photometric and spectroscopic redshifts and (2) numerical simulations of the effects of photometric error on the photometric redshifts.

3.1. *Photometric redshifts: Successes and failures*

In preparation for the Hubble Deep Field Symposium, we set out in early May, 1997 to compare the available photometric redshifts with the available spectroscopic redshifts, which at the time numbered roughly 80 redshifts obtained with the Keck telescope. To our dismay, we found that a surprisingly large fraction (roughly 15%) of the photometric redshifts were clearly discordant with the spectroscopic redshifts, by amounts that were far too large to be attributed to photometric error. Believing that at the relatively bright magnitudes of the spectroscopic limit the reliability of the photometric redshifts should in principle be much higher than suggested by this comparison, we examined each of the discordant redshift pairs in detail, expecting to learn what caused the photometric redshifts to fail. Instead we found that in nearly every case it was the spectroscopic redshift—not the photometric redshift—that was in error.

That some fraction of the published spectroscopic redshifts are in error is hardly unexpected. After all, determining redshifts of extremely faint galaxies by any means—photometric or spectroscopic—is a difficult process subject to a variety of systematic uncertainties. But an uncritical acceptance of the validity of the spectroscopic redshifts has led some previous authors to incorrectly (and negatively) assess the reliability of the photometric redshifts. For this reason, it is important that any comparison of the photometric and spectroscopic redshifts begin by establishing under what circumstances both the photometric redshifts and the spectroscopic redshifts fail to yield reliable results. Our comparison of the photometric and spectroscopic redshifts in early May, 1997 identified eight spectroscopic redshifts that are ambiguous or in error and three photometric redshifts that are in error. (In two cases, both the spectroscopic and the photometric redshift are in error.) Here we describe each of the discordant cases, identifying galaxies by the Version 2 x and y mosaic pixel coordinates.

3.1.1. *The successes*

Galaxy 1591,3681: This galaxy was assigned a spectroscopic redshift of $z = 0.13$ by Cowie (1997) and a photometric redshift of $z = 1.20$ by our analysis. The galaxy occurs within ≈ 1 arcsec of a slightly brighter galaxy with no available spectroscopic redshift and a photometric redshift of $z = 0.16$. Considering typical ground-based seeing, it is likely that the spectroscopic redshift refers to the nearby galaxy, of which the photometric redshift is in excellent agreement with the spectroscopic redshift. In this case, the spectroscopic redshift is ambiguous or in error due to confusion with an overlapping source.

Object 2449,1574: This galaxy was assigned a spectroscopic redshift of $z = 0.318$ by Cowie (1997) and a photometric redshift of $z = 4.48$ by our analysis. The object occurs within ≈ 2 arcsec of a much brighter galaxy with a spectroscopic redshift of $z = 0.321$ and a photometric redshift of $z = 0.24$. The published spectra of the two objects are virtually identical although the colors of the two galaxies are very dissimilar, suggesting that one or the other or both of the spectra are incorrectly assigned. Subsequent spectroscopic observations of this object have established that it is a star (Cowie 1997, private communication). In this case, the spectroscopic redshift is in error due to operator error. This object is also noted below as a failure, because the photometric redshift is discordant with the actual redshift.

Galaxy 561,305: This galaxy was assigned a spectroscopic redshift of $z = 0.47$ by Cowie (1997) and a photometric redshift of $z = 0.76$ by our analysis. The galaxy occurs within ≈ 1 arcsec of a slightly brighter galaxy with no available spectroscopic redshift and a photometric redshift of $z = 0.48$. As in the case of galaxy 1591,3681, it is likely that the spectroscopic redshift refers to the nearby galaxy, of which the photometric redshift is in excellent agreement with the spectroscopic redshift. In this case, the spectroscopic redshift is ambiguous or in error due to confusion with an overlapping source.

Galaxies 790,1877 and 745,1845: These galaxies were assigned spectroscopic redshifts of $z = 0.511$ (Cowie 1997) and photometric redshifts of $z = 2.20$ (galaxy 790,1877) and $z = 0.56$ (galaxy 745,1845). The galaxies occur within ≈ 3 arcsec of each other. The published spectra of the two galaxies are identical, indicating that one or the other or

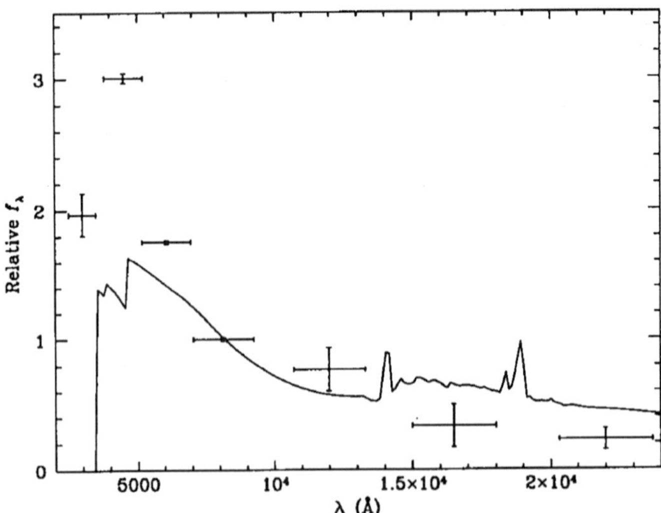

FIGURE 3. Observed photometry of galaxy 1816,1026 compared with model spectrum of an Irr galaxy of redshift $z = 2.775$. Vertical error bars indicate 1σ uncertainties and horizontal error bars indicate filter FWHM. It is very unlikely that the redshift of the galaxy is $z = 2.775$.

both of the spectra are incorrectly assigned. In this case, the spectroscopic redshifts are ambiguous or in error due to operator error.

Galaxy 693,1224: This galaxy was assigned a spectroscopic redshift of $z = 1.231$ by Cowie (1997) and a photometric redshift of $z = 1.08$ by our analysis. The galaxy occurs within ≈ 1 arcsec of a galaxy with no available spectroscopic redshift and a photometric redshift of $z = 1.18$. As in the case of galaxy 1591,3681, it is likely that the spectroscopic redshift refers to the nearby galaxy, of which the photometric redshift is in excellent agreement with the spectroscopic redshift. In this case, the spectroscopic redshift is ambiguous or in error due to confusion with an overlapping source.

Galaxy 1816,1026: This galaxy was assigned a spectroscopic redshift of $z = 2.775$ by Steidel et al. (1996) and a photometric redshift of $z = 1.72$ by our analysis. The observed photometry of this galaxy is compared with the model spectrum of an Irr galaxy of redshift $z = 2.775$ in Figure 3. Results of Figure 3 indicate that the spectroscopic redshift cannot be correct unless our model of intrinsic and intervening neutral hydrogen is grossly incorrect. Furthermore, our examination of the published spectrum of this galaxy (which is of moderate signal-to-noise ratio) fails to discern the claimed absorption features upon which the spectroscopic redshift is based. Subsequent analysis of the spectrum of this galaxy has failed to verity the spectroscopic redshift (Steidel 1997, private communication). In this case, the spectroscopic redshift is ambiguous or in error due to inaccurate identification of spectral features.

Galaxy 1193,3751: This galaxy was assigned a spectroscopic redshift of $z = 2.845$ by Steidel et al. (1996) and a photometric redshift of $z = 0.04$ by our analysis. The observed photometry of this galaxy is compared with the model spectrum of an Irr galaxy of redshift $z = 2.845$ in Figure 4. As in the case of galaxy 1816,1026, results of Figure 4 indicate that the spectroscopic redshift cannot be correct unless our model of intrinsic and intervening neutral hydrogen is grossly incorrect, and our examination of the published spectrum of this galaxy (which is of only moderate signal-to-noise ratio) fails to discern the claimed absorption features upon which the spectroscopic redshift is based.

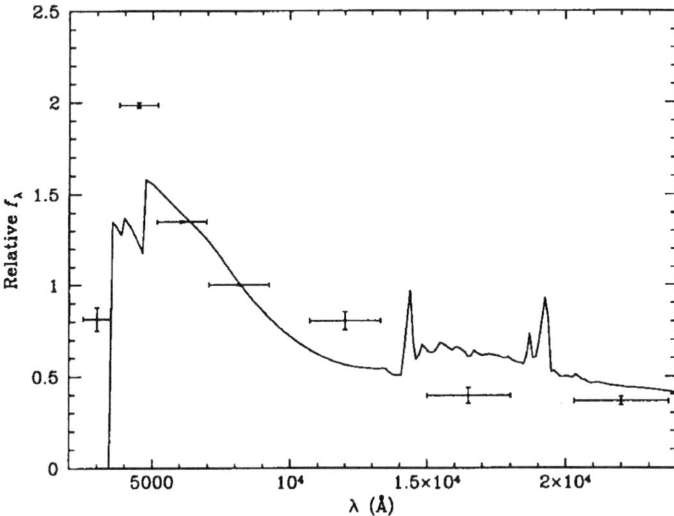

FIGURE 4. Observed photometry of galaxy 1193,3751 compared with model spectrum of an Irr galaxy of redshift $z = 2.845$. Vertical error bars indicate 1σ uncertainties and horizontal error bars indicate filter FWHM. It is very unlikely that the redshift of the galaxy is $z = 2.845$.

Subsequent spectroscopic observations of this galaxy have established a spectroscopic redshift of $z = 2.008$ (Steidel 1997, private communication). In this case, the spectroscopic redshift is in error due to inaccurate identification of spectral features. This galaxy is also noted below as a failure, because the photometric redshift is discordant with the actual redshift.

3.1.2. *The failures*

Object 2449,1574: This object was assigned a photometric redshift of $z = 4.48$ by our analysis and a spectroscopic redshift of $z = 0.318$ by Cowie (1997), although as noted above subsequent spectroscopic observations of this object have established that it is a star (Cowie 1997, private communication). Our analysis does not include spectral templates of stars, which therefore must be removed by hand. In this case, the photometric redshift is in error due to confusion by a star.

Galaxy 1193,3751: This galaxy was assigned a photometric redshift of $z = 0.04$ by our analysis and a spectroscopic redshift of $z = 2.845$ by Steidel et al. (1996), although as noted above subsequent spectroscopic observations of this galaxy have established a spectroscopic redshift of $z = 2.008$ (Steidel 1997, private communication). The redshift likelihood function of this galaxy shows no local maxima near the actual redshift $z \approx 2.008$. In this case, the photometric redshift is in error due to cosmic variance with respect to the spectral templates. This galaxy is also noted above as a success, because the photometric analysis demonstrates that the original spectroscopic redshift cannot be correct.

Galaxy 3242,1972: This galaxy was assigned a photometric redshift of $z = 0.00$ by our analysis and a spectroscopic redshift of $z = 2.591$ by Steidel et al. (1996). The redshift likelihood function of this galaxy shows a local maximum at redshift $z = 2.00$, which has a likelihood 2.9σ less than the global maximum at $z = 0.00$. In this case, the photometric redshift is in error due to cosmic variance with respect to the spectral templates.

Condition	Median Absolute Residual	Clipped RMS Residual	Discordant Number	Total Number	Discordant Fraction
All	0.09	0.17	5	102	0.049
$z < 2$	0.08	0.09	4	78	0.051
$z > 2$	0.32	0.40	2	24	0.083
$z < 2$, "early type"	0.12	0.25	0	10	0.000
$z < 2$, "late type"	0.07	0.09	2	68	0.029

TABLE 3. Comparison of photometric and spectroscopic redshifts

Galaxy 2888,1503: This galaxy was assigned a photometric redshift of $z = 1.32$ by our analysis and a spectroscopic redshift of $z = 2.268$ by Steidel et al. (1996). The redshift likelihood function of this galaxy shows no local maxima near the actual redshift $z \approx 2.268$. In this case, the photometric redshift is in error due to cosmic variance with respect to the spectral templates.

3.1.3. *Summary*

Spectroscopic redshifts of faint galaxies in the HDF are subject to a misidentification rate of $\approx 10\%$, which results from operator error, confusion with overlapping sources, and inaccurate identification of spectral features. Photometric redshifts of faint galaxies in the HDF are subject to a misidentification rate of $\approx 5\%$, which results from cosmic variance with respect to the spectral templates and confusion with stars.

3.2. *Comparison of photometric and spectroscopic redshifts*

Spectroscopic redshift identifications of nearly 120 objects have been obtained with the Keck telescope as of August, 1997. Magnitudes of the spectroscopically identified objects range from $AB(8140) \approx 18$ through $AB(8140) \approx 26$, and redshifts of the spectroscopically identified objects range from $z \approx 0$ through $z \approx 4$. Excluding from consideration (1) spectroscopic redshifts that are described as ambiguous or in error in § 3.1.1 or as uncertain by the spectroscopic observers and (2) spectroscopic redshifts of stars (for which our analysis does not include spectral templates), spectroscopic redshifts of 102 galaxies are available for comparison with the photometric redshifts. Sources of the spectroscopic redshifts are listed in Table 3, and the comparison of photometric and spectroscopic redshifts is shown in Figure 5. Results of Figure 5 indicate that the photometric redshifts are in broad agreement with the spectroscopic redshifts. At redshifts $z < 2$, the photometric redshifts generally trace the spectroscopic redshifts, with no examples of photometric redshifts that are clearly discordant with the spectroscopic redshifts. At redshifts $z > 2$, the photometric redshifts generally trace the spectroscopic redshifts, with two examples of photometric redshifts that are clearly discordant with the spectroscopic redshifts. (The discordant redshifts arise from galaxies 1193,3751 and 3242,1972, which are discussed in § 3.1.2. In both cases, a high-redshift galaxy is incorrectly assigned a low redshift.)

To quantitatively assess the accuracy and reliability of the photometric redshifts by direct comparison of the photometric and spectroscopic redshifts, it is necessary to adopt some robust measure of the distribution of residuals between the photometric and spectroscopic redshifts. This is because (1) the residuals are not drawn from a Gaussian distribution and (2) the RMS residual is dominated by a few highly discordant redshifts, as is evident from Figure 5. Here we consider two such measures of the residuals: First, we consider the median absolute residual between the photometric and spectroscopic

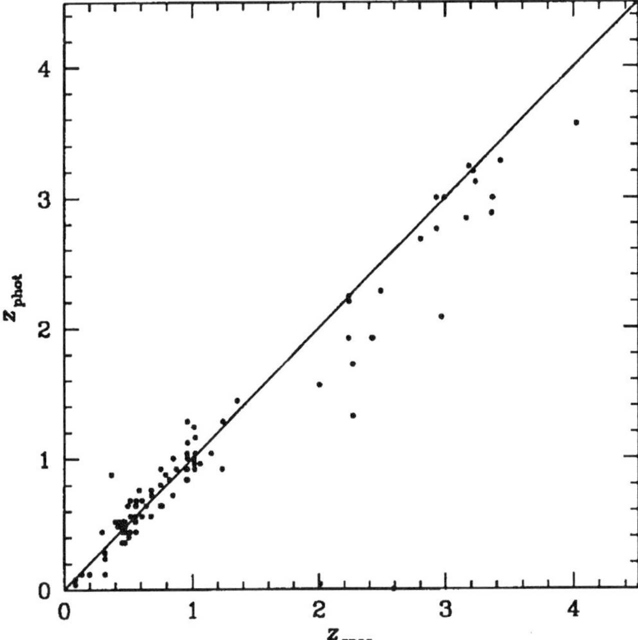

FIGURE 5. Comparison between photometric redshifts $z_{\rm phot}$ and spectroscopic redshifts $z_{\rm spec}$. Solid line shows $z_{\rm phot} = z_{\rm spec}$.

redshifts. Next, we consider the "clipped" RMS residual between the photometric and spectroscopic redshifts. The clipped RMS residual is obtained by applying an iterative "sigma-clipping" algorithm to the residuals, rejecting as discordant photometric redshifts for which the residual exceeds three times the clipped RMS residual. According to this procedure, the "accuracy" of the photometric redshifts is measured by the clipped RMS residual and the "reliability" of the photometric redshifts is measured by the discordant fraction. Results are summarized in Table 3, which lists the sample condition, the median absolute residual, the clipped RMS residual, the number of discordant redshifts, the number of total redshifts, and the discordant fraction.

Several conclusions can be drawn from results of Table 3:

(1) At the redshifts $z \approx 0 - 4$ spanned by the comparison of photometric and spectroscopic redshifts, the accuracy of the photometric redshifts is a few tenths and the reliability of the photometric redshifts approaches 100%. Specifically, the median absolute residual is 0.09, the clipped RMS residual is 0.17, and the discordant fraction is 0.049.

(2) The accuracy of the photometric redshifts is greater at low redshifts than at high redshifts, but the *fractional* accuracy of the photometric redshifts is greater at high redshifts than at low redshifts. At redshifts $z < 2$, the clipped RMS residual is 0.09, which at a representative redshift $z = 0.5$ corresponds to a fractional accuracy of $\approx 18\%$. At redshifts $z > 2$, the clipped RMS residual is 0.40, which at a representative redshift $z = 3$ corresponds to a fractional accuracy of $\approx 13\%$.

(3) The accuracy of the photometric redshifts is greater for late-type galaxies (which do not exhibit prominent broad-band spectral features) than for early-type galaxies (which do exhibit prominent broad-band spectral features). At redshifts $z < 2$ (at which early-

type galaxies are found), the clipped RMS residual for late-type galaxies is 0.09 and the clipped RMS residual for early-type galaxies is 0.25.

There is some evidence that the photometric redshifts of galaxies at redshifts $z \gtrsim 2$ are systematically underestimated by ≈ 0.3, especially at redshifts $z = 2 - 3$. If this systematic offset is accounted for, at redshifts $z > 2$ the clipped RMS residual is decreased to 0.24 and the discordant fraction is increased to 0.115. We have tried but failed to determine the cause of this systematic offset.

3.3. *Photometric error versus cosmic variance*

There are in principle two effects that contribute to residuals between the photometric and spectroscopic redshifts: photometric error and cosmic variance with respect to the spectral templates. Here we evaluate the relative importance of photometric error and cosmic variance by means of a numerical simulation of the effect of photometric error on the photometric redshifts. First, we add random noise (according to the actual flux uncertainties) to the best-fit model fluxes of every object detected in the F814W image. Next, we redetermine the maximum-likelihood redshift estimate of every object detected in the F814W image, using the simulated rather than measured fluxes. Finally, we repeat these steps 100 times in order to determine distributions of residuals between the actual and simulated redshifts. Results are summarized in Figure 6, which shows the distributions of residuals and the median absolute residuals between the actual and simulated redshifts as functions of magnitude and redshift.

Several conclusions can be drawn from results of Figure 6:

(1) Photometric error produces almost no effect on the photometric redshifts at magnitudes $AB(8140) < 25$. At these magnitudes, the median absolute residual is no greater than 0.04, and the distributions of residuals are concentrated into single peaks centered at zero residual.

(2) Photometric error produces a mild effect on the photometric redshifts at magnitudes $AB(8140) = 25 - 26$, which is somewhat more pronounced at redshifts $z > 2$ than at redshifts $z < 2$. Although at these magnitudes the median absolute residual (at all redshifts) is only 0.04 (which is comparable to the median absolute residual at brighter magnitudes), the distributions of residuals at high-redshifts are spread into small secondary peaks centered at large negative residuals in addition to prominent primary peaks centered at zero residual. These secondary peaks—which result from high-redshift galaxies that are incorrectly assigned low redshifts—contain 2.5% of the total at redshifts $z = 2$–3 and 6.1% of the total at redshifts $z = 3$–4.

(3) Photometric error produces an increasingly significant effect on the photometric redshifts with increasing magnitude at magnitudes $AB(8140) > 26$. At redshifts $z = 0$–1, the effect of photometric error at faint magnitudes is to (1) increase the median absolute residuals and (2) produce long tails stretching to large positive residuals on the distributions of residuals. At redshifts $z > 1$, the effect of photometric error at faint magnitudes is to (1) increase the median absolute residuals and (2) produce prominent secondary peaks centered at large negative residuals on the distributions of residuals. At magnitudes $AB(8140) = 26$–27, these secondary peaks—which again result from high-redshift galaxies that are incorrectly assigned low redshifts—contain 9.7% of the total at redshifts $z = 2$–3 and 19.6% of the total at redshifts $z = 3$–4.

(4) At faint magnitudes, it is more likely that high-redshift galaxies will be incorrectly assigned low redshifts than that low-redshift galaxies will be incorrectly assigned high redshifts. For example, at magnitudes $AB(8140) = 27$–28, the long tail stretching to large positive residuals at redshifts $z = 0$–1 contains 27.3% of the total while the secondary peak at large negative residuals at redshifts $z = 3$–4 contains 34.8% of the total.

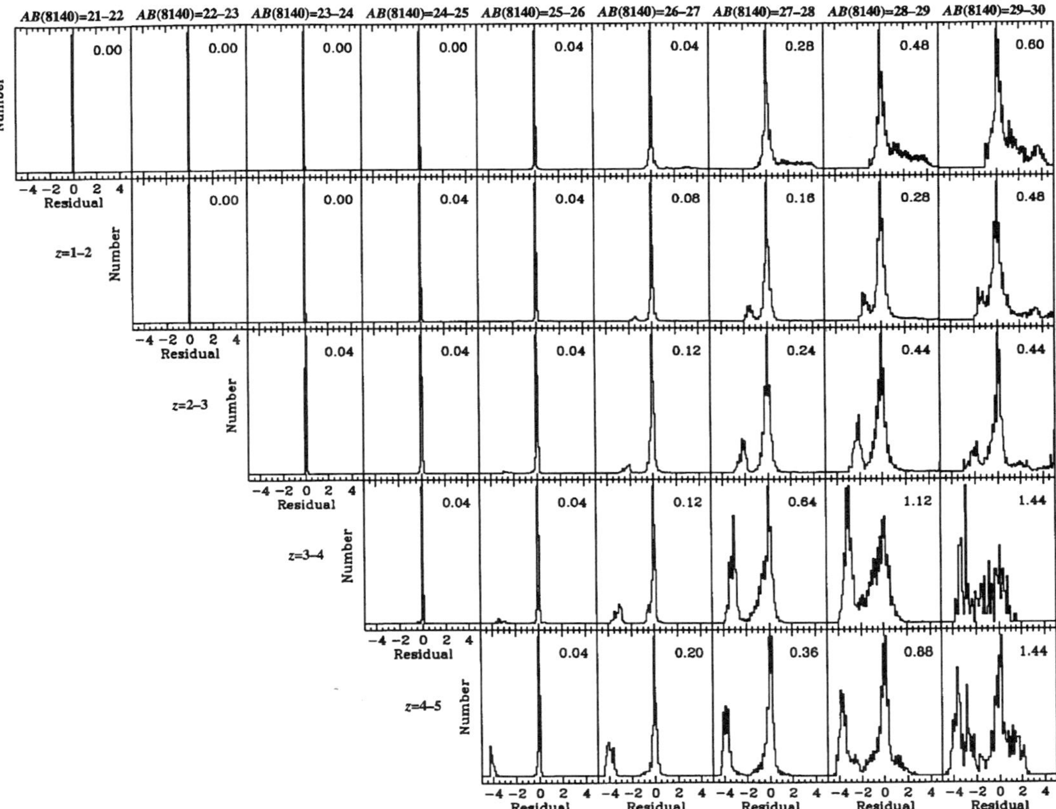

FIGURE 6. Distributions of residuals between best-fit model and simulated redshifts, illustrating the effects of photometric error on the photometric redshifts. Panels are arranged horizontally according to magnitude, from $AB(8140) = 21–22$ through $AB(8140) = 29–30$ in steps of 1 mag, and vertically according to redshift, from $z = 0–1$ through $z = 4–5$ in steps of 1. The median absolute residual is listed at the upper right corner of every panel.

It is possible to drawn general conclusions about the relative importance of photometric error and cosmic variance on the photometric redshifts by combining results of Figure 6 with results of § 3.2. At redshifts $z < 2$, magnitudes of the spectroscopically identified objects range from $AB(8140) = 18.16$ through $AB(8140) = 24.88$, with a median of $AB(8140) = 22.68$. Results of Figure 6 indicate that at these magnitudes photometric error produces almost no effect on the photometric redshifts. At redshifts $z > 2$, magnitudes of the spectroscopically identified objects range from $AB(8140) = 23.40$ through $AB(8140) = 26.75$ with a median of $AB(8140) = 25.08$. Results of Figure 6 indicate that at these magnitudes photometric error produces at most a mild effect on the photometric redshifts. [For example, even at magnitudes as faint as $AB(8140) = 26 - 27$, the median absolute residual produced by cosmic variance is less than the median absolute residual between the photometric and spectroscopic redshifts, although the discordant fraction produced by cosmic variance is consistent with the discordant fraction between the photometric and spectroscopic redshifts.] We therefore conclude that *residuals between the photometric and spectroscopic redshifts are dominated by cosmic variance rather than by photometric error.*

This result establishes the magnitude of the effect of cosmic variance on the photometric redshifts. Specifically, by attributing the entire residual between the photometric and spectroscopic redshifts to the effect of cosmic variance and considering the redshifts $z \approx 0$–4 spanned by the comparison of photometric and spectroscopic redshifts, we conclude that cosmic variance produces a median absolute residual of 0.09, a clipped RMS residual of 0.17, and a discordant fraction of 0.049. As long as the effect of cosmic variance does not increase with decreasing galaxy luminosity—and we can think of no reason that it should—then *these values must apply at all magnitudes*, not just the relatively bright magnitudes of the spectroscopic limit. In practice, this has two important implications: First, at magnitudes brighter than $AB(8140) \approx 26$–27 the accuracy and reliability of the photometric redshifts is limited by cosmic variance and at magnitudes fainter than $AB(8140) \approx 26$–27 the accuracy and reliability of the photometric redshifts is limited by photometric error. Second, the uncertainty associated with any statistical moment of the galaxy distribution—including the effects of sampling error, photometric error, and cosmic variance—may be realistically estimated by means of a "bootstrap" resampling technique, simulating the effect of photometric error by adding random noise to the fluxes and simulating the effect of cosmic variance by adding (rather modest) random noise to the estimated redshifts. We expect that this technique will ultimately play an important role in exploiting the full potential of the broad-band photometric redshift techniques.

4. Galaxies of redshift $z = 5$–7

In this section we describe an application of the broad-band photometric redshift techniques to galaxies of redshift $z = 5 - 7$. A full description of this analysis is reported by Lanzetta, Fernández-Soto, & Yahil (1997).

Results of § 3 demonstrate the accuracy and reliability of the broad-band photometric redshift techniques at the redshifts $z \approx 0$–4 spanned by the comparison of photometric and spectroscopic redshifts. But our analysis identifies galaxies with estimated redshifts as large as $z \approx 5$–7, which are probably inaccessible to ground-based spectroscopy. These galaxies are characterized by strong flux at observed-frame wavelengths $\lambda \approx 8140$ Å and an absence of detectable flux at shorter wavelengths, hence it is not possible by means of optical observations alone to establish continuum spectral energy distributions of the galaxies or to definitively rule out the competing alternatives that (1) the flux detected at observed-frame wavelengths $\lambda \approx 8140$ Å arises due to strong emission lines or (2) the absence of detectable flux at shorter wavelengths arises due to reddening by dust.

To address these issues, we consider the infrared images obtained with the KPNO 4 m telescope by Dickinson et al. (1997, in preparation). Specifically, we concentrate on galaxy 2342,968, which is the brightest galaxy with estimated redshift $z > 5$ (at least at 8140 Å). The infrared photometric analysis applied to galaxy 2342,968 is shown in Figure 7, and the spectral energy distribution of galaxy 2342,968 is shown in Figure 8. Results of Figures 7 and 8 indicate that galaxy 2342,968 is detected at the 0.9σ level of significance at the J band, the 2.7σ level of significance at the H band, and the 4.7σ level of significance at the K band. The primary conclusions of the analysis are two-fold: First, the detection of galaxy 2342,968 at the J, H, and K infrared bands establishes the continuum spectral energy distribution at observed-frame wavelengths spanning $\lambda \approx 8140$ Å through $\lambda \approx 2.2$ μ. This rules out the possibility that galaxy 2342,968 is a lower-redshift galaxy that exhibits a strong emission line at observed-frame wavelength $\lambda \approx 8140$ Å. Second, the detection of galaxy 2342,968 at modest flux levels at the J, H, and K infrared bands limits the permitted reddening. This rules out the possibility that galaxy 2342,968 is a lower-redshift galaxy that is heavily obscured and reddened by dust.

To establish a quantitative measure of the degree to which heavily obscured and reddened lower-redshift galaxies can be ruled out, we repeat the maximum-likelihood redshift determinations of our previous analysis, modifying the galaxy spectral templates by a grid of dust extinction models with color excess $E(B-V)$ ranging from $E(B-V) = 0$ (no dust obscuration) through $E(B-V) = 4$ (very heavy dust obscuration). It is possible to find acceptable high-redshift solutions with very modest amounts of dust extinction, but it is not possible to find acceptable low-redshift solutions with larger amounts of dust extinction. In particular, any solution with $z < 4.4$ is ruled out at more than the 4σ level of significance. The reason is that it is not possible to simultaneously satisfy the sharp discontinuity in flux between the F606W and F814W filters and the absence of strong infrared flux with any low-redshift galaxy spectral template and any amount of dust extinction. A similar analysis applied to the other galaxies with estimated redshift $z > 5$ yields similar results, although with less stringent limits. We conclude that the available evidence supports the photometric redshift identifications and that it is very unlikely that at least the brightest of the galaxies with estimated redshifts $z > 5$ are at substantially lower redshifts.

5. Redshift distribution of faint galaxies

Our analysis determines photometric redshifts of over 2500 galaxies, which together with an assessment of the effects of photometric error and cosmic variance can be used to construct an essentially exhaustive account of statistical properties (and uncertainties) of faint galaxies at magnitudes ranging to $AB(8140) \approx 30$ and redshifts ranging to $z \approx 7$. A complete enumeration of the properties of faint galaxies is beyond the scope of this review, so we instead illustrate the utility of broad-band photometric redshift techniques by presenting the redshift distribution of faint galaxies. The redshift distribution and median redshift of faint galaxies as functions of limiting magnitude threshold are shown in Figure 9.

Several conclusions can be drawn from results of Figure 9:

(1) At the brightest magnitudes $AB(8140) \approx 23$ accessible to the HDF, the galaxy redshift distribution is peaked at low redshifts with a median redshift of $z = 0.64$. This

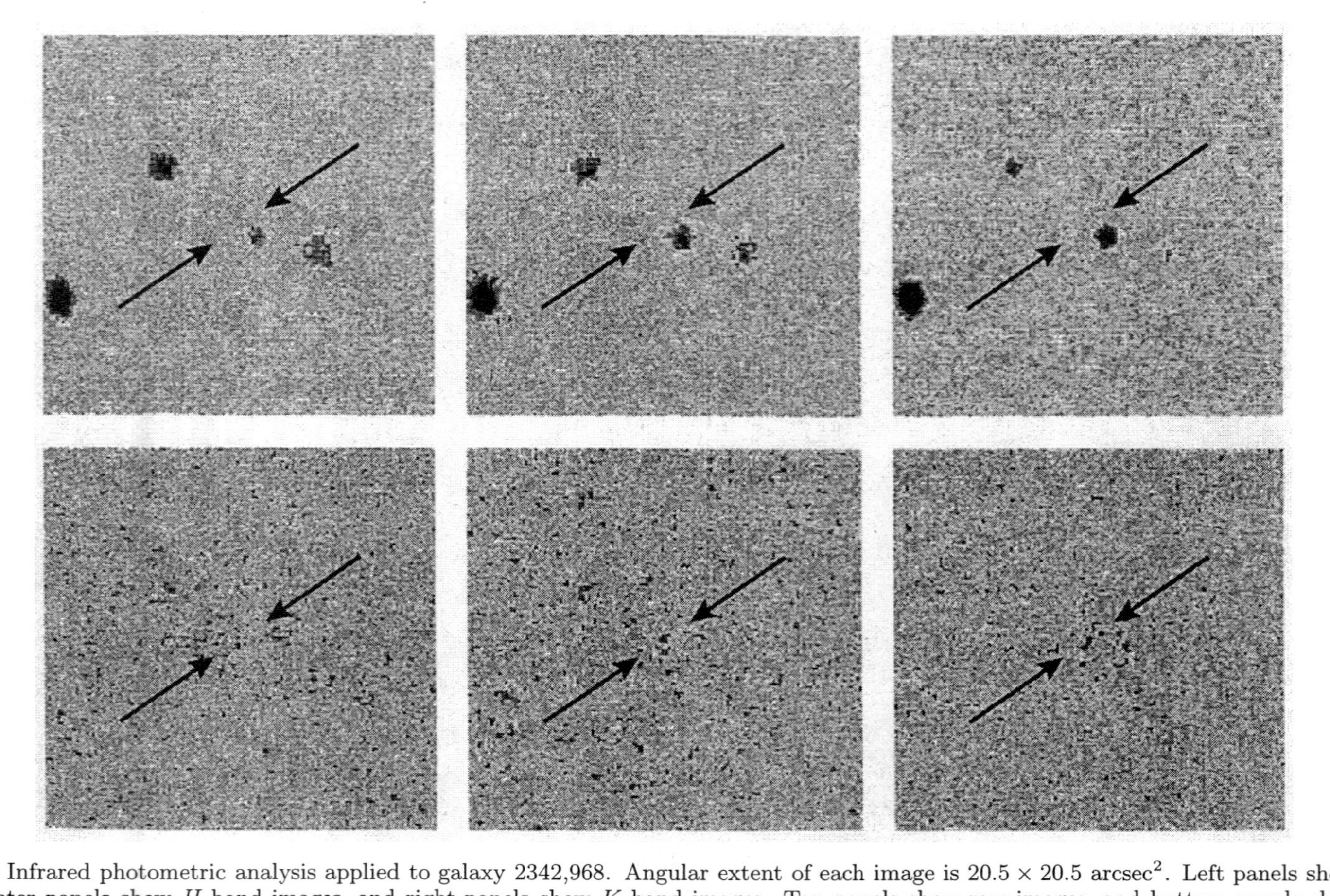

FIGURE 7. Infrared photometric analysis applied to galaxy 2342,968. Angular extent of each image is 20.5×20.5 arcsec^2. Left panels show J-band images, center panels show H-band images, and right panels show K-band images. Top panels show raw images, and bottom panels show images for which models of all objects other than galaxy B have been subtracted. Arrows point to the expected position of galaxy B, as measured from the F814W image.

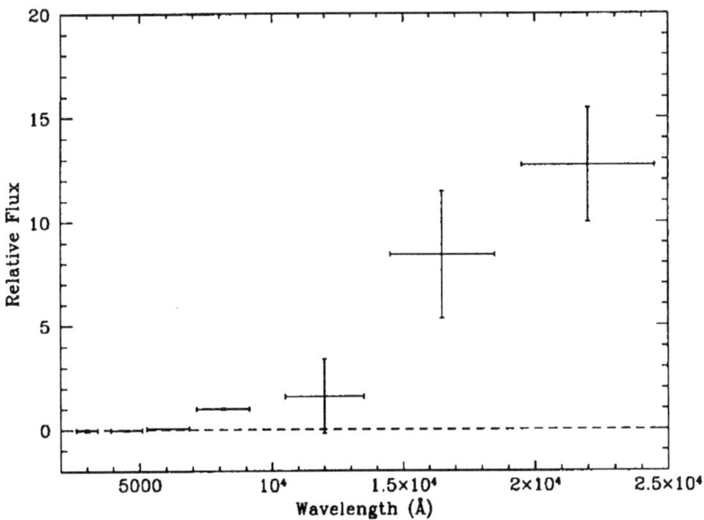

FIGURE 8. Spectral energy distribution of galaxy 2342,968. Vertical error bars indicate flux uncertainties, and horizontal error bars indicate filter FWHM.

agrees with results of deep ground-based galaxy surveys, the deepest of which are sensitive to comparable magnitudes (e.g., Cowie et al. 1996).

(2) High-redshift galaxies with $z > 2$ become prominent at magnitudes $AB(8140) \approx 25$, at which they constitute $\approx 6.1\%$ of the galaxy population. At fainter magnitudes, high-redshift galaxies become increasingly prominent with increasing magnitude threshold. Galaxies of redshift $z > 2$ constitute $\approx 13.9\%$ of the galaxy population at magnitudes $AB(8140) < 26$, 19.5% of the galaxy population at magnitudes $AB(8140) < 27$, and 24.0% of the galaxy population at magnitudes $AB(8140) < 28$.

(3) The median redshift of the galaxy redshift distribution increases monotonically with increasing magnitude threshold. The median redshift of the galaxy redshift distribution grows from med $z = 0.68$ at magnitudes $AB(8140) < 24$, to med $z = 1.04$ at magnitudes $AB(8140) < 26$, to med $z = 1.36$ at magnitudes $AB(8140) < 28$.

(4) The galaxy redshift distribution exhibits a precipitous decline at redshift $z \approx 2$ at all magnitudes $AB(8140) \gtrsim 26$. This is unlikely to be an artifact of the analysis because (1) results of § 3 indicate that redshifts $z > 2$ exhibit a comparable accuracy to redshifts $z < 2$ and (2) there are prominent broad-band spectral features redshifted into the optical filters at redshifts $z \approx 2$.

6. Future directions

Our conclusion is that broad-band photometric redshift techniques can be applied to accurately and reliably estimate redshifts of galaxies that are inaccessible to ground-based spectroscopy. At bright magnitudes the reliability of the photometric redshifts compares favorably with the reliability of the spectroscopic redshifts, and at faint magnitudes the effects of photometric error on the photometric redshifts can be rigorously quantified and accounted for. Yet we suspect that the ability of broad-band photometric techniques to reliably identify galaxies at redshifts beyond the ground-based spectroscopic limit will not become generally accepted until some means of independent verification can be found. The very brightest galaxy at estimated redshifts $z > 5$ (galaxy 2342,968) is of magnitude

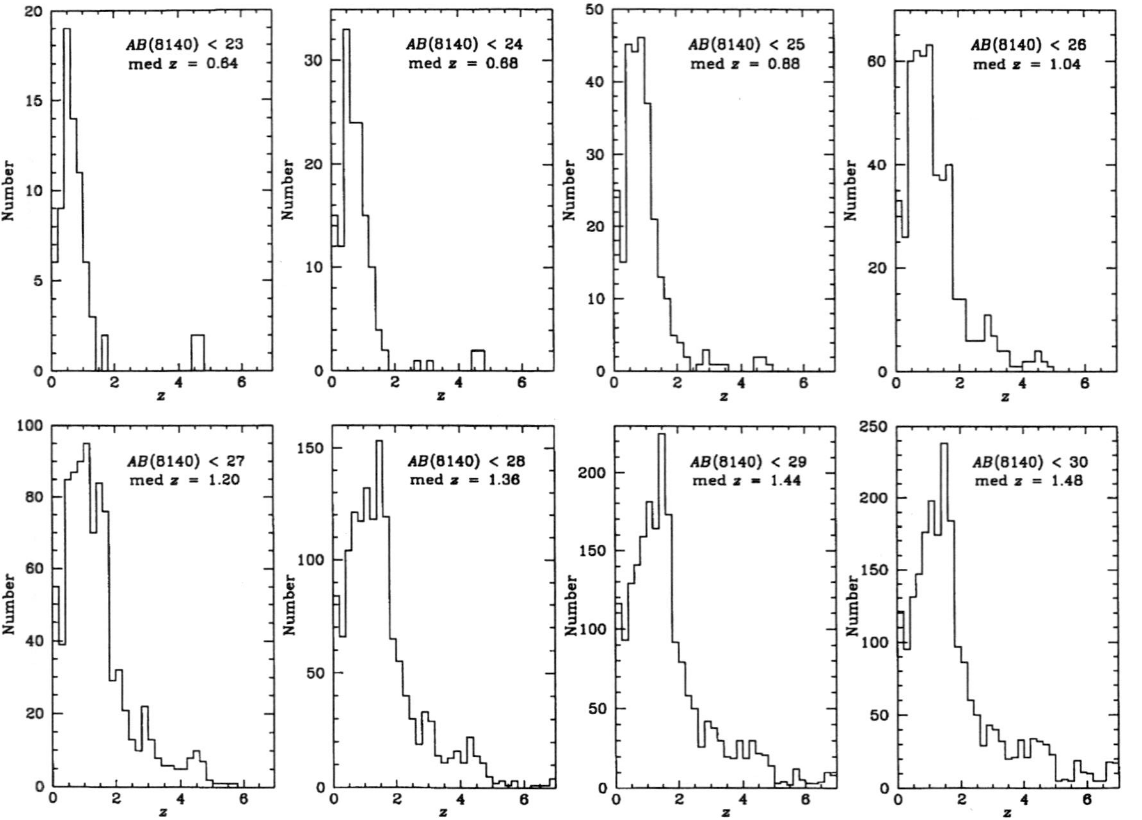

FIGURE 9. Redshift distribution of faint galaxies as a function of limiting magnitude threshold. The limiting magnitude threshold and median redshift is listed in every panel.

$AB(8140) = 26.3$, which is probably inaccessible to Keck spectroscopy. How, then, is progress to be made?

It is easy to show that for background-limited observations of spatially unresolved objects, the ratio of exposure times needed to obtain a given signal-to-noise ratio for space-based versus ground-based observations (assuming identical system throughputs and bandpasses) scales as

$$\frac{t_{\rm ground}}{t_{\rm space}} = \frac{f_\nu({\rm sky})_{\rm ground}}{f_\nu({\rm sky})_{\rm space}} \frac{A_{\rm space}}{A_{\rm ground}} \frac{\Omega({\rm sky})_{\rm ground}}{\Omega({\rm sky})_{\rm space}}, \tag{6.1}$$

where f_ν(sky) is the sky brightness, A is the telescope aperture, and Ω(sky) is the solid angle subtended by a spatial resolution element. At near-infrared wavelengths the first term of this expression evaluates to ≈ 4.4, for HST versus the Keck telescope the second term evaluates to ≈ 0.06, and for space-based seeing of ≈ 0.1 arcsec versus (optimistic) ground-based seeing of ≈ 0.6 the third term evaluates to ≈ 36. Forming the product of these three terms, we conclude that *HST is roughly 10 times more sensitive than the Keck telescope for spectroscopy of faint, spatially unresolved objects*. Although the Keck telescope can take advantage of multi-object spectroscopic capabilities—which HST with STIS probably cannot—this is not an issue for objects at the very limit of detectability or for objects of very low spatial density. We believe that to establish the broad-band photometric techniques at magnitudes and redshifts beyond the spectroscopic limit it will be necessary to use HST with STIS obtain spectroscopic confirmation of a few targets chosen carefully on the basis of broad-band photometric redshifts. These observations will provide the means of establishing the effects of cosmic variance on the photometric redshifts at redshifts $z \gtrsim 4$.

We are grateful to R. Williams and the staff of ST ScI for providing access to the reduced optical images, to M. Dickinson for providing access to the reduced infrared images, and to C. Steidel and M. Dickinson providing access to spectroscopic redshifts in advance of publication. A. F.-S.and K. M. L. were supported by NASA grant NAGW-4422 and NSF grant AST-9624216.

REFERENCES

Bertin, E. & Arnouts, S. 1996 *A&AS* in press.

Bruzual, G., & Charlot, S. 1993 *ApJ* **378**, 47.

Cohen, J. G., Cowie, L. L., Hogg, D. W., Songaila, A., Blandford, R., Hu, E. M., & Shopbell, P. 1996 *ApJ* **471**, L5.

Coleman, G. D., Wu, C. C. & Weedman, D. W. 1980 *ApJS* **43**, 393.

Connolly, A. J., Szalay, A. S., Dickinson, M., Subbarao, M. U., Brunner, R. J. 1997 *ApJ* **486**, L11.

Cowie, L. L. 1997 http://www.ifa.hawaii.edu/ cowie/tts/tts.html.

Cowie, L. L., Songaila, A., Hu, E. M., & Cohen, J. G. 1996 *AJ* **112**, 839.

Ellis, R. S. 1997 *ARAA*, in press.

Gwyn, S. D. J., & Hartwick, F. D. A. 1996 *ApJ* **468**, L77.

Hogg, D. W., Cohen, J. G., Blandford, R., Gwyn, S. D. J., Hartwick, F. D. A., Mobasher, B., Mazzei, P., Sawicki, M., Lin, H., Yee, H. K. C., Connolly, A. J., Brunner, R. J., Csabai, I., Dickinson, M., Subbarao, M. U., Szalay, A. S., Fernández-Soto, A., Lanzetta, K. M., & Yahil, A. 1997, *AJ*, submitted.

Kinney, A. L., Bohlin, R. C., Calzetti, D., Panagia, N. & Wyse, R. F. G. 1993 *ApJS* **86**, 5.

Lanzetta, K. M., Fernández-Soto, A., & Yahil, A. 1997 *ApJ*, submitted.

Lanzetta, K. M., Yahil, A., & Fernández-Soto, A. 1996 *Nature* **381**, 759.

Lowenthal, J. D., Koo, D. C., Guzmán, R., Gallego, J., Phillips, A. C., Faber, S. M., Vogt, N. P., Illingworth, G. D., & Gronwall, C. 1997 *ApJ* **481**, 673.

Madau, P. 1995 *ApJ* **441**, 18.

Mobasher, B., Rowan-Robinson, M., Georgakakis, A., Eaton, N. 1996 *MNRAS* **282**, L7.

Sawicki, M. J., Lin, H., & Yee, H. K. C. 1997 *AJ* **113**, 1.

Steidel, C. C., Giavalisco, M., Dickinson, M., & Adelberger, K. 1996 *AJ* **112**, 352.

Williams, R. E., et al. 1996 *AJ* **112**, 1335.

Zepf, S. E., Moustakas, L. A., & Davis, M. 1997 *ApJ* **474**, L1.

Global evolution of the stellar and interstellar contents of galaxies

By S. MICHAEL FALL

Space Telescope Science Institute, 3700 San Martin Drive, Baltimore, MD 21218

We present a method to combine information about the evolution of galaxies obtained from deep imaging surveys and quasar absorption-line surveys. The connection is made through the global histories of star formation, gas consumption, and metal and dust production in galaxies. We show that the star formation history inferred from imaging surveys, including the Hubble Deep Field, is consistent with the one inferred previously from absorption-line surveys.

---

1. Introduction

The Hubble Deep Field has given us our clearest view yet of the stellar contents of galaxies at high redshifts. This is a remarkable achievement, one that has advanced our knowledge substantially. But even the deepest of imaging surveys can tell us only part of the story of galactic evolution. Quasar absorption-line surveys tell us another, equally important part of the story, the part about the interstellar contents of galaxies. How can we relate the information from these two kinds of surveys? The simplest way to do this is through various average properties of the entire population of galaxies. It is often convenient to express these "global" properties as mean comoving densities and to normalize them to the present closure density. We are particularly interested in the comoving densities of stars, gas, metals, and dust within galaxies, which we denote respectively by Ω_s, Ω_g, Ω_m, and Ω_d. The last three of these are meant to refer to the interstellar media (ISM) of galaxies, exclusive of the intergalactic medium (IGM), although in practice such a distinction may not be very sharp. As defined here, Ω_m includes metals in both the gas and solid (i.e., dust) phases of the ISM. It is often more informative to reexpress Ω_m and Ω_d in terms of the mean metallicity, $Z \equiv \Omega_m/\Omega_g$, and the mean dust-to-gas ratio, $D/G \equiv \Omega_d/\Omega_g$. It is clear that all of these properties are related in the sense that, as new stars form, Ω_s will increase, while in most cases, Ω_g will decrease and Z and D/G will increase. The goal here is to quantify such notions through the equations of "cosmic chemical evolution."

Until recently, there were no emission-based estimates of the global rate of star formation $\dot{\Omega}_s$ at $z \gtrsim 0.3$. The reason for this is that samples of galaxies selected by emission become progressively incomplete and include only brighter objects at higher redshifts. In contrast, samples of galaxies selected by absorption against background quasars do not suffer from this bias. Such observations are exquisitely sensitive to small column densities of absorbing or scattering particles. In principle at least, they enable us to estimate Ω_g, Ω_m, and even Ω_d as functions of redshift. From these and the equations of cosmic chemical evolution, we can then infer the global rate of star formation $\dot{\Omega}_s$. It is amusing to note that this idealistic program does not require the detection of a single stellar photon! We can also combine our estimates of $\dot{\Omega}_s$ with stellar population synthesis models to compute the mean comoving emissivity $\mathcal{E}_\nu$ and the mean intensity of background radiation J_ν. One might then claim to have predicted the "emission history" of the universe from its "absorption history." This article describes a first attempt by Yichuan Pei, Stéphane Charlot, and the author to carry out such a program; a complete account of our work is given elsewhere (Pei & Fall 1995; Fall et al. 1996).

2. Damped Lyman-Alpha galaxies

Before proceeding, it is worth recalling some facts about the statistics of absorption-line systems. Let $f(N_x, z)$ be the column density distribution of particles of any type x that absorb or scatter light. These might, for example, be hydrogen atoms (x = HI), metal ions ($x = m$), or dust grains ($x = d$). By definition, $H_0(1+z)^3|dt/dz|f(N_x, z)dN_x dz$ is the mean number of absorption-line systems with column densities of x between N_x and $N_x + dN_x$ and redshifts between z and $z + dz$ along the lines of sight to randomly selected background quasars. These lines of sight are very narrow (less than a light year across) and pierce the absorption-line systems at random angles and impact parameters. One can show that the mean comoving density of x is given by

$$\Omega_x(z) = \frac{8\pi G m_x}{3cH_0} \int_0^\infty dN_x f(N_x, z) N_x, \tag{2.1}$$

where m_x is the mass of a single particle (atom, ion, or grain). Equation (2.1) plays a central role in this subject. It enables us to estimate the mean comoving densities of many quantities of interest without knowing anything about the structure of the absorption-line systems. In particular, we do not need to know their sizes or shapes, whether they are smooth or clumpy, and so forth. A corollary of equation (2.1) is that the global metallicity, $Z \equiv \Omega_m/\Omega_g$, is given simply by an average over the metallicities of individual absorption-line systems weighted by their gas column densities.

The absorption-line systems of most interest in the present context are the damped Lyα (DLA) systems. It is widely believed that they trace the ISM of galaxies and protogalaxies and are the principal sites of star formation in the universe. There are excellent reasons to adopt this as a working hypothesis. First, the DLA systems have, by definition, $N_{\rm HI} \gtrsim 10^{20}$ cm^{-2}, and this, at least at low redshifts, coincides roughly with a threshold for the onset of star formation (Kennicutt 1989). Second, the DLA systems contain at least 80% of the HI in the universe and appear to be mostly neutral (Lanzetta et al. 1995). The other absorption-line systems, those with $N_{\rm HI} \lesssim 10^{20}$ cm^{-2}, probably contain more gas in total than the DLA systems, but this must be diffuse and mostly ionized. In the following, we regard non-DLA systems as belonging to the IGM, even though some of them might actually be associated with the outer, tenuous parts of galaxies. This distinction—between the mostly-neutral ISM, where stars form, and the mostly-ionized IGM, where they do not—is clearly valid at the present epoch. Thus, it seems appropriate to refer to the DLA systems as DLA galaxies. The precise nature of these objects—whether they large or small, disk or spheroid—remains to be determined, probably by direct imaging. However, as we have already emphasized, the global properties derived from equation (2.1) are not affected by this uncertainty.

The sample of known DLA galaxies now includes about 80 objects (Wolfe et al. 1995). They are distributed over a wide range in redshift, $0 \lesssim z \lesssim 4$, although, as a consequence of selection effects, most of them are confined to the narrower range $2 \lesssim z \lesssim 3$. From observations of DLA galaxies in various subsets of this sample and comparisons with present-day galaxies, the following trends have emerged. The mean comoving density of HI decreases by almost an order of magnitude, from $\Omega_{\rm HI} \approx (1-2) \times 10^{-3} h^{-1}$ at $z \approx 3$ to $\Omega_{\rm HI} \approx 2 \times 10^{-4} h^{-1}$ at $z = 0$, with $h \equiv H_0/(100$ km s^{-1} Mpc$^{-1})$ (Lanzetta et al. 1995; Wolfe et al. 1995; Storrie-Lombardi et al. 1996). The mean metallicity increases by about an order of magnitude, from $Z \approx 0.1 Z_\odot$ or slightly less at $z \approx 2$ to $Z \approx Z_\odot$ at $z = 0$ (Pettini et al. 1994, 1997b; Lu et al. 1996). The mean dust-to-gas ratio increases by a similar factor, while the mean dust-to-metals ratio remains roughly constant at about the present value in the local ISM (Pei et al. 1991; Kulkarni et al. 1997; Pettini et al. 1997a; Welty et al. 1997; Vladilo 1998). The abundances of H_2 and CO generally appear to be lower at $z \gtrsim 2$ than at $z = 0$ (Levshakov et al. 1992; Ge & Bechtold 1997). Unfortunately, the estimates of $\Omega_{\rm HI}$, Z, and D/G are dominated by relatively few systems—those with the highest values of $N_{\rm HI}$, N_m, and N_d. As a result, they are less certain than is sometimes appreciated.

3. Cosmic chemical evolution

The global properties defined above are governed by a set of coupled equations, which are sometimes referred to as the equations of cosmic chemical evolution. In the approximation of instantaneous recycling (and $Z \ll 1$), they take the form

$$\frac{d}{dt}(\Omega_g + \Omega_s) = \dot{\Omega}_f, \tag{3.2}$$

$$\frac{d}{dt}(Z\Omega_g) + (Z - y)\frac{d}{dt}\Omega_s = Z_f\dot{\Omega}_f, \tag{3.3}$$

where y is the IMF-averaged yield. These are the cosmological analogs of the usual equations for the chemical evolution of individual galaxies (Pagel 1997). They are exact in the limit that all galaxies evolve in the same way but otherwise must be regarded as approximations. The "source" terms on the right-hand sides of the equations allow for the exchange of material between the ISM of galaxies and the IGM; they represent the inflow or outflow of gas with metallicity Z_f at a rate $\dot{\Omega}_f$. To illustrate a range of possibilities, we consider three types of evolution: a closed-box model ($\dot{\Omega}_f = 0$), a model with inflow of metal-free gas ($\dot{\Omega}_f = +\nu\dot{\Omega}_s$, $Z_f = 0$), and a model with outflow of metal-enriched gas ($\dot{\Omega}_f = -\nu\dot{\Omega}_s$, $Z_f = Z$). Our inflow and outflow models are the cosmological analogs of the standard models of chemical evolution in the disk and spheroid components of the Milky Way (Larson 1972; Hartwick 1976). We fix the yield y in each model by requiring $Z = Z_\odot$ at $z = 0$. Then the only adjustable parameters are the "initial" comoving density of gas in galaxies $\Omega_{g\infty}$ (in practice, the value of Ω_g at $z \gtrsim 4$) and the relative inflow or outflow rate ν.

To complete the specification of the models, we make two other approximations, both motivated by the observations summarized in the previous section: (1) We neglect any ionized or molecular gas in the ISM of galaxies and set $\Omega_g = 1.3\Omega_{\rm HI}$ (to account for He). (2) We assume that just over half of the metals in the ISM are depleted onto dust grains and set $D/G = 0.6Z$. Our models are designed to reproduce (as input) the observed decrease in the mean comoving density of HI between $z \approx 3$ and $z = 0$. The main complication here is that the observed values of $\Omega_{\rm HI}$ tend to underestimate the true values as a consequence of the obscuration of quasars by dust in foreground galaxies (Fall & Pei 1993). We make a self-consistent correction for this bias in the models by linking the obscuration of quasars to the chemical enrichment of galaxies. It is worth noting that, while this correction has a substantial effect on $\Omega_{\rm HI}$, especially at $z \sim 1$, it does not entail large numbers of "missing" quasars [only ~20% at $z = 2$ and ~ 40% at $z = 4$, consistent with observational constraints (Shaver et al. 1996)]. We neglect another potential bias—that caused by gravitational lensing—because it appears not to be important in the existing samples of DLA galaxies (Le Brun et al. 1997; Perna et al. 1997; Smette et al. 1997). Our models reproduce (as output) the observed increase in the mean metallicity between $z \approx 2$ and $z = 0$ without any fine tuning of the parameters $\Omega_{g\infty}$ and ν. The reason for this is that most of the star formation and hence most of the metal production occur at $z \lesssim 2$.

Figure 1 shows the evolution of the comoving rate of metal production in our models, given by $\dot{\rho}_z = y(3H_0^2/8\pi G)\dot{\Omega}_s$. The predicted rates have maxima at $1 \lesssim z \lesssim 2$ and

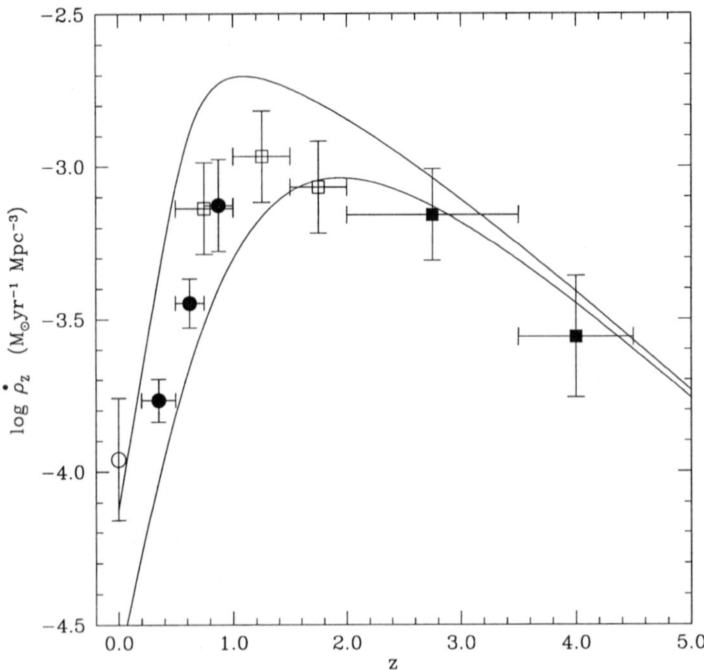

FIGURE 1. Comoving rate of metal production $\dot{\rho}_z$ as a function of redshift z (for $h = 0.5$, $q_0 = 0.5$, and $\Lambda = 0$). The upper curve represents the inflow model and the lower curve represents the closed-box and outflow models, with $\Omega_{g\infty} = 4\times10^{-3}h^{-1}$ and $\nu = 0.5$ [the standard parameters of Pei & Fall (1995)]. The data points represent global Hα and UV emissivities (Gallego et al. 1995; Lilly et al. 1996; Madau et al. 1996, 1998; Connolly et al. 1997).

decline rapidly at lower redshifts (Pei & Fall 1995). Figure 1 also shows emission-based estimates of $\dot{\rho}_z$ from several recent surveys, including the Canada-France Redshift Survey and the Hubble Deep Field (Gallego et al. 1995; Lilly et al. 1996; Madau et al. 1996, 1998; Connolly et al. 1997). These were derived from rest-frame Hα and UV emissivities and the approximate proportionality between UV emission and metal production in stellar populations. The main systematic uncertainties in $\dot{\rho}_z$ stem from corrections for incompleteness at low luminosities (which have been included) and corrections for absorption by dust (which have not been included); as a result, the true uncertainties are probably larger than indicated by the error bars in Figure 1. Evidently, the predicted and observed rates are in broad qualitative, and even some quantitative, agreement (within factors of about two). This is remarkable because our models were constructed only with absorption-line observations in mind, before the emission-based estimates of $\dot{\rho}_z$ were available. We have also combined our chemical evolution models with stellar population synthesis models to compute directly the mean comoving emissivity $\mathcal{E}_\nu$ and, by an integration over redshift, the corresponding mean intensity of background radiation J_ν (Fall et al. 1996). These calculations include a self-consistent treatment of the absorption and reradiation of starlight by dust. The predicted far-IR/sub-mm background is consistent with recent results from the DIRBE and FIRAS experiments on *COBE* (Hauser 1996; Puget et al. 1996).

REFERENCES

Connolly, A J., Szalay, A. S., Dickinson, M., SubbaRao, M. U., & Brunner, R. J. 1997 *ApJ* **486**, L11.

Fall, S. M., Charlot, S., & Pei, Y. C. 1996 *ApJ* **464**, L43.

Fall, S. M., & Pei, Y. C. 1993 *ApJ* **402**, 479.

Gallego, J., Zamorano, J., Aragón-Salamanca, A., & Rego, M. 1995 *ApJ* **455**, L1.

Ge, J., & Bechtold, J. 1997 *ApJ* **477**, L73.

Hartwick, F. D. A. 1976 *ApJ* **209**, 418.

Hauser, M. G. 1996 in *Unveiling the Cosmic Infrared Background*, (ed. E. Dwek), p. 11. American Institute of Physics.

Kennicutt, R. C. 1989 *ApJ* **344**, 685.

Kulkarni, V. P., Fall, S. M., & Truran, J. W. 1997 *ApJ* **484**, L7.

Lanzetta, K. M., Wolfe, A. M., & Turnshek, D. A. 1995 *ApJ* **440**, 435.

Larson, R. B. 1972 *Nature Phys. Sci.* **236**, 7.

Le Brun, V., Bergeron, J., Boissé, P., & Deharveng, J. M. 1997 *A&A* **321**, 733.

Levshakov, S. A., Chaffee, F. H., Foltz, C. B., & Black, J. H. 1992 *A&A* **262**, 385.

Lilly, S. J., Le Fèvre, O., Hammer, F., & Crampton, D. 1996 *ApJ* **460**, L1.

Lu, L., Sargent, W. L. W., Barlow, T. A., Churchill, C. W., & Vogt, S. S. 1996 *ApJS* **107**, 475.

Madau, P., Ferguson, H. C., Dickinson, M. E., Giavalisco, M., Steidel, C. C., & Fruchter, A. 1996 *MNRAS* **283**, 1388.

Madau, P., Pozzetti, L., & Dickinson, M. 1998 *ApJ* in press (astro-ph/9708220).

Pagel, B. E. J. 1997 *Nucleosynthesis and Chemical Evolution of Galaxies.* Cambridge University Press.

Pei, Y. C., & Fall, S. M. 1995 *ApJ* **454**, 69.

Pei, Y. C., Fall, S. M., & Bechtold, J. 1991 *ApJ* **378**, 6.

Perna, R., Loeb, A., & Bartelmann, M. 1997 *ApJ* **488**, 550.

Pettini, M., King, D. L., Smith, L. J., & Hunstead, R. W. 1997a *ApJ* **478**, 536.

Pettini, M., Smith, L. J., Hunstead, R. W., & King, D. L. 1994 *ApJ* **426**, 79.

Pettini, M., Smith, L. J., King, D. L., & Hunstead, R. W. 1997b *ApJ* **486**, 665.

Puget, J.-L., Abergel, A., Bernard, J.-P., Boulanger, F., Burton, W. B., Désert, F.-X., & Hartmann, D. 1996 *A&A* **308**, L5.

Shaver, P. A., Wall, J. V., Kellerman, K. I., Jackson, C. A., & Hawkins, M. R. S. 1996 *Nature* **384**, 439.

Smette, A., Claeskens, J.-F., & Surdej, J. 1997 *New Astronomy* **2**, 53.

Storrie-Lombardi, L. J., McMahon, R. G., & Irwin, M. J. 1996 *MNRAS* **283**, L79.

Vladilo, G. 1998 *ApJ* **493**, 583.

Welty, D. E., Lauroesch, J. T., Blades, J. C., Hobbs, L. M., & York, D. G. 1997 *ApJ* **489**, 672.

Wolfe, A. M., Lanzetta, K. M., Foltz, C. B., & Chaffee, F. H. 1995 *ApJ* **454**, 698.

Model predictions for clustering and morphologies at HDF depths

By MATTHIAS STEINMETZ

Steward Observatory, University of Arizona, Tucson, AZ 85721

The current status of numerical simulations of the formation of galaxies is reviewed. Success and failure of modeling galaxies at low and high redshift is demonstrated using a variety of examples, such as the Tully-Fisher relation, the appearance of high-redshift galaxies and the kinematics of damped Lyα systems. The relationship between the clustering properties of high-z galaxies and the present generation of galaxies is emphasized.

1. Introduction

Hierarchical clustering is at present the most successful paradigm of structure formation in the universe. In this scenario—Cold Dark Matter (CDM) and its variants are perhaps the best known examples—structure grows as systems of progressively larger mass merge and collapse to form newly virialized systems. A large variety of models which are based on the hierarchical clustering hypothesis have been extensively studied using N-body simulations as well as analytical approximations.

Within the last few years, more and more work focussed on smaller structures, and tried to embed galaxy formation into the hierarchical picture. One successful approach has been semi-analytical or phenomenological models (Kauffmann, White & Guiderdoni, 1993; Cole et al. 1994). These models use the extended Press-Schechter theory to predict abundances and merger rates of halos as a function of mass and redshift. Physically motivated recipes are used to model how gas cools, how it settles at the center of dark matter halos and how it is transformed into stars. This phenomenological ansatz provides at comparably low computational cost a very efficient method to predict the formation and evolution of the galaxy population. However, it has only little power in predicting the clustering of galaxies (see, however, Kauffmann, Nusser & Steinmetz, 1997) and it makes no prediction on the detailed formation history of individual galaxies.

Detailed studies of the formation of individual galaxies necessarily require large numerical simulations which include not only gas dynamics, shocks and radiative cooling, but which also incorporate some description of star formation. This article addresses some successes and failures of such simulations. It will focus on issues which seem to be generic to the hypothesis of hierarchical clustering and which depend only little on the details on the underlying cosmogony. Numerical details are discussed in the appendix.

2. Prologue: The overcooling problem

Already the very first quantitative investigations of galaxy formation in hierarchically clustering universes (e.g., White & Rees 1978) exhibited a generic problem of this class of models, nowadays usually referred to as the *overcooling problem.* Since cooling times scale inversely with density, the dissipative collapse of gas must have been more efficient at high redshift because the dark matter halos present at that time (and the universe as a whole) were denser. Cooling is expected to be so efficient at early times that almost all gas within a dark matter halo is able to cool to temperatures of about 10^4 K. Consequently, galaxies tend to be too massive. A related problem is that hierarchical models predict

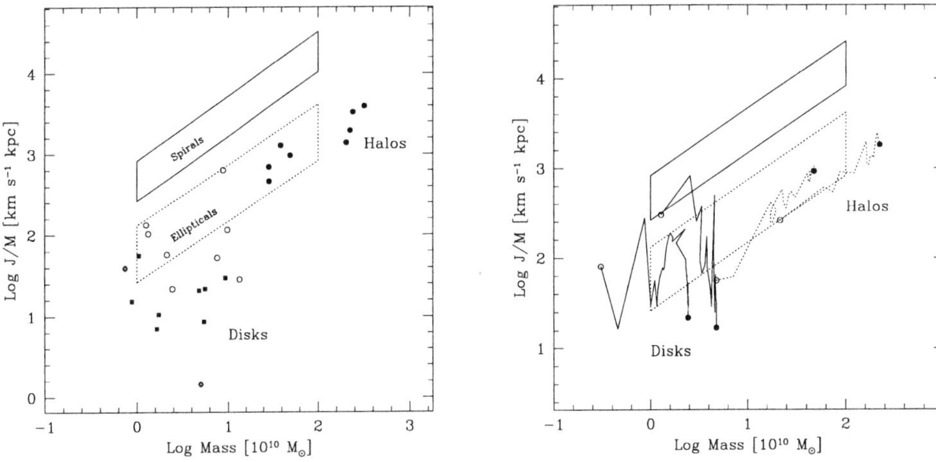

FIGURE 1. Left: The specific angular momentum of dark halos and gaseous disks, as a function of mass. The boxes enclose the region occupied by spiral and elliptical galaxies, as given by Fall (1983). Open circles correspond to runs without UV background; solid squares correspond to a soft UV background ($J_{-21} = 1$, $\alpha = 5$ (equation 2.3)), starred symbols correspond to an extremely energetic background ($J_{-21} = 10$, $\alpha = 1$). Note that the halos' J/M scale approximately as $M^{2/3}$, as expected if all systems had the same value of the rotation parameter λ (see text for a definition). Gaseous disks have much lower angular momenta than observed spirals, a consequence of the role of mergers during the assembly of the disks. Note that the inclusion of a UV radiation field seems to aggravate this problem. Right: Evolution of the dark halo and central gaseous disk in the J/M versus M plane, from $z = 5$ (open circles) to $z = 0$ (solid circles). The mass of the system grows steadily by mergers, which are accompanied by an increase in the spin of the halo and a decrease in the spin of the central disk. The latter results from angular momentum being transferred from the gas to the halo during mergers.

an overabundance of low mass halos with circular velocities below 100 km/s (Cole et al. 1994).

This can be quantified by the following analytical argument (see, e.g., White 1994). At a given redshift z, the mass of a dark matter halo of circular velocity v_c can be written as

$$M_{\rm vir} = 2.6 \times 10^{12}\, {\rm M}_\odot\, \frac{3\zeta}{2}\, h^{-1}\, (1+z)^{-1.5} \left(\frac{v_c}{220\,{\rm km/s}}\right)^3 , \qquad (2.1)$$

the factor ζ representing the age of the universe in units of H_0^{-1}, i.e., for a $\Omega = 1$ universe $\zeta = 2/3$. Following White et al. (1993), the baryon fraction ($M_{\rm bary}/M_{\rm tot}$) within the virial radius should be very similar to the cosmological value $\Omega_{\rm b}/\Omega_0$, with the nucleosynthesis value $\Omega_{\rm b} = 0.0125\, h^{-2}$. The baryonic mass enclosed in a halo of circular velocity v_c is thus given by

$$M_{\rm bary} = \frac{\Omega_{\rm bary}}{\Omega_0} M_{\rm vir} = 3.3 \times 10^{10}\, {\rm M}_\odot\, \frac{3\zeta}{2\Omega_0}\, h^{-3}\, (1+z)^{-1.5} \left(\frac{v_{\rm vir}}{220\,{\rm km/s}}\right)^3 . \qquad (2.2)$$

Let us apply this to the case of the Galaxy. By assuming that the circular velocity of the dark matter halo is similar to the actual rotation velocity of the Galactic disk ($M_{\rm disk} \approx 6 \times 10^{10}\, {\rm M}_\odot$),the following statements can be made:

- In the case of $\Omega_0 = 1$ and $h = 0.5$, the total baryonic mass amounts $2.6 \times 10^{11}\, {\rm M}_\odot$, i.e., 4 times the mass of the galactic disk. For $\Omega = 0.2, \Lambda = 0.8$ the situation is even more extreme, the total baryonic mass is 2.1 $10^{12}\,M_\odot$, i.e., only 3% of the baryonic mass has cooled and settled into the disk.

• The situation is a bit less extreme if a higher h and/or a lower Ω_b is assumed, as, e.g., favored by recent Deuterium measurements by Songaila et al. (1994). Vice versa, even less gas has cooled if a higher Ω_b is assumed, as, e.g., favored by the Deuterium measurements of Tytler, Fan & Burles (1996).

• Since the cooling radius of a Milky Way sized system includes much more mass than the mass of the disk, a very efficient heating mechanism has to be postulated to prevent a large fraction of the baryonic mass from cooling.

• It is of some interest to note that for $\Omega = 1$ and $h = 0.8$ there exist hardly enough baryons to account for the mass observed in the galactic disk.

This simple model nicely reproduces the main feature seen in numerical simulation of galaxy formation (see, e.g., Navarro & Steinmetz 1997), namely that virtually all baryons within a dark matter halo are able to cool resulting in disk galaxies which are too massive.

Numerical simulations also show another shortcoming of the hierarchical clustering hypothesis, which is to some extent related to the overcooling problem: the angular momentum of simulated galaxies is too small. Compared to their observed counterparts galaxies are thus too concentrated. This is shown in Figure 1 which shows the specific angular momentum of dark matter halos and of their central gaseous disks at $z = 0$, as a function of mass. The boxes indicate the loci corresponding to spiral and elliptical galaxies, as compiled by Fall (1983). If, as suggested by Fall & Efstathiou (1980), the collapse of gas would proceed under conservation of angular momentum, the baryonic component would have the same specific angular momentum J/M as the dark matter, however, its corresponding mass would be a factor of 20 smaller (for $\Omega_{\rm bary} = 0.05$, $\Omega_0 = 1$). These disks would be located only slightly below the box for spiral galaxies. However, Figure 1 demonstrates clearly that the spins of gaseous disks are about an order of magnitude lower than that. This is a direct consequence of the formation process of the disks (Navarro, Frenk & White 1995). Most of the disk mass is assembled through mergers between systems whose own gas component had previously collapsed to form centrally concentrated disks. During these mergers, and because of the spatial segregation between gas and dark matter, the gas component transfers most of their orbital angular momentum to the surrounding halos (Frenk et al. 1985; Barnes 1988; Quinn & Zurek 1988).

As already mentioned, efficient feedback processes are required to prevent the gas from excessive cooling. One proposal has been that energy input due to a photoionizing UV background may prevent cooling in low mass halos at higher redshift (Efstathiou 1992). This hypothesis has been tested by numerical simulations but came to a negative result (Quinn, Katz & Efstathiou 1996; Weinberg, Hernquist & Katz 1997; Navarro & Steinmetz 1997). These simulations assumed the presence of a photoionizing background with an energy distribution

$$J(\nu) = J_{-21} \times 10^{-21} \left(\frac{\nu}{\nu_{\rm H}}\right)^{-\alpha} . \qquad (2.3)$$

Though an UV background can delay or even prevent the formation of galaxies with circular velocities of 30–50 km/s and below, its influence on the properties of galaxies with circular velocities exceeding 100 km/s is almost negligible. The amount of cool gas is only moderately reduced by about 10–30%. For extreme assumptions on the UV background it can be reduced by 50%, insufficient to reconcile the observed shape of the galaxy luminosity function. Concerning the angular momentum of gaseous disks, a

UV background even exacerbates the angular momentum problem as shown by the solid squares and starred symbols in Figure 1. This can be easily understood, since gas which falls in late and thus has low densities is most strongly affected by the UV background. Such gas, however, also possesses the highest specific angular momentum.

To solve the overcooling/angular momentum problem thus energy feedback from supernovae has been repeatedly advocated (e.g., Dekel & Silk 1986). The stumbling block for implementing star formation into galaxy formation simulations is certainly the ill-understood physics of star formation and the interaction of evolving stars with the ISM. The negative effect of photoheating on the angular momentum of cold disk also points to an interesting complication which seem to be generic to quite a variety of feedback mechanisms: In order to explain the observed sizes of disk galaxies, the specific angular momentum of the disk must not be much smaller than that of the host dark matter halo. However, it is also gas at large radii which (i) possesses the highest specific angular momentum and which (ii) has the lowest density and for which cooling can be easily suppressed. It is thus a non-trivial problem how the amount of cool gas can be reduced by a large fraction without affecting its specific angular momentum. This problem seems to be especially severe for low-Ω_0 models, where only a small fraction of baryons can be allowed to cool. In the following sections, some results from simulations including star formation and feedback are being discussed. These simulations incorporate supernova feedback due only to thermal energy (see appendix). To some extent they can thus be considered as minimum feedback models.

3. Simulation sample

The galaxy sample which is analyzed in the following sections consists of 21 galaxies with circular velocities between 50 and 250 km/s. This sample has been compiled from a set of 8 high resolution numerical simulations. The mass of a gas particle is between $5 \times 10^6\,\mathrm{M}_\odot$ and $2 \times 10^7\,\mathrm{M}_\odot$, the gravitational softening is 1 kpc. The star formation efficiency has been calibrated so that at $z = 0$ a galaxy with a circular velocity of 200 km/s has a star formation rate of about $1\,\mathrm{M}_\odot/\mathrm{yr}$. Details of the numerical method and the star formation and feedback scheme are presented in the appendix. Each galaxy consists of 10^2–10^4 star particles. Each of these particles represents a population of a few million stars which have been formed in a burst-like manner, i.e., a model galaxy can be considered as a superposition of several thousand mini starbursts. Each star particle is labeled by an age (time since creation) and a metallicity equal to that of its gas progenitor at the time of formation. The luminosity evolution of each of these bursts is then followed by an evolutionary spectral synthesis model (Contardo, Steinmetz & Fritze-von Alvensleben 1998). The model galaxies can thus be "observed" in arbitrary colors. The power of this technique is illustrated in Figure 2, which shows a computer-generated rendition of a CDM-dominated universe observed with the Hubble Space Telescope (HST) with exposures similar to those used for the Hubble Deep Field (HDF). The image contains galaxies of various intrinsic luminosities, placed at different redshifts and taken from the simulation sample so as to reproduce approximately the apparent magnitude and redshift distributions of galaxies in the HDF. The image includes realistic noise levels, and has been processed with a point-spread function (PSF) similar to that of HST. This image illustrates dramatically how simulations can be "observed" and compared directly with high-resolution observations of distant galaxies.

FIGURE 2. Computer-generated synthetic image of a WFPC2/PC field with parameters chosen to match the exposures corresponding to the Hubble Deep Field (HDF). All bands observed in the HDF have been simulated and combined to produce this image. The cosmological model probed is standard CDM.

4. Tully-Fisher relation

The Tully-Fisher (TF) relation in several colors provides an excellent testbed for galaxy formation simulations. The total baryonic mass of a dark-matter halo scales with its circular velocity like $M \propto v^3$ (equation 2.2), and thus already provides a scaling similar to the observed luminosity/velocity relation (see, however, Silk (1997) for a model in which the TF relation arises mainly as a consequence of self-regulated star formation in galactic disks). This basic relation is then modulated by the mass distribution of the gas compared to the dark matter (i.e., the relation between the actual rotation velocity of gas/stars and the virial velocity of the dark-matter halo), the efficiency of transforming gas into stars and the mass-to-light ratios in different bands. All these modulations are likely to depend on the mass (or, equivalently, circular velocity) of the halo and will affect the normalization, scatter, and slope of the TF relation. Hence, the kinematic of galaxies (represented by their rotation velocities) is linked with their current star formation rate (represented by their B-band luminosities), and their star formation history (represented

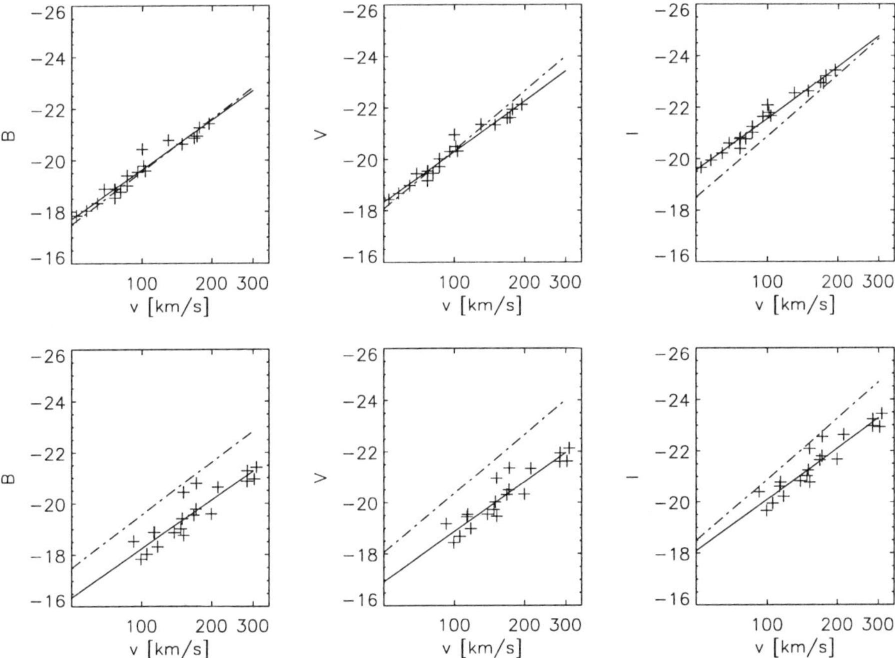

FIGURE 3. Tully-Fisher relation: B (left), V (middle) and I (right) luminosity of numerically simulated galaxies versus their circular velocity at the virial radius (top row) and their maximum rotation velocity (bottom row). The solid lines correspond to a least square fit to the simulated data, the dash-dotted lines correspond to the observed B,V and I Tully-Fisher relation (from Pearce & Tully, 1992).

by their I or K-band luminosities). Finally, these modulations have to preserve the small scatter of about 0.3 mag in the observed TF relation. Further constraints can be derived by comparing the redshift evolution of the TF relation with observations (e.g., Vogt et al. 1997).

Figure 3 shows the B-, V- and I-band TF relation for the numerically simulated galaxies. In order to demonstrate different contributions, the B, V and I luminosities are plotted against the circular velocity of the DM halo and against the maximum velocity of the rotation curve. For comparison a best fit to the simulated data (solid line) and the observed data (dashed line, from Pierce & Tully 1992) is shown. The plot demonstrates that the simulations can qualitatively reproduce many features of the observed data: first of all, a clear steepening of the TF relation from B to I is visible. While the luminosity is proportional to $L_B \propto v^{2.4}$, the I luminosity follows $L_I \propto v^{2.7}$. Also the *rms* scatter is remarkably small, 0.2 mag, if using the virial velocity and 0.4 mag using the maximum rotation velocity. These data are consistent with the observed values of $\Delta M = 0.3$ mag (recall, dark matter halos obey $\Delta M = 0$ mag and $M \propto v^3$). Slope and scatter of the TF relations as well as differences between different bands are a result of the simulations. The calibration of the star formation law only enters by fixing the B luminosity of a galaxy within a halo of circular velocity 200 km/s!

However, a more quantitative comparison shows that the model fails in detail. First of all, the slope of the TF relation is too flat, especially in the I band ($L_I \propto v^{2.7}$ versus observed $L_I \propto v^{3.2}$) which probably indicates that the supernovae feedback is not acting efficiently enough. A more efficient feedback mechanism would deplete the total

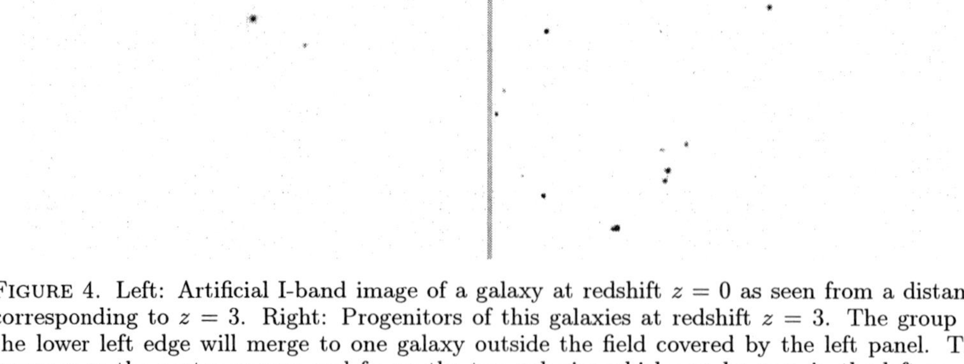

FIGURE 4. Left: Artificial I-band image of a galaxy at redshift $z = 0$ as seen from a distance corresponding to $z = 3$. Right: Progenitors of this galaxies at redshift $z = 3$. The group at the lower left edge will merge to one galaxy outside the field covered by the left panel. The group near the center merges and forms the two galaxies which can be seen in the left panel. Resolution, Noise, PSF and efficiency are taken to match that of the HST WFPC2 camera, exposure time: 123.6 ksec. Each frame has a sidelength of 2.8 Mpc (comoving), corresponding to the area covered by 4 WFPC2 chips.

amount of stars at low circular velocities and thus result in a lower I-band luminosities and a steepening of the I-band TF relation. The systematically low luminosities in the lower row of Figure 3 can be interpreted in two ways: either the luminosities are in fair agreement, but the galaxies are too concentrated (i.e., the maximum of the rotation curve is too high), or the velocities are in fair agreement, but the luminosities are too low. The angular momentum problem mentioned in the prologue supports the first interpretation, and also strengthens the conclusion that feedback has not worked efficiently enough. If, however, the actual rotation velocities of galaxy disks are more similar to the virial velocities of its dark matter halo, the I band TF relation (Figure 3, upper right) indicates, that the total amount of stars in a galaxy of given rotation velocity is too large.

These results can be summarized as follows: the adopted feedback mechanism is too weak to resolve the overcooling/angular momentum problem. The resulting galaxies are too massive and too concentrated compared to their observed counterparts. This effect is especially strong at the low mass end and the slope of the I-band TF relation is too shallow. This discussion, however, also demonstrates how global scaling relations like the TF relation can be used to calibrate models of (large-scale) star formation and feedback in galaxy formation simulations.

5. High redshift galaxies

Although the models still fail to reproduce quantitatively the scaling relations of present day galaxies, the qualitative agreement is probably good enough to take a closer look at the redshift evolution of these galaxies, especially if one concentrates on the high mass end where the influence of feedback processes is weaker.

Figure 4 shows a $z = 0$ galaxy and its progenitor at $z = 3$, a group of protogalactic clumps (PGCs, see also Haehnelt, Steinmetz & Rauch 1996). The baryonic masses of these clumps are only a few times 10^9 $M_\odot$ or even less. Their mutual separation is about a few hundred kpc. All PGCs share virtually the same redshift ($\Delta v \approx 400\,\mathrm{km/s}$). Similar to the galaxy shown in Figure 4, most of the 21 simulated galaxies give rise to a few detectable ($I < 26$) progenitors at $z \approx 2$–3, a behavior which nicely accounts for the increasing evidence for redshift clustering at redshifts above two (Pascarelle et al. 1996; Steidel et al. 1997; Elston & Bechtold, 1998).

The redshift evolution of a subset of galaxies (3 galaxies with $v_c \approx 200\,\mathrm{km/s}$ and 3 galaxies with $v_c \approx 80\,\mathrm{km/s}$) is shown in Figure 5. By following the merging history of a galaxy, the expression "progenitor" is, of course, no longer uniquely determined for redshifts higher than that of last major merging event. Hence, at each branch in the merging tree, the more massive clump is defined as the progenitor. Luminosity, velocity and star formation rate shown in Figure 5 always refers to only one bound clump.

The absolute U magnitude (Figure 5, upper left) brightens with redshift about 1–2 magnitudes between redshift 0 and 2 which indicates a higher star formation rate at higher z. This assumption is confirmed by the lower left plot of Figure 5 which shows the star formation rate of these clumps. At $z = 0$, the star formation rate is a very few $\mathrm{M}_\odot$/yr for 200 km/s galaxies and an order of magnitude lower for 100 km/s galaxies. The star formation rate increases with redshift, however it always stays below a few tens of $\mathrm{M}_\odot$/yr. Note, however, that the simulations do not account for the effects of dust.

At redshift above 2.5 a strong drop in U is visible. This dropout is mainly due to intervening absorption due to neutral hydrogen shortward of 1217 Å(restframe) and, to a smaller extent, absorption within the atmospheres of young massive stars. In R and I, the luminosity is fairly constant. The plots also indicate, that the luminosities of the more massive models are in good agreement with that of high redshift galaxies observed by Windhorst et al. (1994) and Steidel et al. (1996). Also the circular velocity of these clumps is fairly constant. The circular velocity of low v_c objects is even slightly higher at larger redshifts. Also the circular velocity of the 200 km/s galaxies increases slightly with redshift for $z < 1.5$, and it is still above 150 km/s at $z = 3$. So it should not be surprising that galaxies with velocities of a few hundred km/s exist at these redshifts (see also Steinmetz 1997; Baugh et al. 1997).

In summary, the $z \lesssim 3$–4 progenitors of present day galaxies have luminosities and rotation velocities similar to those of their present day counterpart. However, they are much more compact and much less massive.

6. Absorption systems

One very successful application of gasdynamical simulations in cosmology has been the study of the intergalactic medium (IGM), in particular the physical origin of QSO absorption systems. Hydrodynamical simulations similar to those presented here explain the basic properties of QSO absorbers covering many orders of magnitude in HI column density (see, e.g., Cen et al. 1994; Katz et al. 1996; Zhang et al. 1995; Rauch, Haehnelt & Steinmetz 1997). This is also shown in Figure 6 (left). While the lowest column density systems ($\log N(\mathrm{HI}) \approx 12$–14, light gray) arises from gas in voids and sheets of the "cosmic web", systems of higher column density are produced by filaments ($\log N \approx 14$–17, dark gray) or even gas which has cooled and collapsed in virialized halos ($\log N > 17$, black).

So far, numerical simulations have been applied primarily to systems with lower column densities ($\log N \lesssim 17$), corresponding to gas densities below $10^{-2}\,\mathrm{cm}^{-3}$. At such low densities the important physical processes are relatively simple and well understood. Fluctuations are still only mildly non-linear and the gas is essentially in photoionization equilibrium with the UV background. Cooling times are long compared to dynamical time scales.

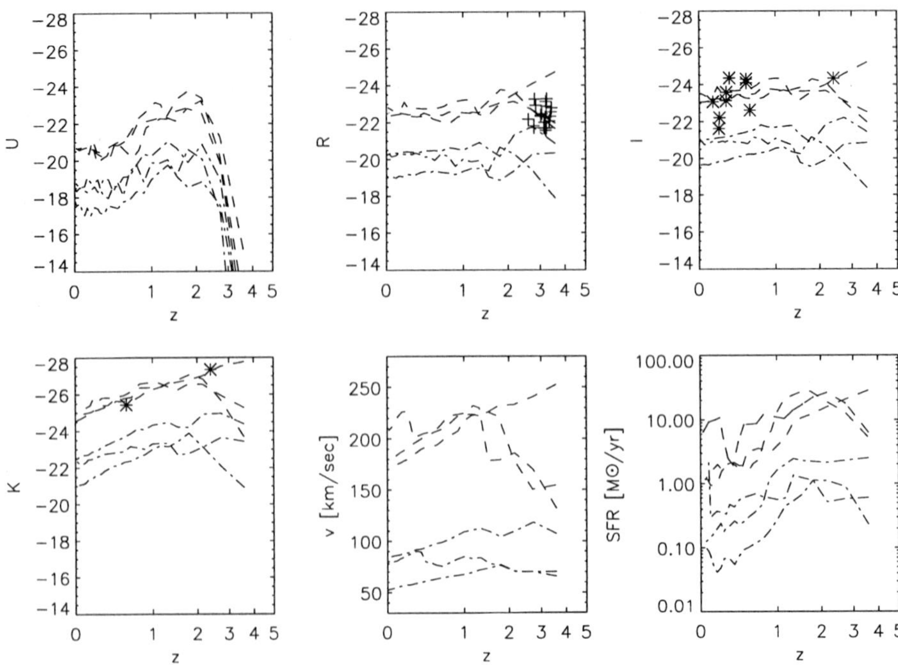

FIGURE 5. Upper row: redshift evolution of the absolute magnitude in U (left) R (middle) and I (right) of three galaxies with $v_c(z = 0) \approx 190\,\mathrm{km/s}$ (dashed) and three galaxies with $v_c(z = 0) \approx 80\,\mathrm{km/s}$ (dashed dotted). For comparison, (+) and (∗) denoted observational data from Steidel et al. (1996) and Windhorst et al. (1994), respectively. Lower row: redshift evolution in K (left), circular velocity as a function of redshift (middle) and star formation rate as a function of redshift (right).

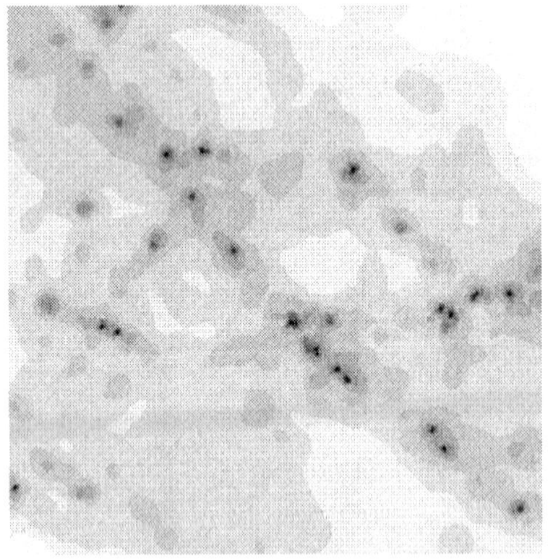

FIGURE 6. Right: HI column density map of a galaxy forming environment at redshift $z = 3$. Light gray correspond to column densities of about $\log N = 13.5$, dark gray to $\log N \approx 15.5$ and black to $\log N > 17.5$. Left: the same simulation shown as an artificial I-band image in an HDF-like exposure. Each frame has a side length of 2.8 Mpc (comoving).

The right panel of Figure 6 shows the corresponding I-band image of the stellar component. The I band image includes noise, PSF and exposure time similar to that of the Hubble Deep Field. The artificial image shows about 8 detectable PGCs. Each of these PGCs is situated close to a region of very high column density ($\log N > 17$). However, there is still a substantial number of Ly-limit and damped Lyα systems, which do not host a stellar PGC.

There has been considerable debate about the physical structures giving rise to damped Lyα absorption systems (DLAS) at high redshift. DLASs have often been interpreted as large, high-redshift progenitors of present-day spirals which have evolved little apart from forming stars (Wolfe 1988). More recently, Prochaska & Wolfe (1997) have argued that only models in which the lines-of-sight (LOS) intersects rapidly rotating thick galactic disks can explain both the large velocity spreads (up to 200 km/s) and the characteristic asymmetries of the observed low ionization species (e.g., SiII) absorption profiles. In particular, they find that if they embed their disk model within a CDM structure formation scenario, the result is inconsistent with the observed velocity widths. However, although Prochaska & Wolfe investigated quite a number of different geometrical and dynamical configurations, all models have in common that the underlying mass distribution is highly symmetric and the models are in dynamical equilibrium.

The importance of asymmetries and non-equilibrium effects has been demonstrated by Haehnelt, Steinmetz & Rauch (1998). Figure 7 shows a typical configuration which gives rise to a high redshift DLAS with an asymmetric SiII absorption profile. The velocity width of about 120 km/s is also quite similar to typical observation. However, no large disk has yet been developed and also the circular velocity of the collapsed object is only 70 km/s. The physical structure which underlies DLASs are turbulent gas flows and inhomogeneous density structures related to the merging of two or more clumps, rather than large rotating disks similar to the Milky Way. Rotational motions of the gas play only a minor role for these absorption profiles. A more detailed analysis also demonstrates that the numerical models easily pass the statistical tests proposed by Prochaska and Wolfe, i.e., hierarchical clustering, in particular the CDM model, is consistent with the kinematics of high-z DLASs. Semianalytical models failed since they are based on the assumption of relaxed disk-like structures rather than the complicated infall pattern seen in the simulations.

7. Summary and conclusion

Numerical simulations of the formation of galaxies through hierarchical clustering have been presented. The simulation outcome has been compared to a large variety of observations for galaxies at low and high redshifts. The main conclusions are the following:

- Hierarchical clustering has proven to be a very successful model for structure formation in the universe. It also provides well defined initial conditions from which the formation of galaxies can be studied. Some observations like, e.g., the increasing evidence for redshift clustering are natural predictions of the hierarchical clustering hypothesis.
- Numerical simulations are an extremely powerful tool to study the formation of galaxies. Only numerical simulation can take full account of the dynamic of the formation process and the complicated interplay between different physical processes such as, e.g., accretion and merging, star formation and feedback, photo heating and radiative cooling.
- Overall, the qualitative picture seems to be fairly consistent, although current models still fail to reproduce the properties of the observed galaxy population in a quantitative manner. Future investigation will have to cope with the effects of (large-scale) star formation and feedback processes.

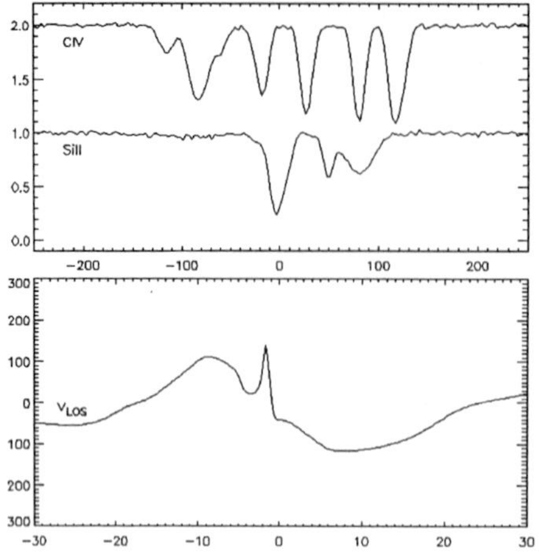

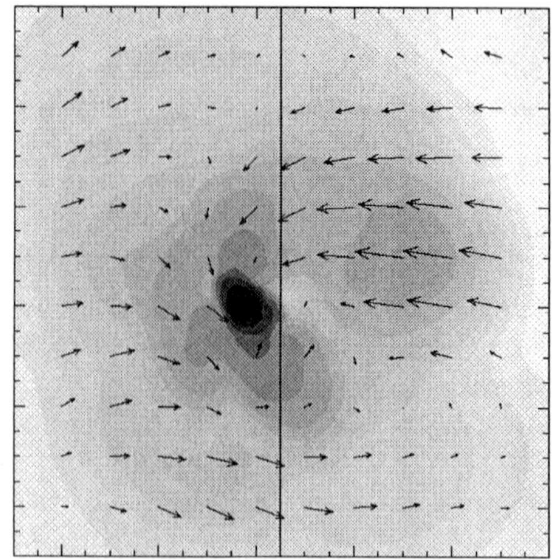

FIGURE 7. Right: Color map of the column density distribution in a 60 kpc around a damped system. Black correspond to HI densities $\log n(\mathrm{HI}) > 1.5$), light grey to $\log n(\mathrm{HI}) \approx -3$). White arrows indicate the velocity field. The white line correspond to the line-of-sight(LOS). In the lower left plot, the velocity field along the LOS is shown. The upper left plot shows the absorption line in CIV 1548 (top) and SiII 1808 (bottom). For readability, CIV has been displaced by 0.5 in flux.

This article includes work from collaborations with G. Contardo, M. Haehnelt, J. Navarro and M. Rauch.

Appendix A. Simulating galaxy formation

A.1. *Numerical method*

The simulation presented in this article have been performed using GrapeSPH (Steinmetz 1996), a particle method that combines the hardware N-body integrator GRAPE (=GRAvity PipE, Sugimoto et al. 1990) with the Smooth Particle Hydrodynamics (SPH) approach to numerical hydrodynamics (Lucy 1977; Gingold & Monaghan 1977). GrapeSPH is fully Lagrangian, free from symmetry restrictions, and highly adaptive in space and time through the use of individual particle timesteps and smoothing lengths. The physical processes implemented in this code include self-gravity, gas pressure, hydrodynamical shocks, radiative cooling and heating by a photoionizing background, and star formation.

A.2. *Star formation*

Star formation is modeled by the creation of collisionless "star" particles in regions where the gas is locally Jeans unstable and where the cooling timescale is shorter than the local dynamical timescale. The local star formation rate per unit volume is assumed to be directly proportional to the gas density and inversely proportional to a local dynamical timescale. A star formation efficiency of 3% has been used (for details see Katz 1992; Navarro & White 1993; Steinmetz & Müller 1994, 1995), a value consistent with that observed in the Milky Way.

The orbits of newly formed stars are subsequently followed in a self-consistent fashion, assuming that they are only affected by gravitational forces. Young star particles devolve energy and metal enriched mass to their surrounding gas, an effect that mimics the energy and mass input by supernovae and evolving stars into the ISM. The supernova energy is added to the thermal energy of the surrounding gas. Input in kinetic energy (Navarro & White 1993) has not been assumed, i.e., this star formation model represents a minimum feedback model.

A.3. *Initial conditions and simulation design*

The large dynamic range needed to resolve the internal structure of galaxies and the full cosmological context of the galaxy formation process is achieved using a two-step procedure. First, galaxy-sized halos are extracted from large cosmological N-body simulations. Second, these systems are resimulated in high-resolution individual runs that use the same initial conditions and tidal fields of the original simulations plus small scale perturbations introduced to account for the increased Nyquist frequency of the second run (Navarro, Frenk & White 1995).

REFERENCES

BARNES, J. 1988 *ApJ* **331**, 699.

BAUGH, C., COLE, S., FRENK, C. S., LACEY, C. 1998 *ApJ*, submitted (astro-ph/9703111).

CEN, R., MIRALDA-ESCUDÉ, J., OSTRIKER, J. P., RAUCH, M. 1994 *ApJ* **437**, L9.

CONTARDO, G., STEINMETZ, M., FRITZE-VON ALVENSLEBEN, U. 1998, in preparation.

COLE, S. M., ARAGÓN-SALAMANCA, A., FRENK, C. S., NAVARRO, J. F., ZEPF, S. E. 1994 *MNRAS* **271**, 781.

DEKEL, A., & SILK, J. 1986 *ApJ* **303**, 39.

EFSTATHIOU, G. P. 1992 *MNRAS* **456**, 43.

ELSTON, R., BECHTOLD, J. 1998, in preparation.

FALL, S. M. 1983 in *Internal Kinematics and Dynamics of Galaxies*, (ed. E. Athanassoula). p. 391. Reidel.

FALL, S. M. & EFSTATHIOU, G. 1980 *MNRAS* **193**, 189.

FRENK, C. S., WHITE, S. D. M., EFSTATHIOU, G. P., AND DAVIS, M. 1985 *Nature* **317**, 595.

GINGOLD, R. A., MONAGHAN, J. J. 1977 *MNRAS* **481**, 375.

HAEHNELT, M., STEINMETZ, M., RAUCH, M. 1996 *ApJ* **465**, L95.

HAEHNELT, M., STEINMETZ, M., RAUCH, M. 1998 *ApJ*, in press (astro-ph/9706201).

KATZ, N. 1992 *ApJ* **391**, 502.

KATZ, N., WEINBERG, D. H., HERNQUIST, L., & MIRALDA-ESCUDÈ J. 1996 *ApJ* **457**, L57.

KAUFFMANN, G., NUSSER, A., STEINMETZ, M. 1997 *MNRAS* **286**, 795.

KAUFFMANN, G., WHITE, S. D. M., & GUIDERDONI, B. 1994 *MNRAS* **267**, 981.

LUCY, L. 1977 *AJ* **82**, 1013.

NAVARRO, J. F., FRENK, C. S., & WHITE, S. D. M. 1995 *MNRAS* **275**, 56.

NAVARRO, J. F., STEINMETZ, M. 1997 *ApJ* **471**, 13.

NAVARRO, J. F., WHITE, S. D. M. 1993 *MNRAS* **265**, 271.

PASCARELLE, S. M., WINDHORST, R. A., KEEL, W. C., ODEWAHN, S. C. **year??** *Nature* **383**, 45.

PIERCE, M. J., TULLY, R. B. 1992 *ApJ* **387**, 47.

PROCHASKA, J. X., & WOLFE, A. M. 1998 *ApJ*, in press, (astro-ph/9704169).

QUINN, P. J. & ZUREK, W. H. 1988 *ApJ* **331**, 1.

QUINN, T., KATZ, N. & EFSTATHIOU, G. 1996 *MNRAS* **278**, L49.

RAUCH, M., HAEHNELT, M. G., STEINMETZ, M. 1997, *ApJ* **481**, 601.

SILK, J. 1997 *ApJ* **481**, 703.

SONGAILA, A., COWIE, L. L., HOGAN, C. J., RUGERS, M. 1994 *Nature* **368**, 599.

STEIDEL, C., GIAVALISCO, M., PETTINI, M., DICKINSON, M., ADELBERGER, K. 1995 *ApJ* **462**, L17.

STEIDEL, C., ADELBERGER, K., DICKINSON, M., GIAVALISCO M., PETTINI, M., KELLOGG, M. 1997 *ApJ*, in press.

STEINMETZ, M. 1996 *MNRAS* **278**, 1005.

STEINMETZ, M. 1997 *Numerical Simulations of Galaxy Formation*, Proc. *Science with the VLT*, p. 156. Springer-Verlag.

STEINMETZ, M., MÜLLER, E. 1994 *A&A* **281**, L97.

STEINMETZ, M., MÜLLER, E. 1995 *MNRAS* **276**, 549.

SUGIMOTO, D., CHIKADA, Y., MAKINO, J., ITO, T., EBISUZAKI, T. & UMEMURA, M. 1990 *Nature* **345**, 33.

TYTLER, D., FAN, X.-M., BURLES, S. 1996 *Nature* **381**, 207.

VOGT, N., ET AL. 1997 *ApJ* **479**, L121.

WEINBERG, D., HERNQUIST, L. & KATZ, N. 1997 *ApJ* **477**, 8.

WINDHORST, R., ET AL. 1994 *ApJ* **435**, 577.

WHITE, S. D. M. 1994 *Les Houches Lectures.*

WHITE, S. D. M., REES, M. J. 1978 *MNRAS* **183**, 341.

WHITE, S. D. M., NAVARRO, J. F., EVRARD, A. E., FRENK, C. S. 1993 *Nature* **366**, 429.

WOLFE, A. M. 1988 in *QSO Absorption Lines: Probing the Universe*, Proc. of the QSO Absorption Line Meeting. Cambridge University Press.

ZHANG, Y., ANNINOS, P., NORMAN, M. L. 1995 *ApJ* **453**, L57.

Selection effects and robust measures of galaxy evolution

By HENRY C. FERGUSON

Space Telescope Science Institute, Baltimore, MD 21218

A variety of subtle, and not-so-subtle selection effects influence the interpretation of galaxy counts, sizes and redshift distributions in the Hubble Deep Field. Comparison of the different HDF catalogs available in the literature and on the world-wide-web reveals generally good agreement, although the effects of different isophotal thresholds and different splitting algorithms are readily apparent. As the basic source detection and photometry algorithms are similar for the different catalogs, the selection effects are likely to affect them all.

Through simulations, we explore the utility of image moments for inferring the true sizes of galaxies. The truncation of galaxy profiles at a fixed isophote has serious consequences, which limit constraints on the size distribution to galaxies with isophotal magnitudes $I_{814} < 27.5$. Present-day L^* spirals would be undetected in the HDF above redshifts $z \approx 1.2$, and present-day ellipticals would disappear at $z \approx 1.8$.

The Lyman break provides a way to identify high-redshift galaxies at very faint magnitudes. However, galaxies that are at redshifts high enough to vanish from the HDF F300W or F450W filters also suffer severely from photometric biases. For example, at fixed total apparent magnitude and physical scale length, a galaxy at $z = 4$ will have a mean surface brightness 1.2 mag fainter than a galaxy at $z = 2.75$. This lower surface brightness will result in an apparent decrease in the number density of objects, and the inferred luminosity density, even for models where there is no intrinsic evolution. We illustrate the effects of these biases on the estimates of the number of Lyman "dropouts" in the HDF and on the luminosity density at $z > 2$.

---

1. Introduction

In trying to decipher the origin of galaxies from a collection of fuzzy patches on the sky, we must keep in mind that even the Hubble Deep Field, with its exquisite depth and resolution, provides a distorted view of the universe at large. This view is distorted by the fact that we are looking only at optical wavelengths, which for the most part probe the rest-frame ultraviolet portion of the spectra of the galaxies of interest. It is distorted by the fact that the background noise of the detector limits detection of galaxies to those that exceed a certain threshold over a certain number of pixels: faint extended objects are extremely difficult to detect; galaxies with multiple peaks in their light distribution may appear as separate objects. It is distorted by the fact that many of the galaxy images overlap, making it difficult to separate one galaxy from the next. Finally, it is distorted by physical effects such as obscuration by dust or gravitational lensing. These effects of dust and lensing are amply discussed in this conference by Madau, Meurer, Rowan-Robinson, Blandford, and others. I will focus on the selection effects, seeking to explore and quantify some of the concerns that have colored many of the discussions of the HDF and other deep galaxy surveys.

2. HDF catalogs

In making the HDF data set non-proprietary, one of the hopes was that different groups would be stimulated to reduce and analyze the data independently with different scientific objectives and different algorithms. To some extent this has happened. The data have been largely reprocessed by several different groups Flynn et al. (1996),

Catalog	Comments
Couch	Sextractor 1683 Objects on WF chips, selected in F814W. Isophotal limit $\mu_{814} = 26.1$ in 0.016 arcsec^2
Lanzetta et al.	Sextractor 1925 Objects on WF chips, selected in separate bands Isophotal limit $\mu_{814} = 26.2$ in 0.048 arcsec^2
Williams et al.	modified FOCAS 3086 Objects selected in F814W+F606W summed image Isophotal limit $\mu_{814} = 27.5$ in 0.04 arcsec^2
Metcalfe et al.	Detection algorithm described in Metcalfe et al. 1991 3700 Objects Thresholds not specified, but probably the deepest.

TABLE 1. Selected HDF catalogs

Ratnatunga (1996), and different techniques have been used for source detection and photometry Williams et al. (1996), Lanzetta et al. (1996), Elson et al. (1996), Couch (1996), Metcalfe et al. (1996). As much of the discussion at this meeting is focused on comparing what we see in the HDF to what we expect from models of galaxy evolution, it is worth examining some of these catalogs to see how well they agree with each other, and how well they reproduce our own preconceptions of how sources should be counted.

For this purpose, I have chosen four catalogs that have been used in HDF publications. All the catalogs have in common the feature that they convolve the data to smooth the noise over a scale relevant to the sources being sought, and then mark as sources those objects that have a certain number of connected pixels more than some multiple of the rms background fluctuations. This procedure is equivalent to Wiener filtering with a Gaussian source model, and is optimal in the least-squares sense for detecting unresolved objects against a random, uncorrelated background. It is not so clearly optimal for finding galaxies in an image where galaxies have a wide range of sizes and have isophotes that in many cases overlap. However, to my knowledge there have not been any HDF catalogs based, for example, on wavelets or on median filtering over a variety of scales, which might have a different set of selection effects.

2.1. *Completeness*

The catalogs are briefly summarized in Table 1, which lists the software package used, the number of objects found, and the rough detection criteria. The total source counts are influenced primarily by the isophotal threshold. To a certain extent it is a matter of taste how deeply to push into the noise. The Metcalfe et al. (1996) catalog is the deepest (or least conservative) while the Lanzetta et al. (1996) catalog is the shallowest (or most conservative). While the total number of sources varies by more than a factor of two between the catalogs, down to $V_{606} = 27.8$ the counts in the different catalogs agree to within 20% of the mean.

2.2. *Object splitting and merging*

A variety of schemes have been developed for splitting up sources that have multiple peaks above the detection isophote into separate "parent" and "daughter" objects. There is no mathematical rigor in these techniques. Identification of daughter objects is typically done by passing successively higher isophotal thresholds over the image and making lists of sources within sources. Different algorithms are then used to "merge"

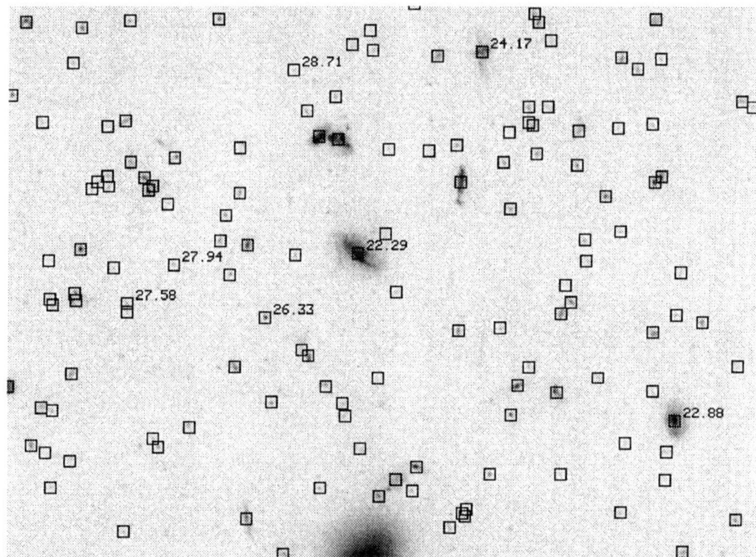

FIGURE 1. A portion of the HDF, with objects brighter than $V_{606} = 28.5$ from the Couch (1996) catalog marked. Note that there are quite a lot of sources clearly visible below this magnitude limit, many of which appear in the catalogs, but we have not marked them on the images to avoid clutter. "Total" F606W AB magnitudes are marked for a few galaxies to give an indication of the agreement between catalogs.

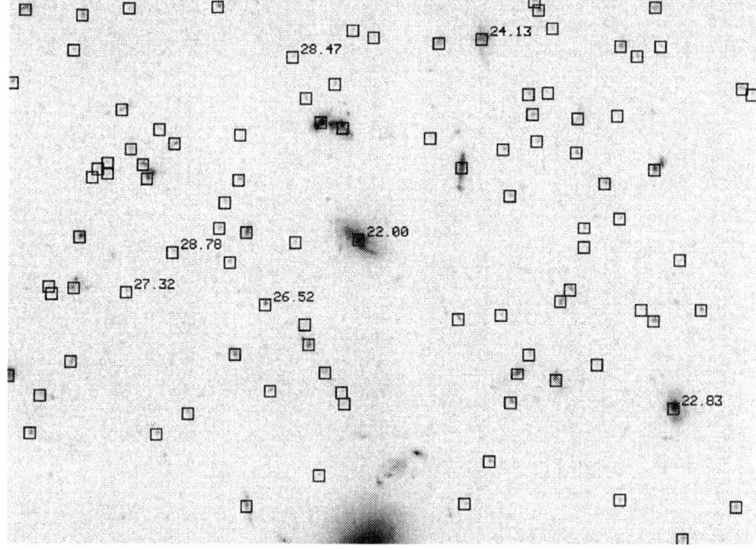

FIGURE 2. Same for the Lanzetta et al. (1996) catalog.

the various pieces back into objects that are likely to be part of the same galaxy. Williams et al. (1996) use color information to help with this merging; in other cases the decisions were made using information from one band.

The differences in how subcomponents are counted are illustrated in the sections of the F606W image shown in Figs. 1–4, where we show the positions of galaxies with total magnitudes brighter than 28.5 from the four different catalogs. Broadly speaking, the Lanzetta et al. (1996) catalog tends to do the least splitting of objects within common

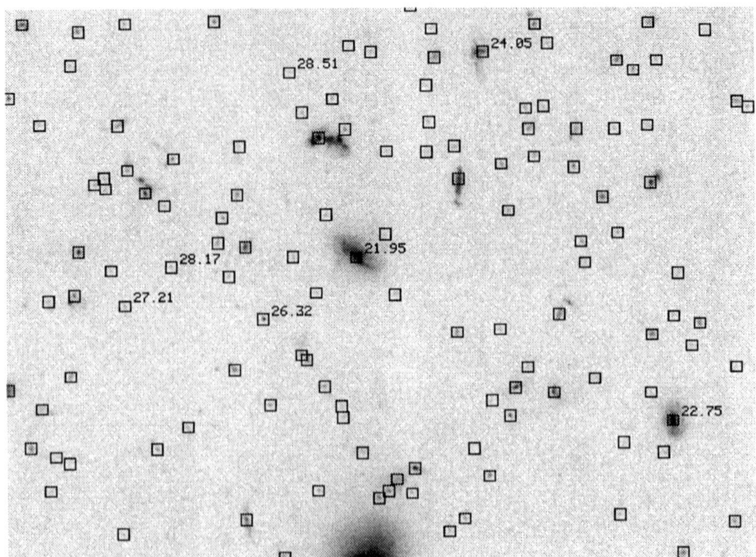

FIGURE 3. Same for the Williams et al. (1996) catalog.

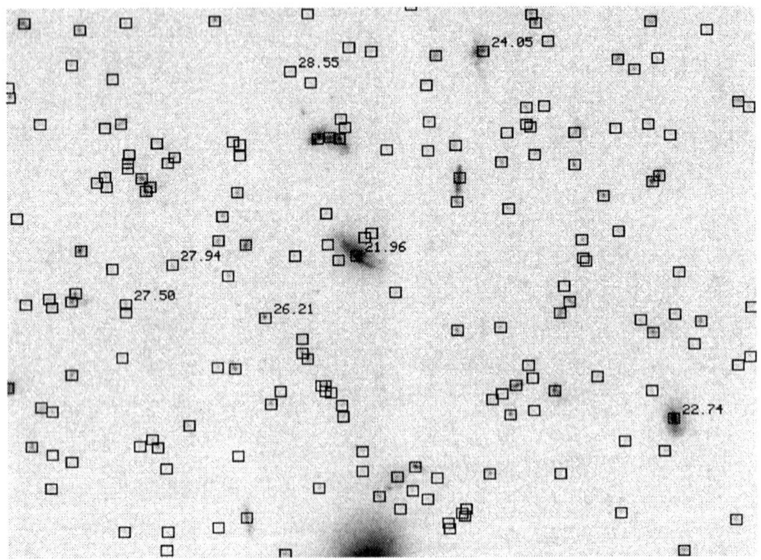

FIGURE 4. Same for the Metcalfe et al. (1996) catalog.

isophotes, while the Metcalfe et al. (1996) catalog does the most splitting. There are many cases, such as the peculiar galaxy just to the right and above the center of the images, where it is not at all obvious how one should count. Even more problematical is the object that looks like a late-type spiral which is superimposed on the outer isophotes of a bright elliptical galaxy near the bottom of the frame. In the Lanzetta et al. (1996) catalog, this object is counted as part of the elliptical. In the Couch (1996) catalog, it is counted as four objects. It is counted as five by Williams et al. (1996) and 4–6 by Metcalfe et al. (1996).

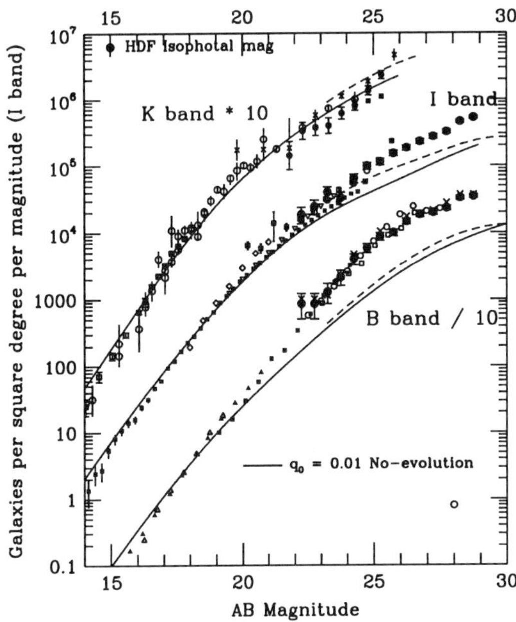

FIGURE 5. HDF counts from Williams et al. (1996), together with a compilation of counts from ground-based surveys. The solid curves show a non-evolving model with $q_0 = 0.01$. The dashed curves show the same model, but with overcounting included. For magnitudes $22 < m_{AB} < 30$, the counts have been multiplied by $1 + (30 - m)/4$, and the magnitudes have been increased by $2.5 \log(1 + (30 - m)/4)$. This illustrates the fairly modest affect that splitting and merging algorithms have on the ability to model galaxy number counts.

From an analysis of the angular correlation function in their HDF catalog (constructed using DAOFIND), Colley et al. (1997) suggest that such difficulties lead to an overestimate in the number of faint galaxies by a factor of 2.5. While this estimate strictly applies only for the subset of 196 color-selected high-redshift candidate objects in their catalog, such overcounting is likely to be true at some level for galaxies at all redshifts in all the catalogs. However, two important facts mitigate the effect of this problem on the interpretation of deep galaxy counts. The first is that the oversplitting of some objects is counteracted by the overmerging of others. That is, many close projections of unrelated objects are counted as one object. This effect can be quantified by simulations. Second, because the artificial splitting of large galaxies into several smaller ones preserves the total luminosity, the effect shifts the number counts both vertically and horizontally in such a way that the predictions of models are not greatly affected. Figure 5 shows the effect of overcounting for a non-evolving model with $q_0 = 0.01$. We have assumed here that galaxies at $I_{814} = 22$ are counted on average as three objects, and that the overcounting decreases with magnitude such by $I_{814} = 30$ objects are counted only once.

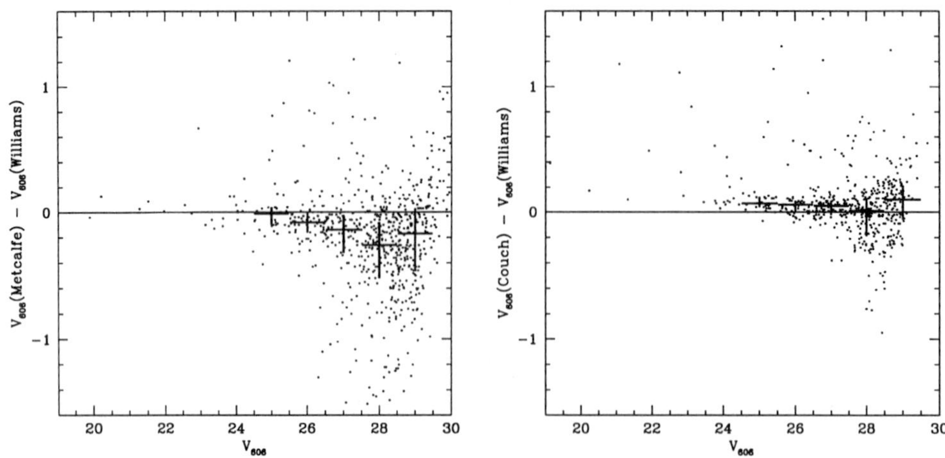

FIGURE 6. The left panel shows a comparison of V_{606} total magnitudes from the Metcalfe et al. and Williams et al. HDF catalogs for galaxies on WF2. The points with error bars show the median and first and third quartiles of the distribution of the residuals in intervals of 1 magnitude. The right panel compares the Williams et al. and the Couch magnitudes.

2.3. *"Total" magnitudes*

Different photometry packages use different techniques to estimate total magnitudes. These estimates are based on extrapolations of galaxy sizes or surface brightnesses below the initial isophotal limits of detection, and face serious difficulties in the particular cases of overlapping galaxies. Even for isolated galaxies, the reliance on the initial estimates for galaxy sizes and/or surface brightness profiles can introduce serious biases for galaxies near the detection threshold. Nevertheless, before addressing the potential biases, it is interesting to compare the results from the different catalogs.

Brighter than $V_{606} = 28$, magnitudes from the different catalogs agree reasonably well. Even if the photometry were in principle perfect, the different splitting algorithms would introduce some scatter at bright magnitudes. Figure 6 shows comparisons between the Williams et al. (1996), Metcalfe et al. (1996), and Couch (1996) magnitudes. The total magnitudes agree to within 0.5 mag rms over the full magnitude range. There is a systematic trend for the Metcalfe et al. magnitudes to be brighter than the Williams et al. mags. The difference increases toward faint mags, and is as much as 0.3 mag at $V_{606} = 28$. The systematic differences between Couch and Williams et al. catalogs are less than 0.15 mag. The good agreement between catalogs suggests that the different photometry packages all do more-or-less the same thing. The discussion of the selection biases below, which focuses specifically on the Williams et al. (1996) catalog, is thus likely to apply to all the catalogs.

3. Galaxy radii

Figure 7 shows the distribution of first-moment radii of galaxies from the Williams et al. HDF catalog and the Medium Deep Survey Ratnatunga et al. (1997).† These isophotal

† The Medium Deep Survey (MDS) catalog is based on observations with the NASA/ESA Hubble Space Telescope, obtained at the Space Telescope Science Institute, which is operated by the Association of Universities for Research in Astronomy, Inc., under NASA contract NAS5-26555. The Medium-Deep Survey is funded by ST ScI grant GO2684.

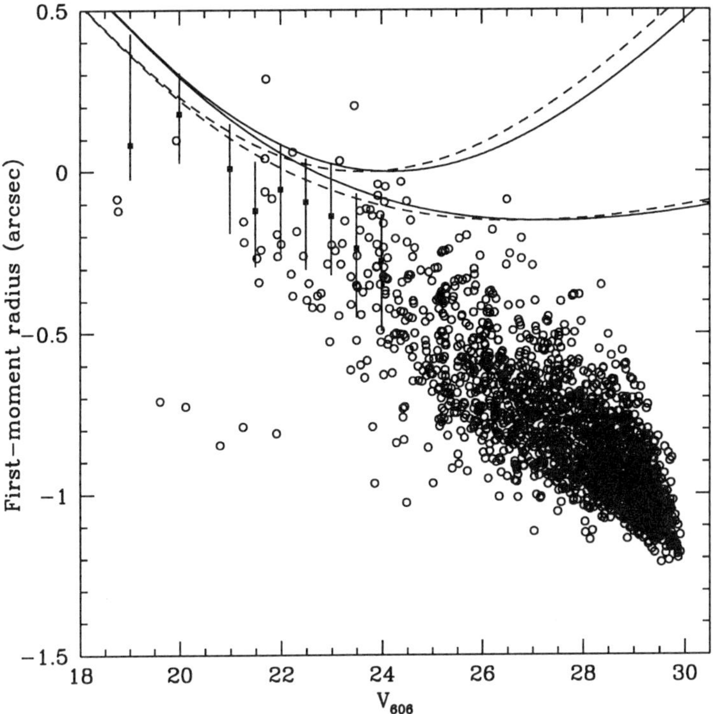

FIGURE 7. Radius-magnitude relations for galaxies in the HDF Williams et al. (1996) catalog (points) and the MDS (points with error bars). The MDS distribution was derived from the catalogs on the MDS web site. The curves show the behavior for non-evolving L^* spirals (solid curves) and ellipticals (dashed curves). The upper pair of curves is for $q_0 = 0.5$; the lower is for $q_0 = 0.01$. *These curves do not take into account selection biases.* The severity of the biases can be seen in the next two figures.

radii keep decreasing right down to the detection limits of the survey. At $V_{606} = 27$ the typical first-moment radius corresponds to less than 2 kpc for *for galaxies at any redshift*, for cosmologies with $q_0 > 0$. The first-moment radius is defined as

$$r_1 = \sum rI(x,y) / \sum I(x,y), \tag{3.1}$$

where $I(x,y)$ is the intensity in each pixel. The quantity $R_k = 2r_1$ is often referred to as the "Kron radius." Kron (1978) showed that in typical ground-based surveys the radius R_k typically encompasses 90% of the light from an object. If the intensity could be measured precisely over the entire galaxy, the relations between first-moment radii and scale radii would be $r_1 = 2\alpha = 1.19r_e$ for spirals, and $r_1 = 2.28r_e$ for ellipticals. The size-magnitude relations for non-evolving L^* ellipticals and spirals are shown in Fig. 7, for two values of q_0.

Taken at face value, the steep radius-magnitude relation implies that faint galaxies are intrinsically very compact. However, r_1 is computed using pixels above the isophotal threshold, and thus is subjected to severe biases as we approach the limiting magnitude of the survey. These biases enter in several ways: (1) because of isophotal selection, at fixed total magnitude larger galaxies will preferentially disappear from the sample; (2) near the detection limit of the survey the first-moment radius r_1 becomes progressively biased toward smaller values; and (3) near the detection limit, total magnitudes computed either from Kron magnitudes or from the flux within some multiple of the isophotal area become progressively biased toward faint values. While effects (2) and (3) can be partially controlled by measuring Petrosian (1976) radii and magnitudes, isophotal selection effects inevitably bias the sample of galaxies selected for such measurements.

We can quantify these selection effects by creating simulated images with the same noise properties as the HDF images and running them through the same source detection and photometry routines. The best way to do this would be to add a small number of galaxies to the HDF image itself and reprocess it hundreds of times. However, for the purpose of this conference I have taken the shortcut of constructing only two simulated images, one with face-on exponential disks and the other with deVaucouleurs profiles. The galaxies have magnitudes between 25 and 30 and a range of scale lengths. The images are somewhat less crowded than the real HDF, so crowding should not affect the results much. In any case, the results here are meant to be indicative rather than definitive. The noise was simulated as described by Ferguson & Babul (1998), and the source detection and photometry were done using the same version of FOCAS with the same parameters used by Williams et al. (1996).

Figure 8 shows the selection boundary for galaxies with exponential profiles (left panel) and deVaucouleurs profiles (right panel). The numbers in the figure give the fraction of the input sample of galaxies recovered as a function half-light radius r_e and total magnitude, I_{814}. The solid curves show the limits of the survey expected from the rms sky noise of the HDF images, assuming FOCAS detects sources with AB magnitude $I_{814} < 30.6$ within a radius of 0.064". This seems to be a reasonably good model of the 50% completeness limit of the Williams et al. catalog. To put this in context, the dashed curves show the size-magnitude relation for non-evolving L^* galaxies. A typical present day spiral would drop below the selection limits of the survey by $z = 1.2$, even though its total magnitude is brighter than $I_{814} = 28$ out to $z = 2.5$. A typical elliptical would be invisible beyond $z = 1.8$.

Figure 8 illustrates the point that the HDF and other deep surveys should not be treated as flux-limited surveys. The sizes of galaxies matter as much as their total luminosities. This fact, although well known, is often ignored in comparing model counts to the data. With the HDF and future surveys it is extremely important to take this into account.

3.1. *Robust constraints on galaxy sizes*

Of course, it would be nice if we could infer something robust about the physical sizes of galaxies in the HDF, independent of any assumed model. To do this, we must translate the selection boundaries in Figure 5 from the theoretical $M_{\rm tot}, r_e$ plane, to the observed $M_{\rm iso}, r_1$ plane. This can be done analytically or numerically for the different profiles, and can also be done empirically from the FOCAS measurements of the simulated galaxies. Figure 9 shows the translated curves from Fig. 8, along with the observed distribution of sizes and magnitudes of galaxies in the HDF. The thin lines show the fiducial galaxies from Fig. 8.

Brighter than $I_{814} = 27.5$, the locus of points clearly peaks at values of r_1 that are well away from the selection boundary. Hence it is probably safe to assume that the relatively compact sizes and high surface brightnesses of the HDF galaxies are a real phenomenon, and not an artifact of galaxy selection. The apparent sizes are smaller than those of present-day L^* disk-dominated spirals, but not significantly smaller than those of luminous ellipticals. Very little can be said about the intrinsic sizes of galaxies fainter than $I_{814} = 27.5$, because it is likely that many galaxies are missed due to low surface brightness. The survey limit determines the upper bound on r_1, and the PSF

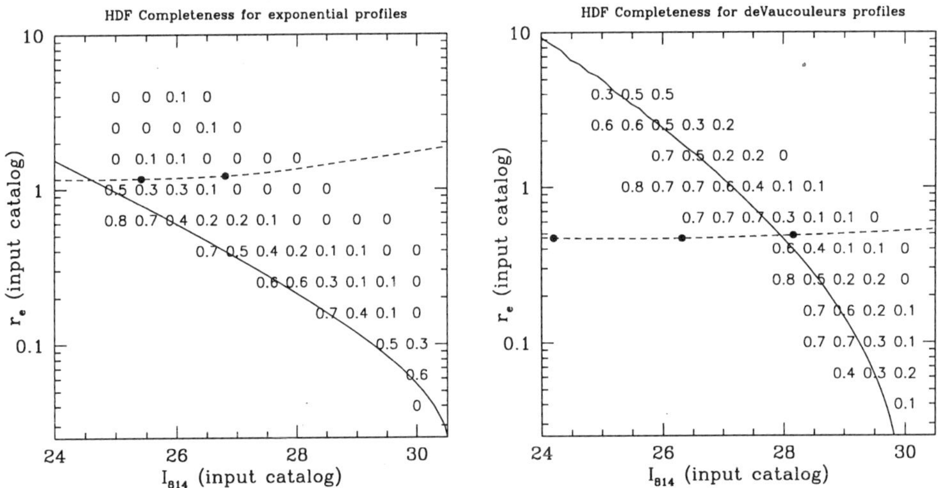

FIGURE 8. Empirical selection boundary for the HDF as a function of half-light radius and total I_{814} magnitude input to the simulations. The left panel shows the limits for face-on galaxies with pure exponential profiles. The right panel shows the same for face-on galaxies with $r^{1/4}$ deVaucouleurs profiles. As there were typically 20 galaxies per bin in the input sample, these estimates are subject to small-number statistics. The solid curves show an analytical model of the FOCAS selection criteria, where we have assumed that a galaxy will be detected if it is brighter than $I_{814} = 30.6$ within a radius of 0.064 arcsec. The dashed curves show the relation between r_e and I_{814} for non-evolving galaxies with $M_B = -21.1$, for $H_0 = 50\,\mathrm{km\,s^{-1}\,Mpc^{-1}}$, and $q_0 = 0.5$. We have assumed a scale-length $\alpha = 6$ kpc for the spiral, and $r_e = 4$ kpc for the elliptical, and have adopted spectral-energy distributions from Coleman, Wu, & Weedman 1980. The solid dots on the curves for both galaxies correspond to $z = 1, 1.5$, and 2 (from left to right); the $z = 1$ point for the spiral is just off the left of the plot at $I_{814} = 23.7$.

determines the lower bound. This truncation of the survey by the isophotal limit means that galaxy counts fainter than $I_{814} = 27.5$ must also be viewed as a lower limit. Also large corrections are required to translate from FOCAS isophotal to true total magnitudes near the survey limit. Such corrections can be derived for models, but cannot be derived with any certainty from the data themselves.

4. Lyman break galaxies

The past two years have brought an explosion in the number of star-forming galaxies identified at redshifts $z > 2$. The HDF has contributed to this by providing a set of robust high-redshift galaxy candidates several magnitudes below the detection limits of current Keck spectroscopy. Madau et al. (1996) used the statistics of Lyman-break objects in the HDF to estimate the luminosity-density in redshift intervals centered at $< z >= 2.75$ and $< z >= 4$. Together with constraints from the CFRS survey Lilly et al. (1995) and a local Hα survey Gallego et al. (1995), this analysis provides an indication of the metal-formation rate and integrated star-formation rate as a function of redshift. Because it puts the observations in a physical context, the "Madau diagram" has become a popular foil for discussing and testing models of galaxy formation, and has been subject to a fair amount of scrutiny. Much of the debate centers on how much dust there is in star-forming galaxies at high redshift, and on what corrections are needed for extinction and for galaxies that are too dusty to detect (see contributions by Madau, Rowan-Robinson, and Meurer in this conference). The luminosities of the $< z >= 2.75$ and the $< z >= 4$ samples are computed at essentially the same rest-frame wavelengths (~ 1500 Å, and

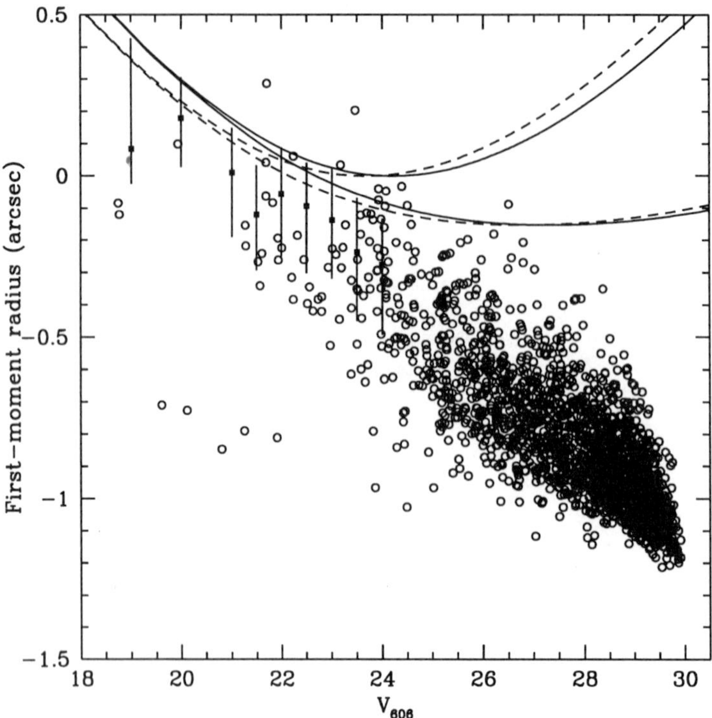

FIGURE 9. HDF angular-size—magnitude relation. Sizes are the first-moment radii and magnitudes are FOCAS I_{814} isophotal magnitudes. The thick lines show the selection boundaries for spirals (solid line) and ellipticals (dashed line), with the same assumptions used for Figure 4. The thin lines show the fiducial non-evolving L^* spiral and elliptical galaxies. The solid dots on the curves for both galaxies correspond to $z = 1, 1.5$, and 2 (from left to right; the first dot for the spiral is off the plot).

thus are probably subject to the similar amounts of extinction. If anything the extinction should be less on average for the $< z >= 4$ sample, since there is likely to be less dust at high redshift. This suggests that the decreasing luminosity density from $z = 2.75$ to $z = 4$ is a real effect, indicative of a change in the overall star-formation rate of the universe.

However, it is important to keep in mind that the Lyman break objects, like all other galaxies in the HDF, are selected above a fixed isophote, and are subject to the same selection effects. For example, at fixed total apparent magnitude and physical scale length, a galaxy at $z = 4$ will have a mean surface brightness 1.25 mag fainter than a galaxy at $z = 2.75$. If the Lyman-break objects fall near the selection boundary, this change in surface brightness could in principle translate to a 0.5 dex change in the inferred luminosity density, which is roughly what is observed between these two redshifts.

To explore this effect, we have examined the statistics of Lyman Break galaxies in the models of Ferguson & Babul (1998), comparing what would be derived from an ideal survey, to what we derive from simulated images of the HDF. Figure 10 shows position of model galaxies in Madau et al. (1996) color-color plot. For this figure, the total magnitudes of the galaxies from the models are used, and we have truncated the sample at $I_{814} = 27.6$, which corresponds to the $\sim 10\sigma$ detection limit of the HDF. The Madau et al. (1996) selection criteria applied to this sample yields a total of 62

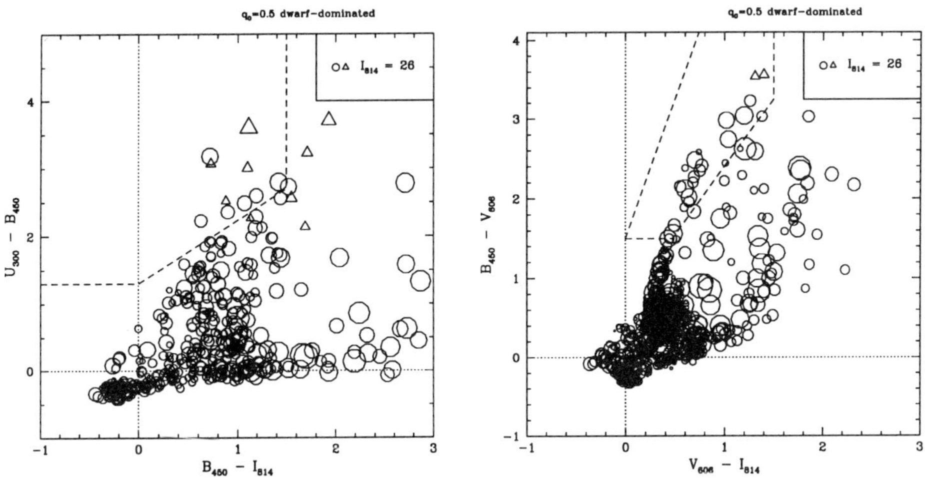

FIGURE 10. Color-color diagrams fashioned after Madau et al. for galaxies in the $q_0 = 0.5$ dwarf-dominated model of Ferguson & Babul 1998. The positions of the galaxies in this diagram are determined from the *true total magnitudes* of the input model. The number of Lyman-break objects within the dashed boundaries should be compared to the next figure, where the same galaxies have been placed in simulated images and measured using FOCAS with the Williams et al. (1996) parameters.

Data set	Number of U dropouts	Total V_{606}	$\dot{\rho}_Z$	Number of B dropouts	Total I_{814}	$\dot{\rho}_Z$
Input model	62	21.15	4.9×10^{-4}	232	19.14	50×10^{-4}
FOCAS	10	21.58	3.3×10^{-4}	21	20.83	10×10^{-4}
HDF	69	21.48	3.6×10^{-4}	14	23.29	1.1×10^{-4}

TABLE 2. Model vs. data for HDF Lyman break objects

"U dropouts" and 232 "B dropouts" (compared to 69 and 14, respectively, in the real HDF).

Figure 11 shows FOCAS measurements *of the same model* from simulated images. This figure should be directly comparable to those in Madau et al. (1996). The color selection criteria yield 10 U dropouts and 21 B dropouts. This illustrates the potential severity of the selection effects on the statistics of Lyman break objects. The number density of Lyman-break objects in the input model differs from the number recovered by FOCAS by roughly one order of magnitude!

The effect on the inferred luminosity density is much less severe, but is still quite significant, at least for the models we have tested. Table 2 shows the total magnitude (from the summed flux of all the detected galaxies) and derived metal-formation rates $\dot{\rho}_Z$ (in units of $M_\odot\,yr^{-1}\,Mpc^{-3}$) for Lyman-break galaxies selected directly from the model, from the simulated images, and from the real HDF. Isophotal selection decreases the inferred metal-formation rate by a factor of 1.5 for the U dropouts and 4.7 for the B dropouts. However, the high metal-formation rate at $z > 3.5$ for this model is still at odds with the data (see Ferguson & Babul (1998)). Tests with another model show similar results, although the correction factors to go from the FOCAS-measured luminosity density to the luminosity density in the underlying model are not the same.

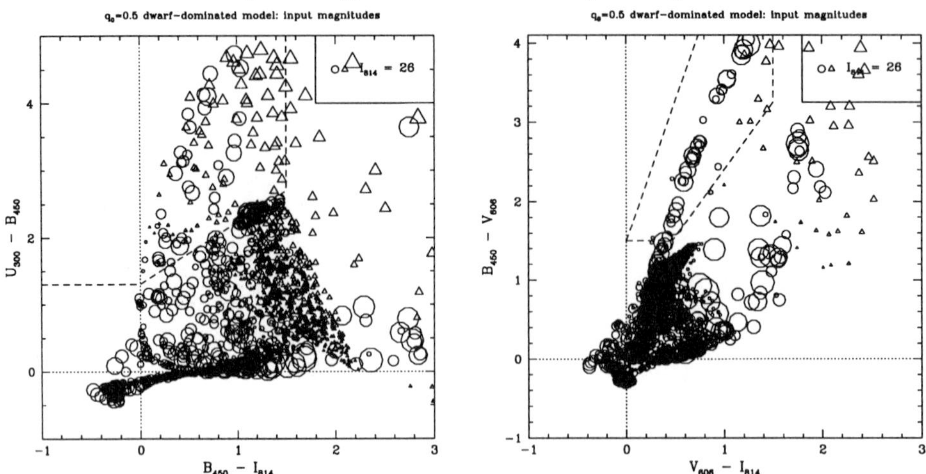

FIGURE 11. Same as the previous figure, but for galaxies measured using FOCAS (with the same parameters used by Williams et al.) in simulated HDF images. Only about 10% of the Lyman-break objects in the input model are recovered by FOCAS. There are various reasons for this. Many of the galaxies that look bright in Fig. 10 have large scale lengths, and hence have FOCAS isophotal magnitudes that are much fainter than their model total magnitudes. In other cases, the FOCS photometric errors in change the colors enough to move the galaxy outside the selection boundary.

5. Summary

I have tried in this presentation to give some quantitative insight into how selection effects influence the interpretation of the HDF data. In general, the treatment of the HDF as a flux limited survey is a bad approximation. Magnitudes and radii of galaxies are strongly affected by the measurement techniques (which vary from catalog to catalog) and by the detection algorithm (which is virtually the same in all catalogs generated to date). Faintward of $I_{814} = 27.5$, the angular size distribution of galaxies in the HDF is severely truncated by the survey selection boundaries. Brighter than this, it appears that the HDF galaxies are in general more compact than L^* spirals, but not necessarily more compact that luminous present-day ellipticals. The angular-size distribution is reasonably well matched by a low-q_0 pure-luminosity evolution model Ferguson & Babul (1998), although such a model does not reproduce the statistics of Lyman-break objects, even when selection effects are accounted for.

The survey selection criteria can have a very strong effect on the statistics of Lyman-break objects, potentially influencing the source counts by an order of magnitude, and the inferred luminosity density by up to a factor of 5. The corrections factors depend on the input model.

While this is a fairly discouraging view of quality of information on faint galaxies that can be gleaned from the HDF, nevertheless I think it is fairly accurate. Approximate corrections for some of these selection effects can be derived from the HDF data themselves, but for the most part the magnitude of the selection bias depends on the intrinsic distributions of galaxy sizes and luminosities, which are not well constrained at faint magnitudes. Fortunately, the selection effects are fairly straightforward to model. Comparisons of galaxy evolution models to the HDF really must involve this step.

This work was supported by NASA grant AR-06337, awarded by the Space Telescope Science Institute, which is operated by the Association of Universities for Research in Astronomy, Inc., for NASA under contract NAS5-26555.

REFERENCES

COLEMAN, G. D., WU, C.-C., & WEEDMAN, D. W. 1980 *ApJS* **43**, 393.

COLLEY, W., GNEDIN, O., OSTRIKER, J. P., & RHOADS, J. E. 1997 *ApJ* **488**, 579.

COUCH, W. J. 1996 http://ecf.hq.eso.org/hdf/catalogs/

ELSON, R. A. W., SANTIAGO, B. X., & GILMORE, G. F. 1996 *New Astron.* **1**, 1.

FERGUSON, H. C., & BABUL, A. 1998 *MNRAS* in press.

FLYNN, C., GOULD, A., & BAHCALL, J. N. 1996 *ApJ* **466**, L55.

GALLEGO, J., ZAMORANO, J., ARAGON-SALAMANCA, A., & REGO, M. 1995 *ApJ* **455**, L1.

KRON, R. G. 1978 PhD Thesis, University of California at Berkeley.

LANZETTA, K. M., YAHIL, A., & FERNANDEZ-SOTO, A. 1996 *Nature* **381**, 759.

LILLY, S. J., LE FEVRE, O., CRAMPTON, D., HAMMER, F., & TRESSE, L. 1995 *ApJ* **455**, 50.

MADAU, P., FERGUSON, H. C., DICKINSON, M., GIAVALISCO, M., STEIDEL, C. C., & FRUCHTER, A. S. 1996 *MNRAS* **283**, 1388.

METCALFE, N., SHANKS, T., CAMPOS, A., FONG, R., & GARDNER, J. P. 1996 *Nature* **383**, 236

PETROSIAN, V. 1976 *ApJ* **209**, L1.

RATNATUNGA, K. 1996, private communication.

RATNATUNGA, K., GRIFFITHS, R., & OSTRANDER, E. 1997, in preparation.

WILLIAMS, R. E., ET AL. 1996 *AJ* **112**, 1335.

Disk galaxy evolution

By JOSEPH SILK

Departments of Astronomy and Physics, and Center for Particle Astrophysics, University of California, Berkeley, CA 94720

1. Introduction

One can attempt to form galaxies either from first principles, or via deconstruction of nearby examples. The *ab initio* approach is philosophically appealing, and has achieved extraordinary success in the realm of large-scale structure. Robust predictions include the correlation functions that describe galaxy clustering, the mass function of galaxy clusters, and the rotation curves of luminous galaxies.

A spectacular recent success has been in predicting the properties of the Lyman alpha clouds, intergalactic gas seen in absorption towards distant quasars. With input of the ionizing radiation field inferred from the observed quasar luminosity function into a cold dark matter large-scale simulation that incorporates baryons, one can reproduce the distribution of observed column densities and line profiles. One infers the baryon abundance, and concludes that most of the baryons inferred from primordial nucleosynthesis are observed in the Lyman alpha clouds.

However, diffuse gas clouds are relatively primitive objects. Relatively crude, but large, simulations suffice. One does not have to be overly concerned about the fate of cooling gas. Galaxy formation has proved to represent a far more difficult challenge for any large-scale structure theory that commences in the very early universe.

2. Forwards approach to galaxy formation

The distribution of galaxy luminosities is described by a luminosity function $\phi(L)$ that has three parameters: number density normalization ϕ_*, characteristic luminosity L_*, and asymptotic slope α at luminosity $L \ll L_*$ so that $\phi(L) = \phi_*(L_*/L)^\alpha \exp(-L/L_*)$. Analytic theory provides the mass function of the first nonlinear objects to form by gravitational instability of a local Gaussian density fluctuation field. The resulting mass function has an exponential cut-off at M_* and $\alpha \approx 2$. The mass M_* corresponds to that of objects currently going nonlinear. These are groups of galaxies. To derive the mass function of galaxies, one requires the baryons to have undergone cooling. This effectively sets a redshift threshold, and objects that satisfy the cooling condition have masses of the order of those of the most massive galaxies.

The superficial success in deriving a characteristic luminosity has not held up. The deeper is the cold dark matter potential well, the greater is the mass of cooled gas. Cooling must be terminated. The predicted slope of $\alpha = 2$ is steeper than the observed slope of $\alpha \approx 1$. This generally results in a ratio of mass-to-luminosity that is excessive, relative to the mass-luminosity ratios inferred for spirals or ellipticals from the normalization of the Tully-Fisher and fundamental plane correlations, respectively. In order to hide this discrepancy in the properties of dwarf galaxies, which are known to have low surface brightnesses and high mass-to-light ratios, appeal must be sought to feedback from star formation and death that effectively makes star formation progressively less efficient in lower mass galaxies. Following this prescription, star formation algorithms have been applied that account for the observed luminosity function, but at the price of excessive mass-to-luminosity ratios for bright galaxies. Conversely, models that satisfy the observed mass-to-luminosity ratios cannot easily be reconciled with the galaxy luminosity function and the abundance of luminous galaxies (Navarro, Frenk and White 1996).

Another difficulty with galaxy formation models has arisen with the size of galaxy disks (Navarro, Frenk and White 1995). Clumpy collapse and baryonic cooling result in effective dynamical friction on the dense baryonic clumps and efficient angular momentum transfer. The resulting disk size is found from simulations to be about 20 percent of that predicted for a homogenous collapse of baryons in an initially slowly rotating dark halo, with initial rotation characterized by the dimensionless angular momentum parameter λ predicted from tidal torque theory to be about 0.07. A disk half-mass radius $\sim 0.2\lambda\, R_i$, where R_i is the dark halo radius of $\sim$ 50–100 kpc, is far smaller than the $\sim$ 5–10 kpc measured for typical disks. Presumably, feedback associated with star formation will help resolve the disk crisis, but a plausible prescription for this remains to be developed.

Another difficulty that has taxed the ingenuity of the cold dark matter simulations concerns the rotation curves of dwarf galaxies. A dwarf spiral such as DDO 154 has the best measured of any rotation curve, to about 21 disk exponential scale lengths. Moreover such dwarf galaxies are everywhere dominated by dark matter, and provide an attractive laboratory for testing the simulations of cold dark matter rotation curves. The simulated density profiles invariably have a central r^{-1} cusp which is not seen in the data. Some new physics, for example considering an additional component of dark (baryonic or non-baryonic) matter appears to be needed (Burkert and Silk 1997).

Baryons appear to provide both the problem and the potential solution, provided that a suitable prescription for star formation and associated feedback can be implemented. A fundamental theory of star formation is lacking. We do not understand star formation in Orion, let alone in the early universe or in the extreme conditions of merging galaxies. Star formation in protogalaxies is especially challenging. The first stars could have had a mass distribution very different from the present day initial mass function. One could argue equally convincingly that the primordial initial mass function was biased towards massive stars, or to low mass stars, or indeed that the first stars formed in highly exotic conditions in the vicinity of supermassive black holes. The best approach to cosmology appears to be to implement a theoretical prescription for star formation that is closely coupled to observational data on nearby star forming galaxies, and that is in essence semi-phenomenological.

3. Semi-phenomenological galaxy evolution

Observations that help guide a theoretical model for star formation center on studies of the Milky Way and of nearby star-forming spiral galaxies. From the iron distribution in nearby F and G stars, and the observed Fe-metallicity-age relation, one can deduce the star formation history appropriate to a region of the Milky Way that extends for several kpc around the Sun. H_2 studies of nearby disk galaxies, combined with H I and CO maps, give the dependence of star formation rate on total gas surface density. Theory of a cold self-gravitating gas-rich disk shows that it is unstable to the growth of small fluctuations at growth rate $Q^{-1}\Omega$, where $\Omega(r)$ is the differential rotation rate, provided that the Toomre parameter $Q \approx \kappa\sigma_g/\pi G\mu_{gas}f \lesssim 1$. Here κ is the epicyclic frequency (equal to $2^{1/2}\Omega$ if the rotation curve is flat), σ_g is the gas velocity dispersion, μ_{gas} is the gas surface density, and f is a factor (of order unity if the gas fraction exceeds $\sim$ 20 percent) that allows for the effect of the stellar component. Density fluctuation growth develops in the non-linear regime to form molecular clouds which in turn eventually become sufficiently unstable to collapse and form stars. The connection between clouds and star formation is a complex issue that involves the intermediary of magnetic fields, which stabilize the less massive clouds, and feedback from star formation and death, which help maintain the cloud velocity dispersion, both extrinsic and intrinsic, as well as the pressure of the interstellar medium.

A semi-phenomenological approach allows one to bypass much of the theoretical uncertainty, while retaining key theoretical concepts, and test and calibrate a model against the observational data. The star formation rate can be described by

$$\mathrm{SFR} = \epsilon \mu_{gas}\, \Omega(r)\ |(1 - Q^2)/Q|$$

together with early metal-poor gas infall, where ϵ is effectively a measure of efficiency of star formation (in mass fraction of gas that forms stars per dynamical time) (Wang and Silk 1994; Prantzos and Aubert 1995). The outer parts of disks, where the gas surface density has declined such that $Q < 1$, form stars at a greatly reduced rate, at least within giant HII regions, and verify that the threshold dependence, predicted from linear theory, is at least a crude approximation to reality. This expression clearly neglects such important aspects of disk star formation as the role of density waves in stimulating cloud formation and growth, a necessary precursor to star formation.

Despite the theoretical oversimplification of the star formation rate *ansatz*, its application to the Milky Way enables several independent types of data to be modelled in a relatively robust manner. A reasonably generic model for global star formation incorporates, and accounts for, the radial profiles of total gas, star, and star formation rate surface densities as well as for disk chemical evolution. Disk chemical properties include the distribution of stellar metallicities, the age-metallicity relation, and the metallicity gradient towards the center of the galaxy. All of this data can be fit, provided that the star formation efficiency parameter is small: $\epsilon \approx 0.05$.

Star formation must eventually self-regulate so that the star formation efficiency is low. The empirical clue as to how this occurs comes from the observation that throughout the star-forming region of the disk, $Q \sim 1$, in other words, the disk is marginally unstable. One can imagine this situation arises because of feedback from star formation, and in particular from dying stars. Supernova remnants are an important source of momentum input into interstellar gas, which loses energy radiatively as interstellar clouds collide. In a cold disk, in the absence of star formation, Q would initially decrease. Once $Q < 1$, the disk is gravitationally unstable. There are various consequences: the non-linear instabilities form clouds and drive disk motions, and star formation erupts, with ensuing supernova production, as massive stars die, that accelerates the gas and provides a source of cloud random motion. However the result of increasing the gas velocity dispersion is that Q increases, in part since the random motions of the old stars respond dynamically, providing a negative feedback that regulates the instabilities. The star formation efficiency must be low if supernovae dominate feedback, since the efficiency is of order the ratio of the specific momentum of the gas to that injected by supernovae, a ratio that is typically a few percent.

The detailed operation of negative feedback requires consideration of the multiphase structure of the interstellar medium. Supernova remnants generate hot bubbles of $\sim 10^6$ K gas. Multiple supernovae operate collectively, since a typical OB association will contain dozens of OB stars that evolve into supernovae. A typical disk, containing thousands of OB associations, can be visualized as a network of hot bubbles that compress and pressurize the cold (HI and H_2) gas clouds. If the hot bubbles overlap, and the volume fraction of the hot phase is more than ~ 50 percent, the supernova injection of momentum controls the pressure of the cold gas, and also can drive a galactic wind. The hot phase dominance is described by the porosity parameter P, defined to be the product of remnant 4-volume at maximum size and rate of production of remnants. Using analytic fits to simulations of supernova-driven shells, one can show that

$$P \propto \dot{\rho}_* p_{gas}^{-1.36} \rho_{gas}^{-2.72} ,$$

where $\dot{\rho}_*$ is the star formation rate and p_{gas} is the interstellar medium pressure. The pressure is mostly turbulent, so that $p_{gas} \approx \rho_{gas}\sigma_{gas}^2$.

If one now assumes that star formation rate scales as a power of the gas density, $\dot{\rho}_* \propto \rho_{gas}^n$, it follows that

$$P \propto \rho_{gas}^{n-1.47} \sigma_{gas}^{-2.72} .$$

Suppose that $n \approx 3/2$, as expected in the generic model (since SFR $\propto \mu_{gas}\Omega \overset{\propto}{\sim} p_{gas}^{3/2}$). If the porosity is large, the clouds are compressed and are likely to have induced star formation. However the resulting supernovae drive further remnants and turbulence, enhancing σ_{gas} and thereby providing regulative feedback. At high σ_{gas}, the maximum bubble radius, and thus the porosity, is reduced. Moreover, as σ_{gas} increases, Q rises, stabilizing the disk and quenching the cloud formation rate.

The Tully-Fisher relation provides an example of how self-regulation controls the global star formation rate in protodisks. Let us simplify the star formation rate to

$$\dot{\mu}_* = \epsilon \, \mu_{gas} \, \Omega ,$$

and introduce $Q = \Omega\sigma_{gas}/\pi G\mu_{gas}$ as well as $P \propto \dot{\rho}_*\sigma_g^{-2.7}\rho_{gas}^{-1.5}$ to eliminate μ_{gas} and σ_{gas} (for disks of scale length $H \approx \sigma_g^2/G\mu_{gas}$). The efficiency can also be eliminated by using the cloud acceleration condition. This leads (Silk 1997) to a Tully-Fisher-like relation for the total star formation rate

$$\dot{M}_* \propto V_{rot}^{5/2} \, P^{-0.9} \, Q^{-3/2} .$$

The dispersion is expected to be narrow because of the interplay between P and Q: as Q decreases, the enhanced supernova rate initially must in turn enhance porosity (by driving bubbles), although as the turbulence increases the porosity must eventually decrease, thereby providing negative feedback and self-regulation.

Star formation is dominated by blue light, and the expression for the star formation rate can be interpreted as a derivation of the blue Tully-Fisher relation. An empirical relation between star formation rate, as measured by Hα luminosity, and I surface brightness can account for the steeper slope of the Tully-Fisher relation observed in the I-band. The most intriguing aspect of the Tully-Fisher relation is its low dispersion of about 15 percent. Perhaps the complex interplay between P and Q could account for the low dispersion in the Tully-Fisher relation.

In a more violent and chaotic situation, such as could be induced by a merger, the velocity dispersion of the gas is limited by the gravitational potential well depth, rather than by feedback via disk stability. Hence porosity attains a (limiting) value if $n \approx 1.5$. If $n > 1.5$, and in a merger one might expect $n \approx 2$ or 2.5, the porosity increases as the gas density increases towards the central kpc. Any dependence on σ_{gas} has saturated. There is no effective feedback and star formation is likely to be highly efficient.

A protogalactic starburst could occur if two conditions are satisfied. The porosity must be large. Also to form stars, the interstellar medium must be able to radiate away the energy injected from supernovae. This latter condition is $\dot{n}_{SM} \, E_{SM} = \rho_{gas}\sigma_g^2/2t_{cool}$, and can be reexpressed as an upper bound on σ of about 100 km s^{-1}. More generally there is approximate balance between heating and cooling

$$\sigma \approx 100 \, \frac{\text{km}}{\text{s}} \, \left[n_{gas}^{0.1} \, P^{-0.3} (Z/Z_0)^{0.2} \right] .$$

This suggests that the fact that most galaxies occupy a narrow range in σ of around 50–300 km s^{-1} can be understood as a consequence of self-regulation which leads to P in the range 0.1–1.

4. Backwards galaxy formation: Predictions

The generic model of disk formation can be summarized, over the approximately flat region of the rotation curve, as

$$\text{SFR} \propto \mu_{gas}\,\Omega + \text{metal-poor gas infall.}$$

This implies that disks were smaller in the past. The effect should be observable at $z \gtrsim 1$, but is confused by the bulge components and the spread in morphological types. The Hubble Deep Field has provided an attractive source of data with which one can search for morphological evolution. In practice, only a marginal signal is found in the case that galaxies grow by infall relative to the no-infall case (Bouwens, Cayon and Silk 1997).

Another prediction that be quantified by tracing the backward evolution of disks is enrichment of the gaseous protodisk. From the star formation rate, one can infer the enrichment history for some specified gas infall rate. This can be translated into metallicity versus redshift. Comparison with observations of the abundances in damped Lyman alpha absorption line systems shows that the mean trend and dispersion in undepleted elements such as zinc can be understood in terms of protodisk chemical evolution, as viewed at a range of galactocentric distances and inclinations.

The Milky Way, and in particular the nearby 2–3 kpc, provides a template for chemical evolution from the metallicity distribution of old stars. Combination of the metallicity distribution with the age-metallicity relation for nearby old stars enable one to reconstruct the local star formation history. Star formation is found to have undergone a pronounced peak, by about a factor of $6-8$ relative to the present day value, some 4–5 Gyr ago. Translating the star formation history to a dependence on redshift shows that the steep enhancement in the past is equivalent in shape to the star formation rate peak reported by Madau et al. (1996) between $z = 1$ and 2. Moreover, if the star formation rate history is normalized to the redshift peak, one can infer the instantaneous gas fraction at any redshift, if the star formation rate is proportional to a power (typically $1 \lesssim n < 2$) of the gas density.

The inferred gas density is several times the value seen directly in the damped Lyman alpha clouds. This is consistent with the inference from modeling chemical abundances of the clouds with a simple chemical evolution model that a substantial gas reservoir is required at high redshift. According to one interpretation, the additional gas is associated with absorbing gas clouds towards quasars that are not observed in magnitude-limited surveys because of dust extinction along the line of sight.

5. Conclusions

Galaxy formation is a subject where fundamental issues are plagued by ignorance of many key parameters. *Ab initio* theory is compelling in principle, but hard to implement in practice. The backwards approach to galaxy formation is semi-phenomenological and relatively robust. Inside-out formation of disks is the key prediction. Disks are expected to be smaller in the past. Disk abundance gradients grow with time. Star formation history peaked several Gyr ago. It is tempting to speculate that such a peak coincided with an infall event or merger of a satellite galaxy, for the typical protodisk counterpart of a Milky Way-type galaxy.

Spheroid formation remains a murkier area. The backwards approach to forming spheroids is best exemplified by studying ultraluminous starbursts. These are invariably associated with, and presumably triggered by, galaxy mergers. Gas rich mergers of protodisks are the hallmark of protospheroid formation, and provide an environment where efficient star formation can produce a dense spheroidal stellar system. Ultraluminous starbursts display high efficiency of star formation, as well as evidence for spheroid formation inferred from the luminosity profile at infrared wavelengths. Unfortunately, we lack a detailed model for star formation in protospheroids that can account for the high efficiency of star formation that is observed (starbursts) or inferred (spheroids). In the spirit of semi-phenomenological modeling, one can take a starburst as described by an exponentially declining burst of star formation, and incorporate this into a merger prescription for spheroid formation.

The gas reservoir is the key to such an approach. Clearly, gas infall occurs for field galaxies, predominantly gas-rich disks, but was presumably aborted in rich clusters, which are dominated by old spheroid components. Understanding the availability and longevity of the gas reservoir requires knowledge of feedback from ongoing star formation. Feedback in disks must be very different, and much stronger, than feedback in protospheroids, in order to account for the low star formation efficiency in disks compared to the high star formation efficiency that characterized spheroid formation. Eventually, one may hope to combine the semi-phenomenological approach, commencing with present epoch star formation, with *ab initio* star formation that commences within gaseous protogalaxies. At present, however, it is apparent that observations, both of young and evolved galaxies, are driving theoretical modeling of galaxy formation.

This research has been supported in part by grants from NASA and NSF. Some of the results described here have been obtained in collaboration with R. Bouwens and L. Cayon. I acknowledge with gratitude the hospitality of the Institut d'Astrophysique de Paris as a Blaise-Pascale Visiting Professor, and the Institute of Astronomy at Cambridge as a Sackler Visiting Astronomer.

REFERENCES

Burkert, A. and Silk, J. 1997 *ApJL*, in press.

Bouwens, R., Cayon, L. and Silk, J. 1997 *ApJL*, in press.

Navarro, J. F., Frenk, C. S. and White, S. D. M. 1995 *MNRAS* **275**, 56.

Navarro, J. F., Frenk, C. S. and White, S. D. M. 1996 *ApJ* **462**, 563.

Madau, P. et al. 1996 *MNRAS* **283**, 1388.

Prantzos, N. and Aubert, O. 1995 *A&A* **302**, 69.

Silk, J. 1997 *ApJ* **481**, 703.

Wang, B. and Silk, J. 1994 *ApJ* **427**, 759.

The evolution of luminous matter in the universe

By PIERO MADAU

Space Telescope Science Institute, 3700 San Martin Drive, Baltimore, MD 21218

I review a technique for interpreting faint galaxy data which traces the evolution with cosmic time of the galaxy luminosity density, as determined from several deep spectroscopic samples and the *Hubble Deep Field* imaging survey. The method relies on the rest-frame UV and near-IR continua of galaxies as indicators, for a given initial mass function (IMF) and dust content, of their instantaneous star formation rate (SFR) and total stellar mass, and offers the prospect of addressing in a coherent framework an important set of subjects: cosmic star formation history, dust in primeval galaxies, shape of the IMF, stellar mass-to-light ratios of present-day galaxies, extragalactic background light, Type II supernovae and heavy element enrichment history of the universe. The global spectrophotometric properties of field galaxies are well fit by a simple stellar evolution model, defined by a time-dependent SFR per unit comoving volume, a universal IMF which is relatively rich in massive stars, and a modest amount of dust reddening. The model is able to account for the entire background light recorded in the galaxy counts down to the very faint magnitude levels probed by the HDF, and produces visible mass-to-light ratios at the present epoch which are consistent with the values observed in nearby galaxies of various morphological types. The bulk ($\gtrsim 60\%$) of the stars present today formed relatively recently ($z \lesssim 1.5$), consistently with the expectations from a broad class of hierarchical clustering cosmologies, and in good agreement with the low level of metal enrichment observed at high redshifts in the damped Lyman-α systems. Throughout this review I emphasize how the poorly constrained amount of starlight that was absorbed by dust and reradiated in the far-IR at early epochs represents one of the biggest uncertainties in our understanding of the evolution of luminous matter in the universe. A "monolithic collapse" model, where half of the present-day stars formed at $z > 2.5$ and were enshrouded by dust, can be made consistent with the global history of light, but overpredicts the metal mass density at high redshifts as sampled by QSO absorbers.

---

1. Introduction

As the best view to date of the optical sky at faint flux levels, the *Hubble Deep Field* (HDF) has offered to many astronomers the opportunity to study the galaxy population in unprecedented details. In particular, the deep HDF images have rapidly become a key testing ground for the two competing scenarios that have been widely used in the past few years to interpret the observed properties of galaxies. In what may be called the "traditional" scheme, one starts from the local measurements of the distribution of galaxies as a function of luminosity and Hubble type and models their photometric evolution assuming a well defined collapse epoch, pure-luminosity evolution thereafter, and a set of parameterized star formation histories (Tinsley 1980; Bruzual & Kron 1980; Koo 1985; Pozzetti, Bruzual, & Zamorani 1996). These, together with an initial mass function (IMF) and a world model, are then adjusted to match the observed number counts, colors, and redshift distributions. Beyond its intrinsic simplicity, the main advantage of this kind of approach is that it can be made consistent with the classical view that ellipticals and spiral galaxy bulges formed early in a single burst of duration 1 Gyr or less (e.g., Bower et al. 1992; Ortolani et al. 1995). Because in this "monolithic" collapse scenario for spheroids much of the action happens at high-z, however, these models predict, in the absence of a significant amount of dust obscuration, far more Lyman-break "blue dropouts" than are seen in the HDF (Pozzetti et al. 1997; Ferguson & Babul 1997). Moreover, they cannot reproduce the rapid evolution—largely driven by late-type galaxies—of the optical luminosity density with lookback time observed by Lilly et al. (1996) and Ellis et al. (1996).

A more physically motivated way to interpret the observations is to construct semi-analytic hierarchical models of galaxy formation and evolution (White & Frenk 1991; Kauffmann, White, & Guiderdoni 1993; Cole et al. 1994; Baugh et al. 1997). Here, one starts from a power spectrum of primordial density fluctuations, follows the growth of dark matter halos by accretion and mergers, and adopts various prescriptions for gas cooling, star formation, feedback from supernovae and stellar winds, and dynamical friction. These are tuned to match the statistical properties of both nearby and distant galaxies. In this scenario, there is no period when bulges and ellipticals form rapidly as single units and are very bright: rather, small objects form first, eventually settling into disks, and merge continually over a range of redshifts to make ellipticals (see reviews by G. Kauffmann and S. White in these proceedings). While reasonably successful in recovering the counts, colors, and redshift distributions of faint galaxies, a generic difficulty of such models is the inability to simultaneously reproduce the observed local luminosity density and the zero-point of the Tully-Fisher relation (White & Frenk 1991; Cole et al. 1994).

In this talk I will describe an alternative method, which focuses on the emission properties of the galaxy population *as a whole*. It traces the cosmic evolution with redshift of the galaxy luminosity density—as determined from several deep spectroscopic samples and the HDF imaging survey—and offers the prospect of an empirical determination of the global star formation history of the universe and initial mass function of stars independently, e.g., of the merging histories and complex evolutionary phases of individual galaxies. The technique relies on two basic properties of stellar populations: a) the UV-continuum emission in all but the oldest galaxies is dominated by short-lived massive stars, and is therefore a direct measure, for a given IMF and dust content, of the instantaneous star formation rate (SFR); and b) the rest-frame near-IR light is dominated by near-solar mass evolved stars that make up the bulk of a galaxy's visible mass, and can then be used as a tracer of the integrated stellar mass density. By modeling the "emission history" of the universe at ultraviolet, optical, and near-infrared wavelengths from the present epoch to $z \approx 4$, I will try to shed some light on what I believe are key questions in structure formation and evolution studies, and provide a first glimpse to the history of the conversion of neutral gas into stars within galaxies.

The initial applications of this novel technique have been presented by Lilly et al. (1996), Madau et al. (1996, hereafter M96), and Madau, Pozzetti, & Dickinson (1997). A complementary effort—which starts instead from the analysis of the evolving gas content and metallicity of the universe—can be found in Fall, Charlot, and Pei (1996). A flat cosmology with $q_0 = 0.5$ and $H_0 = 50\,h_{50}\,\mathrm{km\,s^{-1}\,Mpc^{-1}}$ will be adopted in this review.

2. Indicators of past and present star formation activity

Stellar population synthesis has become a standard technique to study the spectrophotometric properties of galaxies. In the following, I shall make extensive use of the latest version of Bruzual & Charlot (1993) isochrone synthesis code, optimized with an updated library of stellar spectra (Bruzual & Charlot 1997), to predict the time change of the spectral energy distribution of a stellar population. I will consider here two possibilities for the IMF, a Salpeter (1955) function and a Scalo (1986) IMF, which is flatter for

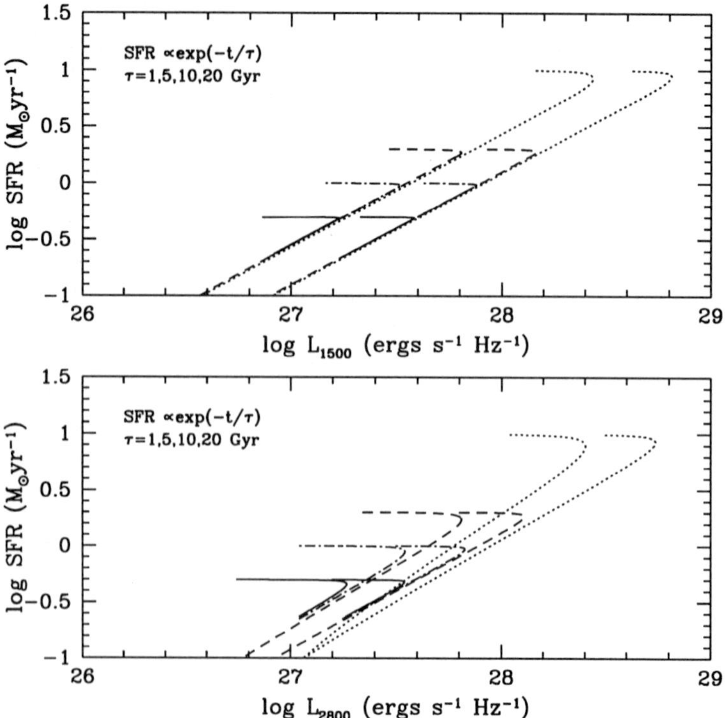

FIGURE 1. SFR-UV relation for models with various exponentially declining star formation rates at ages between 0.01 and 15 Gyr. *Solid lines:* $\tau = 20$ Gyr. *Dot-dashed lines:* $\tau = 10$ Gyr. *Dashed lines:* $\tau = 5$ Gyr. *Dotted lines:* $\tau = 1$ Gyr. The set of curves on the left-hand side of the plot assume a Scalo IMF, the ones on the right-hand side a Salpeter function.

low-mass stars and significantly less rich in massive stars than Salpeter. In all models the metallicity is fixed to solar values and the IMF is truncated at 0.1 and 125 $\mathrm{M}_\odot$.

2.1. *Birthrate—Ultraviolet relation*

The UV continuum emission from a galaxy with significant ongoing star formation is entirely dominated by late-O/early-B stars on the main sequence. As these have masses $\gtrsim 10\,\mathrm{M}_\odot$ and lifetimes $t_{MS} \lesssim 2 \times 10^7$ yr, the measured luminosity becomes proportional to the stellar birthrate and independent of the galaxy history for $t \gg t_{MS}$. This is depicted in Figure 1, where the power radiated at 1500 Å and 2800 Å is plotted against the instantaneous SFR for a model stellar population with different star formation laws, SFR $\propto \exp(-t/\tau)$, where τ is the duration of the burst. After an initial transient phase where the UV flux rises rapidly and the turnoff mass drops below $10\,\mathrm{M}_\odot$, a steady state is reached where one can write

$$L_{UV} = \text{const} \times \frac{\text{SFR}}{\mathrm{M}_\odot\,\mathrm{yr}^{-1}}\ \mathrm{ergs\,s^{-1}\,Hz^{-1}}, \tag{2.1}$$

with const= $(3.5\times10^{27}, 5.0\times10^{27})$ at (1500 Å, 2800 Å) for a Scalo IMF, and const= 8.0×10^{27} in the same wavelength range for a Salpeter IMF, quite insensitive to the details of the past star formation history. Note how, for burst durations $\lesssim 1$ Gyr and a Scalo IMF, the luminosity at 2800 Å becomes a poor SFR indicator after a few e-folding times, when the contribution of intermediate-mass stars becomes significant. After averaging over the whole galaxy population, however, we will find that the (unreddened) UV continuum is always a good tracer of the instantaneous rate of conversion of cold gas into stars.

2.2. *Birthrate—Hα relation*

In the optical wavelength range, the Hα(λ6563) luminosity from a galaxy is another probe of its current star formation activity (Kennicutt 1983), as, for ionization bounded H II regions, the integrated line emission scales directly with the Lyman-continuum flux of the embedded OB stars. Assuming case-B recombination theory and a Salpeter function, the Hα luminosity can be related to the local SFR according to

$$L_{\mathrm{H}\alpha} = 1.5 \times 10^{41} \left(\frac{\mathrm{SFR}}{\mathrm{M}_\odot\,\mathrm{yr}^{-1}} \right) \mathrm{ergs\,s}^{-1}. \tag{2.2}$$

The coefficient for a Scalo IMF is approximately four times smaller. Note that, shortward of the Lyman edge, the differences in the predicted ionizing radiation from model atmospheres of hot stars can be quite large (Charlot 1996a), and must be taken into account when interpreting the results of surveys for Hα-emitting galaxies (Gallego et al. 1995).

2.3. *Supernova frequency*

The frequency of supernovae (SNe) is also intimately related, for a given IMF, to the stellar birthrate. This is true, in particular, for "core-collapse supernovae", SN II and SN Ib/c, which have massive, short-lived ($t_{MS} \lesssim 2 \times 10^7$ yr) progenitors. It might also be approximately true for Type Ia SNe—believed to result from the thermonuclear disruption of accreting CO white dwarfs in binary systems (for a recent review see Ruiz-Lapuente, Canal, & Burkert 1997)—if the favorite route follows a fast evolutionary track, and the deflagration occurs within few Gyr of (binary) stellar birth. Since the Hubble time is equal to 3 Gyr at $z = 1.7$, it would then be possible to neglect the delay between stellar birth and the SN Ia it eventually yields at all epochs $z \ll 1.7$ (but see Yungelson & Livio 1997).

For a Salpeter IMF and a lower mass cutoff for the progenitor star of $10\,\mathrm{M}_\odot$, the core-collapse supernova rate (SNR) can be related to the stellar birthrate according to

$$\mathrm{SNR} = 0.0055 \times \left(\frac{\mathrm{SFR}}{\mathrm{M}_\odot\,\mathrm{yr}^{-1}} \right) \mathrm{yr}^{-1}. \tag{2.3}$$

The coefficient for a Scalo IMF is 2.6 times smaller. I will show later how SNe may provide an independent test of the global star formation history of the universe at low redshifts.

2.4. *Mass—Infrared relation*

If we assume a time-independent IMF, we can use the results of stellar population synthesis modeling, together with the observed UV emissivity, to infer the evolution of the star formation activity in the universe (M96). The biggest uncertainty in this procedure is due to dust reddening, as newly formed stars which are completely hidden by dust would not contribute to the UV luminosity. The effect is potentially more serious at high redshifts, as for a fixed observer-frame bandpass, one is looking further in the ultraviolet with increasing lookback time. For example, SMC-type foreground dust with $E(B-V) \gtrsim 0.1$ would produce an extinction at 1500 Å greater than 1.3 magnitudes. On the other hand, it should be possible to test the hypothesis that star formation regions remain largely unobscured by dust throughout much of galaxy evolution by looking at the near-infrared light density. This will be affected by dust only in the most extreme,

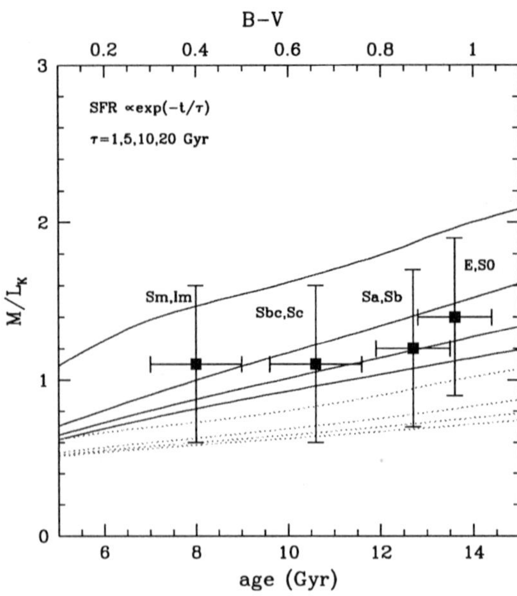

FIGURE 2. Total (processed gas+stars) mass-to-K band light ratio versus age for models with various exponentially declining star formation rates. *Solid lines:* Salpeter IMF. *Dotted lines:* Scalo IMF. From top to bottom, each set of curves depict the values for $\tau = 1$, 5, 10, and 20 Gyr, respectively. The data points show the luminous mass-to-infrared light ratio versus $B-V$ color (top axis) observed in nearby galaxies of various morphological types (see Charlot 1996b). The observations refer to the mass within the galaxy Holmberg radius, where the contribution by the dark matter halo is expected to be small.

rare cases, as it takes an $E(B-V) > 4$ mag to produce an optical depth of unity at 2.2 μm.

Although different types of stars—such as supergiants, AGB, and red giants—dominate the K-band emission at different ages in an evolving stellar population, the mass-to-infrared light ratio is relatively insensitive to the star formation history (Charlot 1996b). Figure 2 shows M/L_K (in solar units) as a function of age for models with various exponentially declining SFR compared to the values observed in nearby galaxies of early to late morphological types. As the stellar population ages, the mass-to-infrared light ratio remains very close to unity, independent of the galaxy color and Hubble type. We can use this interesting property to estimate the visible mass in galaxies from the local K−band luminosity density, $\log \rho_K(0) = 27.05 \pm 0.1\,\mathrm{h}_{50}\,\mathrm{ergs\,s^{-1}\,Hz^{-1}\,Mpc^{-3}}$ (Gardner et al. 1997). The observed range $0.6\,h_{50} \lesssim M/L_K \lesssim 1.9\,h_{50}$ translates into a mass density of stars+gas at the present day of

$$2 \times 10^8 \lesssim \rho_{s+g}(0) \lesssim 6 \times 10^8\,\mathrm{h}_{50}^2\,\mathrm{M}_\odot\,\mathrm{Mpc}^{-3} \tag{2.4}$$

($0.003 \lesssim \Omega_{s+g} \lesssim 0.009$). I will show later how the observed integrated UV emission, with the addition of some modest amount of reddening, may account for the bulk of the baryons traced by the K-band light, and how initial mass functions with relatively few high-mass stars (such as the Scalo IMF), or models with a large amount of dust extinction at all epochs will tend to overproduce the near-infrared emissivity.

3. Galaxy emissivity as a function of redshift

The integrated light radiated per unit volume from the entire galaxy population is an average over cosmic time of the stochastic, possibly short-lived star formation episodes of individual galaxies, and should follow a relatively simple dependence on redshift. In the UV—where it is proportional to the global SFR—its evolution should provide information on the mechanisms which may prevent the gas within virialized dark matter halos from radiatively cooling and turning into stars at early times, or on the epoch when galaxies exhausted their reservoirs of cold gas. From a comparison between different wavebands it should be possible to set constraints on the average IMF and dust content of galaxies.

The observed continuum emissivity, $\rho_\nu(z)$, from the present epoch to $z \approx 4$ is shown in Figure 3 in six broad passbands centered around 0.15, 0.2, 0.28, 0.44, 1.0, and 2.2 μm. The data are taken from the K-selected redshift survey of Gardner et al. (1997) and Cowie et al. (1996), the I-selected CFRS (Lilly et al. 1996) and B-selected Autofib (Ellis et al. 1996) samples, a redshift survey of galaxies imaged in the rest-frame ultraviolet at 2000 Å with the FOCA balloon-borne camera (Treyer et al. 1997), the photometric redshift catalog for the HDF of Connolly et al. (1997)—which take advantage of deep infrared observations by Dickinson et al. (1997)—and the color-selected UV and blue "dropouts" of M96 (see also Madau 1997).They have all (except for the Treyer et al. value) been corrected for incompleteness by integrating over the best-fit Schechter function in each redshift bin,

$$\rho_\nu(z) = \int L_\nu \phi(L_\nu, z) dL_\nu = \Gamma(2+\alpha)\phi_* L_*. \tag{3.5}$$

As it is not possible to reliably determine the faint end slope of the luminosity function from the Connolly et al. (1997) and M96 data sets, a value of $\alpha = -1.3$ has been assumed at each redshift interval for comparison with the CFRS sample (Lilly et al. 1995). The error bars are typically less than 0.2 in the log, and reflect the uncertainties present in these corrections and, in the HDF $z > 2$ sample, in the volume normalization and color-selection region.

Despite the obvious caveats due to the likely incompleteness in the data sets, different selection criteria, and existence of systematic uncertainties in the photometric redshift technique, the spectroscopic, photometric, and Lyman-break galaxy samples appear to provide a remarkably consistent picture of the emission history of field galaxies.† This points to a rapid drop in the volume-averaged SFR in the last 8–10 Gyr, and to a redshift range where the bulk of the stellar population was actually assembled: $1 \lesssim z \lesssim 2$.

4. Population synthesis

It is interesting to see now whether a simple stellar evolution model, defined by a time-dependent SFR per unit volume and a universal IMF, may reproduce the global UV, optical, and near-IR photometric properties of galaxies. In a stellar system with arbitrary star formation rate, the luminosity density at time t is given by the convolution integral

$$\rho_\nu(t) = \int_0^t L_\nu(\tau) \times \mathrm{SFR}(t-\tau) d\tau, \tag{4.6}$$

† While there is no evidence for a gross mismatch at the $z \approx 2$ transition between the photometric redshift sample of Connolly et al. (1997) and the M96 UV dropout sample, one should note that only one spectroscopic confirmation has been obtained so far at $z \approx 4$ (Dickinson 1997).

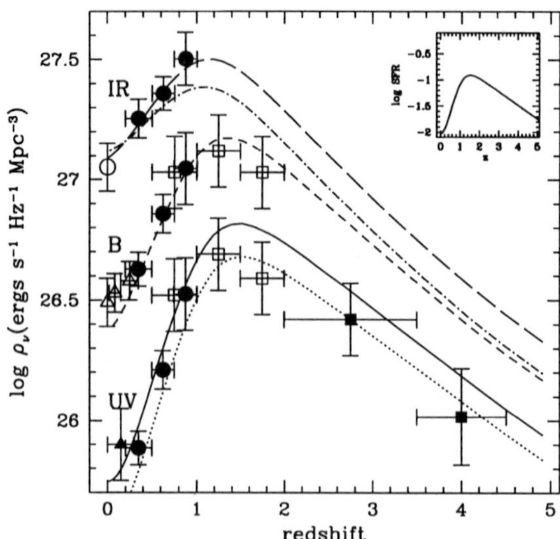

FIGURE 3. Evolution of the comoving luminosity density at rest-frame wavelengths of 0.15 (*dotted line*), 0.28 (*solid line*), 0.44 (*short-dashed line*), 1.0 (*long-dashed line*), and 2.2 (*dot-dashed line*) μm. The data points with error bars are taken from Lilly et al. (1996) (*filled dots*), Connolly et al. (1997) (*empty squares*), Madau et al. (1996) and Madau (1997) (*filled squares*), Ellis et al. (1996) (*empty triangles*), Treyer et al. (1997) (*filled triangle*), and Gardner et al. (1997) (*empty dot*). The inset in the upper-right corner of the plot shows the SFR density ($M_\odot$ yr^{-1} Mpc^{-3}) versus redshift which was used as input to the population synthesis code. The model assumes a Salpeter IMF, SMC-type dust in a foreground screen, and a universal $E(B-V) = 0.1$.

where $L_\nu(\tau)$ is the specific luminosity radiated per unit initial mass by a generation of stars with age τ. After relating the observed UV continuum emissivity to a SFR density, one can then use, e.g., Bruzual-Charlot's synthesis code to predict the time evolution of the spectrophotometric properties of a stellar population in a comoving volume large enough to be representative of the universe as a whole. In doing so, we will bypass all the ambiguities associated with the study of morphologically distinct samples, but, at the same time, we will not be able to specifically address the evolution of particular subclasses of objects, like the oldest ellipticals, whose star formation histories may have differed significantly from the global average. It is fair at this stage to point out two other significant limitations of this approach: a) It focuses on the emission properties of "normal", optically-selected field galaxies which are only moderately affected by dust—a typical spiral emits 30% of its energy in the far-infrared region (Saunders et al. 1990). Starlight which is completely blocked from view even in the near-IR by a large optical depth in dust will not be recorded by this technique, and the associated baryonic mass and metals missed from our census. The contribution of infrared-selected dusty starbursts to the present-day total stellar mass density cannot be very large, however, for otherwise the current limits to the energy density of the mid- and far-infrared background would be violated (Puget et al. 1996; Kashlinsky, Mather, & Odenwald 1996; Fall et al. 1996; Guiderdoni et al. 1997). Locally, infrared luminous galaxies are known to produce only a small fraction of the IR luminosity of the universe (Soifer & Neugebauer 1991); and b) It does not include the effects of cosmic chemical evolution on the predicted galaxy colors.

All the population synthesis models assume solar metallicity, and thus will generate colors that are slightly too red for objects with low metallicity, e.g., truly primeval galaxies.

4.1. *A fiducial model: Salpeter IMF*

Figure 3 shows the model predictions for the evolution of ρ_ν at rest-frame ultraviolet to near-infrared frequencies for a Salpeter IMF. In the absence of dust reddening, this relatively flat IMF generates spectra that are slightly too blue to reproduce the observed mean (luminosity-weighted over the entire population) galaxy colors. The effect of dust attenuation can be included by multiplying equation (4.6) by $p_{\rm esc}$, a redshift-independent term equal to the fraction of emitted photons which are not absorbed by dust. For purposes of illustration, I will assume a foreground screen model, $p_{\rm esc} = \exp(-\tau_\nu)$, and SMC-type dust.† This should only be regarded as a crude approximation, since hot stars can be heavily embedded in dust within star-forming regions, there will be variety of extinction laws, and the dust content of galaxies will evolve with redshift. While the existing data are too sparse to warrant a more elaborate analysis, this simple prescription will highlight the main features and assumptions of the model. The shape of the predicted and observed $\rho_\nu(z)$ relations is then found to agree better to within the uncertainties if a modest amount of dust extinction, $E(B-V) = 0.1$, is included. In this case, the observed UV luminosities must be corrected upwards by a factor of 1.4 at 2800 Å, and 2.1 at 1500 Å.

As expected, while the ultraviolet emissivity traces remarkably well the rise, peak, and sharp drop in the instantaneous star formation rate (the smooth function shown in the inset on the upper-right corner of the figure), an increasingly large component of the longer wavelengths light reflects the past star formation history. The peak in the luminosity density at 1.0 and 2.2 μm occurs then at later epochs, while the decline from $z \approx 1$ to $z = 0$ is more gentle than observed at shorter wavelengths. In the instantaneous recycling approximation (Tinsley 1980), the total stellar mass density produced at time t is

$$\rho_s(t) = (1-R)\int_0^t {\rm SFR}(t)dt, \tag{4.7}$$

where R is the mass fraction of a generation of stars that is returned to the interstellar medium, $R \approx 0.3$ for a Salpeter IMF (R is closer to 0.2 for a Scalo function). The total stellar mass density at $z = 0$ is then $\rho_s(0) = 3.7 \times 10^8\, {\rm M}_\odot\, {\rm Mpc}^{-3}$, with a fraction close to 65% being produced at $z > 1$, and only 20% at $z > 2$. In the assumed cosmology, about half of the stars observed today are more than 9 Gyr old, and only 20% are younger than 5 Gyr.

4.2. *A case with a Scalo IMF*

Figure 4 shows the model predictions for a Scalo IMF. The fit to the data is now much poorer, since this IMF generates spectra that are too red to reproduce the observed mean galaxy colors, as first noted by Lilly et al. (1996). Because of the relatively large number of solar mass stars formed, it produces too much long-wavelength light by the present epoch. The addition of dust reddening would obviously make the fit even worse. The total stellar mass density produced is similar to the Salpeter IMF case.

† Since what is relevant here is the absorption opacity, I have multiplied the extinction optical depth by a factor of 0.6, as the albedo of dust grains is known to approach asymptotically 0.4–0.5 at ultraviolet wavelengths (see, e.g., Pei 1992).

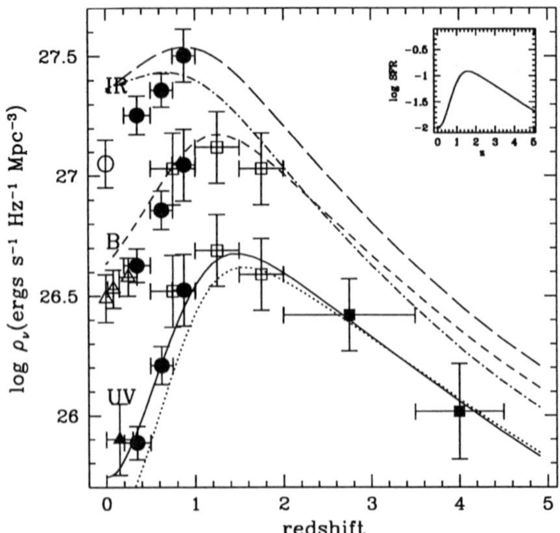

FIGURE 4. Same as in Figure 3, but assuming a Scalo IMF and no dust extinction. This model overproduces the local K-band emissivity by a factor of 2.

5. Clues to galaxy formation and evolution

The results shown in the previous section have significant implications for our understanding of the global history of star and structure formation. Here I discuss a few key issues which will assist in interpreting the evolution of luminous matter in the universe.

5.1. *The brightness of the night sky*

An important check on the inferred emission history of field galaxies comes from a study of the extragalactic background light (EBL), an indicator of the total optical luminosity of the universe. The contribution of known galaxies to the EBL can be calculated directly by integrating the emitted flux times the differential galaxy number counts down to the detection threshold. I have used a compilation of ground-based and HDF data down to very faint magnitudes (Pozzetti et al. 1997; Williams et al. 1996) to compute the mean surface brightness of the night sky between 0.35 and 2.2 μm. The results are plotted in Figure 5, along with the EBL spectrum predicted by our modeling of the galaxy luminosity density,

$$J_\nu = \frac{1}{4\pi}\int_0^\infty dz\frac{dl}{dz}\rho_{\nu'}(z) \tag{5.8}$$

where $\nu' = \nu(1+z)$ and dl/dz is the cosmological line element. The overall agreement is remarkably good, with the model spectrum being only slightly bluer, by about 20–30%, than the observed EBL. The straightforward conclusion of this exercise is that *the star formation history depicted in Figure 3 appears able to account for the entire background light recorded in the galaxy counts down to the very faint magnitudes probed by the HDF.*

5.2. *Luminous mass-to-light ratios of present-day galaxies*

The fiducial Salpeter IMF model generates a present-day stellar mass density of $\Omega_s h_{50}^2 \approx 0.004$, about 10% of the nucleosynthesis constrained baryon density, $\Omega_b h_{50}^2 \approx 0.05 \pm 0.01$ (Walker et al. 1991). The (luminosity-weighted) visible mass-to-light ratios range from

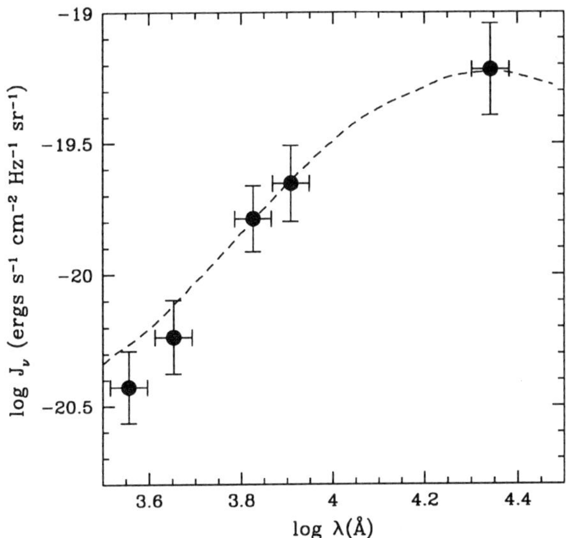

FIGURE 5. Spectrum of the extragalactic background light as derived from a compilation of ground-based and HDF galaxy counts (see Pozzetti et al. 1997). The 2σ error bars arise mostly from field-to-field variations. *Dashed line:* Model predictions for the fiducial model (star formation history of Figure 3).

about 4 in the B-band to 1 in K, consistent with the values observed in nearby galaxies of various morphological types (see, e.g., Persic & Salucci 1992 and references therein). Note, however, that the predicted M/L are quite sensitive to the lower-mass cutoff of the IMF, as very-low mass stars can contribute significantly to the mass but not to the integrated light of the whole stellar population. For example, a lower cutoff of $0.2\,M_\odot$ instead of the $0.1\,M_\odot$ adopted would decrease the mass-to-light ratio by a factor of 1.3. Although one could in principle reduce the inferred star formation density by adopting a top-heavy IMF, richer in massive UV-producing stars, in practice a significant amount of dust reddening—hence of "hidden" star formation—would then be required to match the observed galaxy colors. The net effect of this operation would be a baryonic mass comparable to the estimate above and a large infrared background.

5.3. *Evolution of the supernova rate with redshift*

In recent years there has been a renewed interest in the search for supernovae. While SN Ia may provide one of the best distance indicators at high redshifts (Kim et al. 1997), a direct measurement of the rate of Type II SNe could be used as an independent test for the star formation and heavy element enrichment history of the universe. Figure 6 shows the normalized rate of core-collapse SNe predicted by our best-fit model as a function of cosmic time. The frequency is computed in the rest-frame of the supernovae, and is expressed in SNu (SNe per $10^{10}\,L_\odot(B)$ per century). Unlike the rate per unit comoving volume, which will trace the rise, peak, and drop of the star formation density, the blue luminosity-weighted frequency is a monotonic increasing function of redshift. The predicted rate is in good agreement (to within 0.2 in the log) with the local values estimated by van den Bergh & Tammann (1991), and Cappellaro et al. (1997). In the near future, ongoing searches for distant SNe should provide enough data to constrain

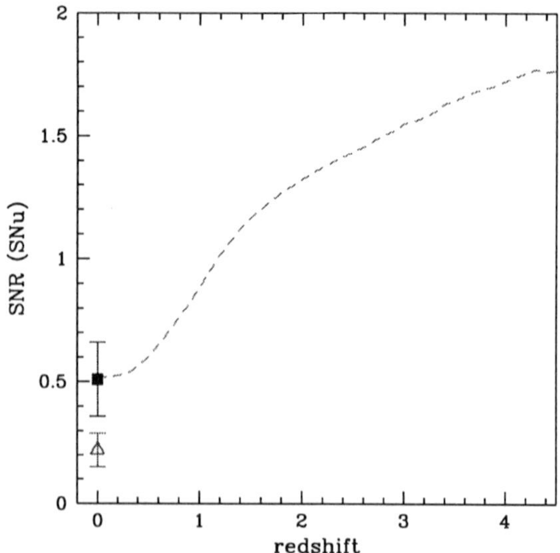

FIGURE 6. Rest-frame rate of core-collapse supernovae (Type II+Ib/c) in SNu (SNe per 10^{10} $L_\odot(B)$ per century) versus redshift. The data points with error bars are taken from van den Bergh & Tammann (1991) (*filled square*), and Cappellaro et al. (1997) (*empty triangle*). They have been normalized according to the local blue luminosity function by spectral type of Heyl et al. (1997). The *dashed line* depicts the rate predicted from our Salpeter IMF, $E(B-V) = 0.1$ fiducial model, assuming a lower mass cutoff for the progenitors of 10 $M_\odot$.

the star formation history of the universe at intermediate redshifts, $z \approx 0.5$–1 (Della Valle & Madau 1997).

5.4. *Star formation at high redshifts: Monolithic collapse versus hierarchical clustering models*

The biggest uncertainty present in our estimates of the star formation density at $z > 2$ is probably associated with dust reddening, but, as the color-selected HDF sample includes only the most actively star-forming young objects, one could also imagine the existence of a large population of relatively old or faint galaxies still undetected at high-z. The issue of the amount of star formation at early epochs is a non trivial one, as the two competing models, monolithic collapse versus hierarchical clustering, make very different predictions in this regard. From stellar population studies we know in fact that about half of the present-day stars are contained in spheroidal systems, i.e., elliptical galaxies and spiral galaxy bulges (Schechter & Dressler 1987). In the monolithic scenario these formed early and rapidly, experiencing a bright starburst phase at high-z (Eggen, Lynden-Bell, & Sandage 1962; Tinsley & Gunn 1976). In hierarchical clustering theory instead ellipticals form continuously by the merger of disk/bulge systems (Kauffman et al. 1993), and most galaxies never experience star formation rates in excess of a few solar masses per year (Baugh et al. 1997). Figure 7 shows the star formation history predicted by a fiducial CDM model (normalized to reproduce the abundance of rich clusters, Cole et al. 1994) and compared to observational estimates. Overall, the agreement between theoretical predictions and data is quite good. Both appear to produce only a small fraction, about 15–20%, of the current stellar content of galaxies at $z \gtrsim 2$–2.5. In fact, the tendency to form the bulk of the stars at relatively low redshifts is a generic feature not only

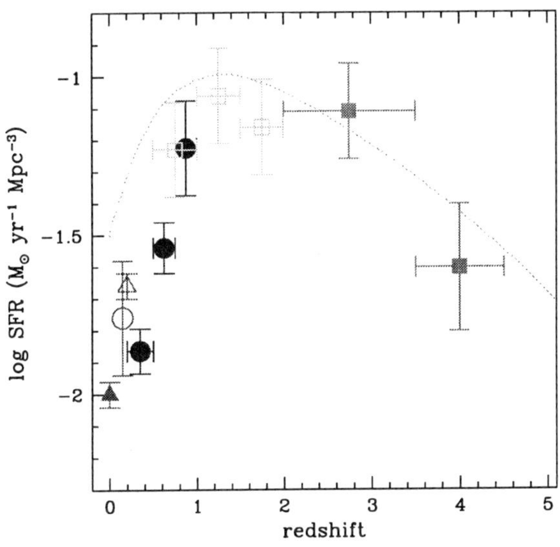

FIGURE 7. Cosmic star formation history. The data points with error bars are taken from the UV broadband measurements of Lilly et al. (1996) (*filled dots*), Connolly et al. (1997) (*empty squares*), Madau et al. (1996) and Madau (1997) (*filled squares*), Treyer et al. (1997) (*empty dot*). Also plotted are the determinations from the Hα luminosity functions of Gallego et al. (1995) (*filled triangle*) and Tresse & Maddox (1997) (*empty triangle*). The data points have been converted to total SFR assuming a Salpeter IMF and a universal $E(B-V)=0.1$. Because of dust extinction, the observed rest-frame UV fluxes have been corrected upwards by a factor of 1.4 at 2800 Å and 2.1 at 1500 Å. The *dotted line* depicts the SFR per comoving volume predicted by a fiducial CDM model in which structure forms hierarchically (from Cole et al. 1994).

of the $\Omega_0 = 1$ CDM cosmology, but also of successful low-density CDM models (Cole et al. 1994; Baugh et al. 1997). While uncertainties still remain in this comparison—e.g., because of the poorly known effects of dust obscuration and possible incompleteness in the high-z sample—one should note the tendency of the theoretical curve to sit above the data points at all epochs, thereby predicting luminosity-weighted mass-to-light ratios at the present time which are higher than observed.

It is of interest then to ask how much larger could the volume-averaged SFR at high-z be before its fossil records—in the form of long-lived, near solar-mass stars—became easily detectable as an excess of K-band light at late epochs. In particular, is it possible to envisage a model where 50% of the present-day stars formed at $z > 2.5$ and were shrouded by dust? The predicted emission history from such a model is depicted in Figure 8. To minimize the long-wavelength emissivity associated with the radiated ultraviolet light, a Salpeter IMF has been adopted. Consistency with the HDF data has been obtained assuming a dust extinction which increases rapidly with redshift, $E(B-V) = 0.011(1+z)^{2.2}$. This results in a correction to the rate of star formation of a factor ~ 5 at $z = 3$ and ~ 15 at $z = 4$. The total stellar mass density today is $\rho_s(0) = 5.0 \times 10^8\ \mathrm{M}_\odot\ \mathrm{Mpc}^{-3}$ ($\Omega_s h_{50}^2 = 0.007$).

Overall, the fit to the data is still acceptable, showing how the blue and near-IR light at $z < 1$ are *relatively poor indicators of the star formation history at early epochs*. The reason for this is the short timescale available at $z \gtrsim 2$, which makes the present-day stellar mass density rather insensitive to a significant boost of the stellar birthrate at high

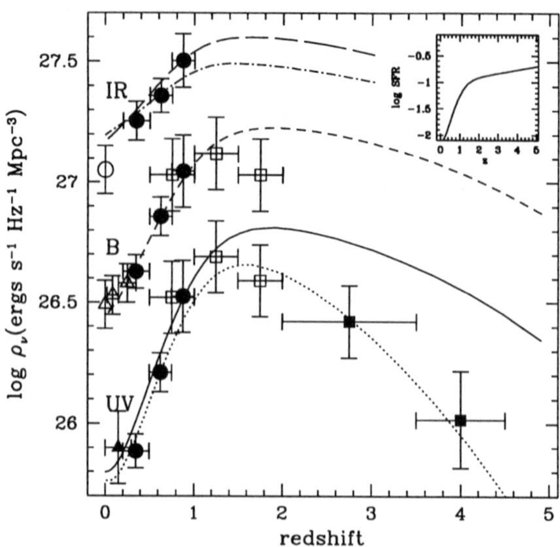

FIGURE 8. Test case with a much larger star formation density at high redshift than indicated by the HDF dropout analysis. The model—designed to mimic a "monolithic collapse" scenario—assumes a Salpeter IMF and a dust opacity which increases rapidly with redshift, $E(B-V) = 0.011(1+z)^{2.2}$. Notation is the same as in Figure 3.

redshifts. By contrast, variations in the global SFR around $z \sim 1.5$, where the bulk of the stellar population was assembled, have a much larger impact. The adopted extinction-redshift relation, in fact, implies negligible reddening at $z \lesssim 1$. Relaxing this—likely unphysical—assumption would cause the model to significantly overproduce the K-band local luminosity density. I have also checked that an even larger amount of hidden star formation at early epochs, as recently advocated by Rowan-Robinson et al. (1997) and Meurer et al. (1997), would generate too much blue, 1 μm and 2.2 μm light to be still consistent with the observations. An IMF which is less rich in massive stars would only exacerbate the discrepancy.

5.5. *The ultraviolet colors of Lyman-break galaxies*

Figure 9 shows a comparison between the HDF data and the model predictions for the evolution of galaxies in the $U-B$ vs. $V-I$ color-color plane according to the star formation history of Figure 3. The HDF ultraviolet passband—which is bluer than the standard ground-based U filter—permits the identification of star-forming galaxies in the interval $2 \lesssim z \lesssim 3.5$. Galaxies in this redshift range predominantly occupy the top left portion of the $U-B$ vs. $V-I$ color-color diagram because of the attenuation by the intergalactic medium and intrinsic absorption (M96). Galaxies at lower redshift can have similar $U-B$ colors, but are typically either old or dusty, and are therefore red in $V-I$ as well.

It is clear that the Salpeter IMF, $E(B-V) = 0.1$ model reproduces quite well the rest-frame UV colors of high-z objects in HDF. This may suggest that interstellar dust is already present in these young galaxies and that it attenuates their 1500 Å luminosities by a factor of ~ 2. A UV extinction of about 1 mag is also indicated by a comparison between UV and Hβ luminosities in three bright UV dropouts (Pettini et al. 1997). One should note that the prescription for a "correct" de-reddening of the Lyman-break galaxies at

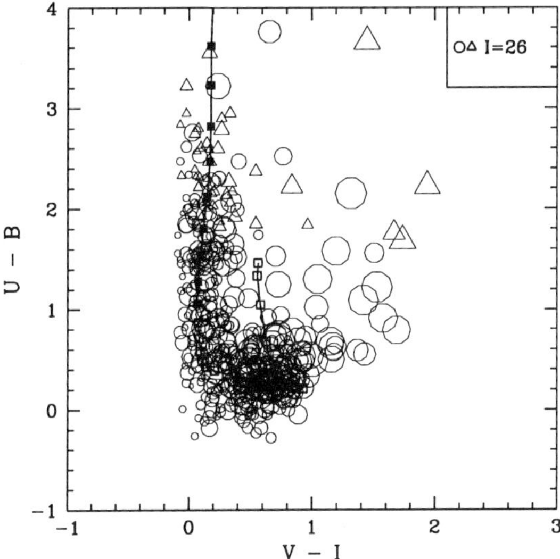

FIGURE 9. *Solid line:* model predictions for the color evolution of galaxies according to the star formation history of Figure 3. The points (*filled squares* for $z > 2$ and *empty squares* for $z < 2$) are plotted at redshift interval $\Delta z = 0.1$. *Empty circles* and *triangles:* colors of galaxies in the HDF with $22 < B < 27$. Objects undetected in U (with $S/N < 1$) are plotted as triangles at the 1σ lower limits to their $U - B$ colors. Symbols size scales with the I mag of the object, and all magnitudes are given in the AB system. The "plume" of Lyman-break galaxies is clearly seen in the data.

$z \sim 3$ is the subject of an ongoing debate (e.g., Meurer et al. 1997; Dickinson et al. 1997; Pettini et al. 1997). Adopting the greyer extinction law deduced by Calzetti et al. (1994) from the integrated spectra of nearby starbursts would require larger corrections to the SFR at high-z in order to match the observed colors. The consequence of this, however, would be the overproduction of red light at low redshifts, as noted in § 5.4. Redder spectra can also result from an aging population or an IMF which is, at early epochs, less rich in massive stars than the adopted ones.

5.6. *Constraints from the mid- and far-infrared background*

Ultimately, it should be possible to set some constraints on the total amount of star formation that is hidden by dust over the entire history of the universe by looking at the cosmic infrared background (CIB) (Fall et al. 1996; Burigana et al. 1997; Guiderdoni et al. 1997). Studies of the CIB provide information which is complementary to that given by optical observations. If most of the star formation activity takes place within dusty gas clouds, the starlight which is absorbed by various dust components will be reradiated thermally at longer wavelengths according to characteristic IR spectra. The energy in the CIB would then exceed by far the entire background optical light which is recorded in the galaxy counts.

From an analysis of the smoothness of the *COBE* DIRBE maps, Kashlinsky et al. (1996) have recently set an upper limit to the CIB of 10–15 nW m^{-2} sr^{-1} at $\lambda = 10$–$100\ \mu m$ assuming clustered sources which evolve according to typical scenarios. An analysis using data from *COBE* FIRAS by Puget et al. (1996) (see also Fixsen et al. 1996) has revealed an isotropic residual at a level of $3.4\ (\lambda/400\ \mu m)^{-3}$ nW m^{-2} sr^{-1} in the 400–1000 μm range, which could be the long-searched CIB. The detection, recently revisited by Guiderdoni et al. (1997), should be regarded as uncertain since it depends critically on the subtraction of foreground emission by interstellar dust.

By comparison, the total amount of starlight that is absorbed by dust and reprocessed in the infrared is 7.5 nW m^{-2} sr^{-1} in the model depicted in Figure 3, about 30% of the total radiated flux. The monolithic collapse scenario of Figure 8 generates 6.5 nW m^{-2} sr^{-1} instead. The resulting CIB spectrum is expected to be rather flat because of the spread in the dust temperatures—cool dust will likely dominate the long wavelength emission, warm small grains will radiate mostly at shorter wavelengths—and the distribution in redshift. While both models appear then to be consistent with the data (given the large uncertainties associated with the removal of foreground emission and with the observed and predicted spectral shape of the CIB), it is clear that too much infrared light would be generated by scenarios that have significantly larger amount of hidden star formation at early and late epochs.

5.7. *Chemical enrichment*

We may at this stage use our set of models to establish a cosmic timetable for the production of heavy elements (with atomic number $Z \geq 6$) in relatively bright field galaxies (see M96). What we are interested in here is the universal rate of ejection of newly synthesized material. In the approximation of instantaneous recycling, the metal ejection rate per unit comoving volume can be written as

$$\dot{\rho}_Z = y(1-R) \times \mathrm{SFR}, \tag{5.9}$$

where the *net*, IMF-averaged yield of returned metals is

$$y = \frac{\int m p_{\mathrm{zm}} \phi(m) dm}{(1-R) \int m\phi(m) dm}, \tag{5.10}$$

p_{zm} is the stellar yield, i.e., the mass fraction of a star of mass m that is converted to metals and ejected, and the dot denotes differentiation with respect to cosmic time.

The predicted end-products of stellar evolution, particularly from massive stars, are subject to significant uncertainties. These are mainly due to the effects of initial chemical composition, mass-loss history, the mechanisms of supernova explosions, and the critical mass, M_{BH}, above which stars collapse to black holes without ejecting heavy elements into space (Maeder 1992; Woosley and Weaver 1995). The IMF-averaged yield is also very sensitive to the choice of the IMF slope and lower-mass cutoff. For a Scalo IMF in the assumed mass range ($0.1 < M < 125\,\mathrm{M}_\odot$), for example, the net yield is typically a factor of 3.3 lower than Salpeter. At the same time, a lower cutoff of $0.5\,\mathrm{M}_\odot$ would boost the net yield by a factor of 1.9 for Salpeter and 1.7 for Scalo. Note that some of these ambiguities partially cancel out when computing the total metal ejection rate, as the product $y \times \mathrm{SFR}$ is less sensitive to the slope of the IMF than the yield or the rate of star formation, and is insensitive to the lower mass cutoff. Observationally, the best-fit "effective yield" (derived assuming a closed box model) is 0.025 $Z_\odot$ for Galactic halo clusters, 0.3 $Z_\odot$ for disk clusters, 0.4 $Z_\odot$ for the solar neighborhood, and 1.8 $Z_\odot$ for the Galactic bulge (Pagel 1987). The last value may represent the universal true yield, while the lower effective yields found in the other cases may be due, e.g., to the loss of enriched material in galactic winds.

Figure 10 shows the total mass of metals ever ejected, ρ_Z, versus redshift, i.e., the sum of the heavy elements stored in stars and in the gas phase as given by the integral of equation (5.9) over cosmic time. The values plotted have been computed from the star formation histories depicted in Figures 3 and 8, and have been normalized to $y\rho_s(0)$, the

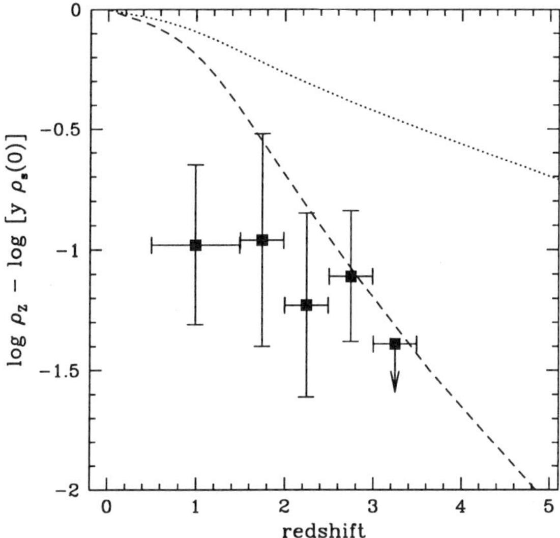

FIGURE 10. Total mass of heavy elements ever ejected versus redshift for the Salpeter IMF model of Figure 3 (*dashed line*) and the monolithic collapse model of Figure 8 (*dotted line*), normalized to $y\rho_s(0)$, the total mass density of metals at the present epoch. *Filled squares:* column density-weighted metallicities (in units of solar) as derived from observations of the damped Lyman-α systems (Pettini et al. 1997).

mass density of metals at the present epoch according to each model. A characteristic feature of the two competing scenarios is the rather different average metallicity expected at high redshift. For comparison, I have also plotted the *gas metallicity*, $Z_{\rm DLA}/Z_\odot$, as deduced from observations by Pettini et al. (1997) of the damped Lyman-α systems (DLAs). At early epochs, when the gas consumption into stars is still low, the metal mass density predicted from these models gives, in a closed box model, a measurement of the metallicity of the gas phase. If DLAs and star-forming field galaxies have the same level of heavy element enrichment, then one would expect a rough agreement between $Z_{\rm DLA}$ and the model predictions at $z \gtrsim 3$. This is not true at $z \lesssim 2$, when a significant fraction of heavy elements is locked into stars.† From Figure 10, it does appear that the monolithic collapse model overpredicts the cosmic metallicity at high redshifts as sampled by the DLAs. In order for such a model to be acceptable, the gas traced by the DLAs would have to be physically distinct from the luminous star formation regions observed in the Lyman-break galaxies, and to be substantially under-enriched in metals compared to the cosmic mean.

5.8. *The cluster-field analogy*

It has been recently pointed out by Renzini (1997) and Mushotzky & Loewenstein (1997) that, in the absence of any systematic cluster/field differences, clusters of galaxies may also provide an indication of the metal formation history of the universe. In the fiducial model of Figure 3, the global mean metallicity of the local universe is $y\Omega_s/\Omega_b \sim 0.1y/Z_\odot$ solar, to be compared with the overall cluster metal abundance, $\sim 1/3$ solar. If $y \sim Z_\odot$,

† More complex chemical evolution models which reproduce the evolving gas content and the metal enrichment history of the DLAs have been developed by Lanzetta, Wolfe, & Turnshek (1995) and Pei & Fall (1995).

the efficiency of metal production must have been larger in clusters than in the field, in spite of both having a similar baryon-to-star conversion efficiency, $\Omega_s/\Omega_b \sim 10\%$. Cluster-related processes, like ram-pressure stripping, would then be responsible for enriching the intracluster medium. Alternatively, a larger IMF-averaged metal yield, $y \sim 3\ Z_\odot$, may solve the apparent discrepancy. In this case, field galaxies would have to have ejected most of the heavy elements they produce, and there should be a comparable share of metals in the intergalactic medium (IGM) as there is in the intracluster gas. The characteristic metal ejection rate per unit comoving volume associated with such a pollution level would be large,

$$\dot{\rho}_{Z,\mathrm{IGM}} \approx (0.013\,\mathrm{M}_\odot\,\mathrm{yr}^{-1}\,\mathrm{Mpc}^{-3}) \left(\frac{\Omega_{\mathrm{IGM}}h_{50}^2}{0.05}\right) \left(\frac{3Z_{\mathrm{IGM}}}{Z_\odot}\right) \left(\frac{f_{\mathrm{inj}}}{0.7}\right)^{-1} \left(\frac{\Delta t}{2.5\,\mathrm{Gyr}}\right)^{-1}, \tag{5.11}$$

about ten times larger than derived at $z = 2$ from our modeling of the galaxy emission history (with $y \sim Z_\odot$). Here, Ω_{IGM} is the baryonic density parameter of the IGM phase and f_{inj} is the fraction of heavy elements injected into the IGM during a timescale Δt. It is hard to see, however, how massive galaxies with deep potential wells could be responsible for large outflows of metal-enriched gas. Recent observations of metal lines in the Lyα forest clouds at $z \sim 3$, while pointing towards some widespread chemical enrichment at early epochs, suggest typical metallicities of only 0.003 to 0.01 solar (Cowie et al. 1995). Similar low values have been inferred in local Lyα absorbers (Shull et al. 1997).

It is a pleasure to thank the hospitality of the European Southern Observatory, Garching, where this review was largely written. I have benefited from many useful discussions on various topics related to this talk with C. Baugh, G. Bruzual, S. Charlot, A. Connolly, M. Della Valle, M. Pettini, A. Renzini, M. Treyer, and my collaborators, L. Pozzetti and M. Dickinson. Support for this work was provided by NASA through grant AR-06337.10-94A from the Space Telescope Science Institute, which is operated by the Association of Universities for Research in Astronomy, Inc., under NASA contract NAS5-26555.

REFERENCES

Baugh, C. M., Cole, S., Frenk, C. S., & Lacey, C. G. 1997 *ApJ*, submitted.

Bower, R. G., Lucey, J. R., & Ellis, R. S. 1992 *MNRAS* **254**, 589.

Bruzual, A. G., & Charlot, S. 1993 *ApJ* **405**, 538.

Bruzual, A. G., & Charlot, S. 1997, in preparation.

Bruzual A. G., & Kron, R. G. 1980 *ApJ* **241**, 25.

Burigana, C., Danese, L., De Zotti, G., Franceschini, A., Mazzei, P., & Toffolatti, L. 1997 *MNRAS* **287**, L17.

Calzetti, D., Kinney, A. L., & Storchi-Bergmann, T. 1994 *ApJ* **429**, 582.

Cappellaro, E., Turatto, M., Tsvetkov, D. Yu., Bartunov, O. S., Pollas, C., Evans, R., & Hamuy, M. 1997 *A&A* **322**, 431.

Charlot, S. 1996a in *From Stars to Galaxies* (ed. C. Leitherer & U. Fritze-von Alvensleben). ASP Conference Series.

Charlot, S. 1996b in *The Universe at High-z, Large Scale Structure, and the Cosmic Microwave Background* (ed. E. Martinez-Gonzalez & J. L. Sanz). p. 53. Springer.

Cole, S., Aragón-Salamanca, A., Frenk, C. S., Navarro, J. F., & Zepf, S. E. 1994 *MNRAS* **271**, 781.

Connolly, A. J., Szalay, A. S., Dickinson, M. E., SubbaRao, M. U., & Brunner, R. J. 1997 *ApJ*, in press.

Cowie, L. L., Songaila, A., Hu, E. M., & Cohen, J. G. 1996 *AJ* **112**, 839.

Cowie, L. L., Songaila, A., Kim, T.-S., & Hu, E. M. 1995 *AJ* **109**, 1522.

Della Valle, M., & Madau, P. 1997, in preparation.

Dickinson, M. E. 1997, private communication.

Dickinson, M. E., et al. 1997, in preparation.

Eggen, O. J., Lynden-Bell, D., & Sandage, A. R. 1962 *ApJ* **136**, 748.

Ellis, R. S., Colless, M., Broadhurst, T., Heyl, J., & Glazebrook, K. 1996 *MNRAS* **280**, 235.

Fall, S. M., Charlot, S., & Pei, Y. C. 1996 *ApJ* **464**, L43.

Ferguson, H. C., & Babul, A. 1997 *MNRAS*, in press.

Fixsen, D. J., Cheng, E. S., Gales, J. M., Mather, J. C., Shafer, R. A., Wright, E. L. 1996 *ApJ* **473**, 576.

Gallego, J., Zamorano, J., Arag'on-Salamanca, A., & Rego, M. 1995 *ApJ* **455**, L1.

Gardner, J. P., Sharples, R. M., Frenk, C. S., & Carrasco, B. E. 1997 *ApJ* **480**, L99.

Guiderdoni, B., Bouchet, F. R., Puget, J.-L., Lagache, G., & Hivon, E. 1997 *Nature*, in press.

Heyl, J., Colless, M., Ellis, R. S., & Broadhurst, T. 1997 *MNRAS* **285**, 613.

Kashlinsky, A., Mather, J. C., & Odenwald, S. 1996 *ApJ* **473**, L9.

Kauffmann, G., White, S. D. M., & Guiderdoni, B. 1993 *MNRAS* **264**, 201.

Kennicutt, R. C. 1983 *ApJ* 272, 54.

Kim, A. G., et al. 1997 *ApJ* **476**, L63.

Koo, D. C. 1985 *AJ* **90**, 418.

Lanzetta, K. M., Wolfe, A. M., & Turnshek, D. A. *ApJ* **440**, 435.

Lilly, S. J., Le Févre, O., Hammer, F., & Crampton, D. 1996 *ApJ* **460**, L1.

Lilly, S. J., Tresse, L., Hammer, F., Crampton, D., & Le Févre, O. 1995 *ApJ* **455**, 108.

Madau, P. 1995 *ApJ* **441**, 18.

Madau, P. 1997, in *Star Formation Near and Far* (ed. S. S. Holt & G. L. Mundy). p. 481. AIP Press.

Madau, P., Ferguson, H. C., Dickinson, M. E., Giavalisco, M., Steidel, C. C., & Fruchter, A. 1996 *MNRAS* **283**, 1388 (M96).

Madau, P., Pozzetti, L., & Dickinson, M. E. 1997 *ApJ*, in press.

Maeder, A. 1992 *A&A* **264**, 105.

Meurer, G. R., Heckman, T. M., Lehnert, M. D., Leitherer, C., & Lowenthal, J. 1997 *AJ* **114**, 54.

Mushotzky, R. F., & Loewenstein, M. 1997 *ApJ* **481**, L63.

Ortolani, S., Renzini, A., Gilmozzi, R., Marconi, G., Barbuy, B., Bica, E., & Rich, M. R. 1995 *Nature* **377**, 701.

Pagel, B. E. J. 1987 in *The Galaxy* (ed. G. Gilmore & B. Carswell). p. 341. Reidel.

Pei, Y. C. 1992 *ApJ* **395**, 130.

Pei, Y. C., & Fall, S. M. 1995 *ApJ* **454**, 69.

Persic, M., & Salucci, P. 1992 *MNRAS* **258**, 14P.

Pettini, M., Smith, L. J., King, D. L., & Hunstead, R. W. 1997 *ApJ*, in press.

Pettini, M., Steidel, C. C., Dickinson, M., Kellogg, M., Giavalisco, M., & Adelberger, K. L. 1997 in *The Ultraviolet Universe at Low and High Redshift* (ed. W. Waller). AIP Press.

Pozzetti, L., Bruzual, G. A., & Zamorani, G. 1996 *MNRAS* **281**, 953.

Pozzetti, L., Madau, P., Zamorani, G., Ferguson, H. C., & Bruzual, G. A. 1997 *MNRAS*, submitted.

Puget, J.-L., Abergel, A., Bernard, J.-P., Boulanger, F., Burton, W. B., Desert,

F.-X., & Hartmann, D. 1996 *A&A* **308**, L5.

Renzini, A. 1997 *ApJ*, in press.

Rowan-Robinson, M., et al. 1997 *MNRAS* **289**, 490.

Ruiz-Lapuente, P., Canal, R., & Burkert, A. 1997 in *Thermonuclear Supernovae* (ed. P. Ruiz-Lapuente, R. Canal, & J. Isern). Kluwer.

Salpeter, E. E. 1955 *ApJ* **121**, 161.

Saunders, W., Rowan-Robinson, M., Lawrence, A., Efstathiou, G., Kaiser, N., Ellis, R. S., & Frenk, C. S. 1990 *MNRAS* **242**, 318.

Scalo, J. N. 1986 *Fundam. Cosmic Phys.* **11**, 1.

Schechter, P. L., & Dressler, A. 1987 *AJ* **94**, 56.

Shull, J. M., Penton, S., Stocke, J. T., Giroux, M. L., van Gorkom, J. H., & Carilli, C. 1997 in preparation.

Soifer, B. T., & Neugebauer, G. 1991 *AJ* **101**, 354.

Tinsley, B. M. 1980 *Fundam. Cosmic Phys.* **5**, 287.

Tinsley, B. M., & Gunn, J. E. 1976 *ApJ* **203**, 52.

Tresse, L., & Maddox, S. J. 1997 preprint.

Treyer, M. A., Ellis, R. S., Milliard, B., & Donas, J. 1997 in *The Ultraviolet Universe at Low and High Redshift* (ed. W. Waller). AIP Press).

van den Bergh, S., & Tammann, G. A. 1991 *ARA&A* **29**, 363.

Walker, T. P., Steigman, G., Schramm, D. N., Olive, K. A., & Kang, H. 1991 *ApJ* **376**, 51.

White, S. D. M., & Frenk, C. S. 1991 *ApJ* **379**, 25.

Williams, R. E., et al. 1996 *AJ* **112**, 1335.

Woosley, S. E., & Weaver, T. A. 1995 *ApJS* **101**, 181.

Yungelson, L., & Livio, M. 1997 *ApJ*, submitted.

Color-selected high redshift galaxies and the HDF

By MARK DICKINSON

The Johns Hopkins University and Space Telescope Science Institute

The quality, depth, and multi-color nature of the Hubble Deep Field images makes them an excellent resource for studying galaxies at $z > 2$ using selection techniques based on the presence of the 912 Å Lyman break. I present a descriptive review of this method and of the properties of the objects which it identifies, and summarize spectroscopic progress on galaxies with $2 < z < 4$ in the HDF. Using ground-based and HDF samples of Lyman break galaxies I discuss the luminosity function of galaxies at $z \approx 3$, and consider the effects of extinction on the star formation rates that are derived from the UV luminosity information. Infrared observations of the HDF provide data on the rest-frame optical properties of $z \approx 3$ galaxies, which are briefly described.

---

1. Introduction

Although the study of galaxies at high redshift neither begins nor ends with the Hubble Deep Field (HDF), this conference has demonstrated the ways in which the HDF has served to focus the attention of the community on the properties of galaxies at $z > 2$. In part this is because the HDF imaging data was obtained through several filters, permitting the use of color selection techniques to isolate and study populations of galaxies at various redshifts. Because the HDF images are so deep, colors can be measured with unusually high precision (and in a spatially resolved fashion *within* individual galaxies) for objects which ordinarily would be considered very faint for ground-based telescopes. Additionally, the HDF images easily detect galaxies at magnitudes well beyond the spectroscopic limits of even the largest telescopes. The desire to understand their nature requires *some* idea of their distances, and it is therefore tempting to look for photometric means of estimating redshifts without the benefit of ordinary spectroscopy.

For these reasons, many presentations at this symposium have considered various applications of "photometric redshift" estimation in the HDF and in other data sets. One such method takes advantage of the ubiquitous 912 Å Lyman limit discontinuity, which is redshifted into the HDF bandpasses at $z \gtrsim 2$. In recent years, color selection based on the Lyman limit has developed into a highly successful means of detecting galaxies at large redshifts, as I will review below. In designing the HDF observations, our working group at ST ScI incorporated F300W imaging into the four-filter scheme for two reasons, one scientific and one purely practical. First, such data offers the potential for Lyman break selection of high redshift galaxies. Second, we wished to take advantage of the "bright" portions of the orbit during Continuous View Zone visibility, when scattered earthlight severely impacts WFPC2 imaging at redder wavelengths. The reduced amplitude of the scattered background at UV wavelengths, and the fact that F300W imaging with WFPC2 is not normally background limited anyway, made bright-time observing through that bandpass a suitable use for these otherwise disadvantageous observing intervals.

I begin with a descriptive review of the Lyman break technique, illustrated using data from the HDF. Although the HDF Lyman break galaxy sample is much smaller than that which has been identified from ground based imaging studies (which cover much larger solid angles), the excellent photometric precision of the HDF data and the large percentage of Lyman break galaxies present at the fainter magnitude limits which it probes makes it quite useful for illustrating the principles of the method. In §2.1 I summarize the current status of spectroscopy on HDF Lyman break galaxies. In §3 I discuss some of the statistical properties of Lyman break galaxies, concentrating on their numbers, luminosities, and colors, and inferences that can be derived from these measurements. Finally, §4 discusses the rest-frame optical properties of $z > 2$ galaxies in the HDF using deep infrared imaging data. In his contribution to this volume, Mauro Giavalisco discusses our ground-based Lyman break galaxy sample in greater detail, and addresses the spectroscopic, morphological, and clustering properties of Lyman break galaxies, which I will largely neglect here.

In this paper, HDF galaxy selection and object names are based on the catalog of Williams et al. (1996). Optical photometry is reported on the AB magnitude system, with the WFPC2 bandpasses indicated by U_{300}, B_{450}, V_{606} and I_{814}. In general I have used new, revised photometry of the Williams et al. catalog objects based on optimized apertures which maximize signal-to-noise for color measurement. This has the advantage of allowing robust Lyman break galaxy selection to somewhat fainter magnitudes than was previously possible.

2. Lyman break selected galaxies in the HDF

Figure 1 illustrates the principle of the Lyman break color selection technique using a galaxy from the HDF as an example. In the absence of dust extinction, an actively star-forming galaxy should have a blue continuum at rest-frame ultraviolet wavelengths, nearly flat in f_ν units. Blueward of the 912 Å Lyman limit, however, photoelectric absorption by intervening sources of neutral Hydrogen will sharply truncate the spectrum. This hydrogen may be located in the photospheres of the UV-emitting stars themselves, in the interstellar medium of the distant galaxy, or along the intergalactic sightline between us and the object. When observed at some large redshift, the rest-frame Lyman limit of a galaxy shifts between some pair of bandpasses (e.g., the WFPC2 U_{300} and B_{450} filters in figure 1), and the galaxy "drops out" when viewed through the bluer filter because of the suppression of its flux. In addition to the Lyman continuum absorption, the cumulative effect of the Lyman α forest lines introduces an additional spectral break shortward of Lyman α at the emission redshift of the galaxy. This flux suppression is increasingly strong at higher redshifts as the forest thickens, and introduces its own color effects, particularly for galaxies at $z > 3$.

Color selection based on the effects of the Lyman limit and Lyman α forest has been used for many years in surveys for distant QSOs (e.g., Warren et al. 1987). The method was applied to the study of distant galaxies by Guhathakurta et al. (1990) and Songaila, Cowie & Lilly (1990), who used it set limits on the number of star-forming galaxies at $z \approx 3$ in faint galaxy samples. Steidel & Hamilton (1992) and Steidel, Hamilton & Pettini (1995) reported the detection of significant numbers of high redshift galaxy candidates using this method. Spectroscopic confirmation of their redshifts was first presented by Steidel et al. (1996a), and WFPC2 images of select examples were published by Giavalisco et al. (1996b). To date, the majority of Lyman break selected galaxies come from the $U_n G\mathcal{R}$ survey of Steidel et al., which has identified more than a thousand candidates and has spectroscopically confirmed (as of this writing) more than 400 galaxies at $z \approx 3$. This survey is discussed in greater detail by Giavalisco in this volume, and in the present paper I will often refer to it as the "ground-based sample" to distinguish it from HDF-selected objects.

Although Lyman break galaxies in the HDF are fewer in number than those in the ground-based sample, the quality and depth of the HDF imaging data offer a number of

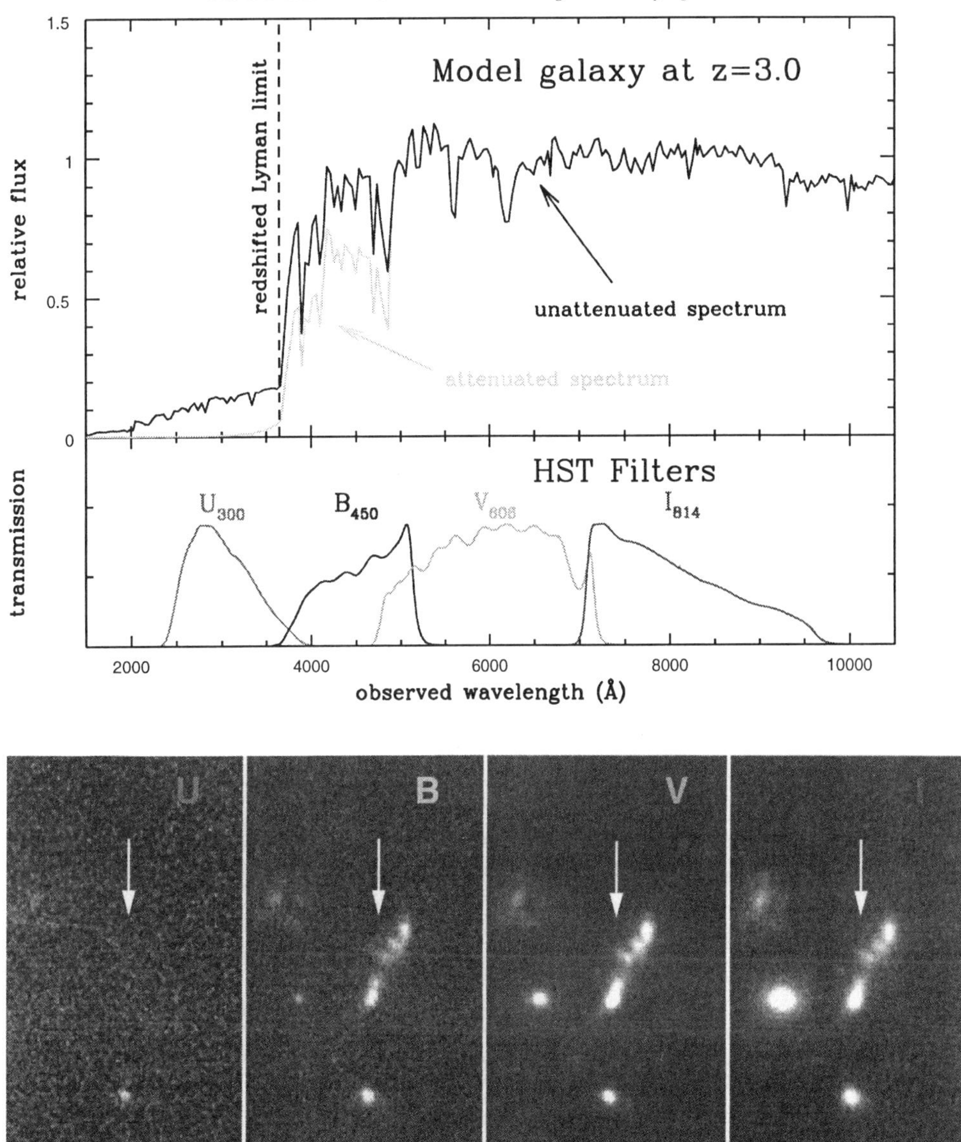

FIGURE 1. Illustration of the Lyman break technique as applied to the Hubble Deep Field. The upper panel shows a model spectrum of a star forming galaxy observed at $z = 3$. Its flat UV continuum is truncated by the 912 Å Lyman limit, which is redshifted between the U_{300} and B_{450} filters (WFPC2 bandpasses shown below spectrum). In addition to photospheric absorption in the UV-emitting stars, the effects of intergalactic neutral hydrogen further suppress the continuum in the U_{300} and B_{450} bands. At bottom, an HDF galaxy is shown in the four WFPC2 bandpasses. Clearly visible at I_{814}, V_{606} and B_{450}, it vanishes in the U_{300} image. This galaxy has been spectroscopically confirmed to have $z = 2.8$.

advantages. The HDF can be used to detect Lyman break galaxies at fainter apparent magnitudes than has been achieved in ground-based data, and the precision of the B_{450}, V_{606} and I_{814} photometry ensures small random errors on color measurements. Moreover, the depth and resolution of the WFPC2 imaging permits detailed morphological study of these objects. The primary *disadvantage* of the HDF is its small field of view, and

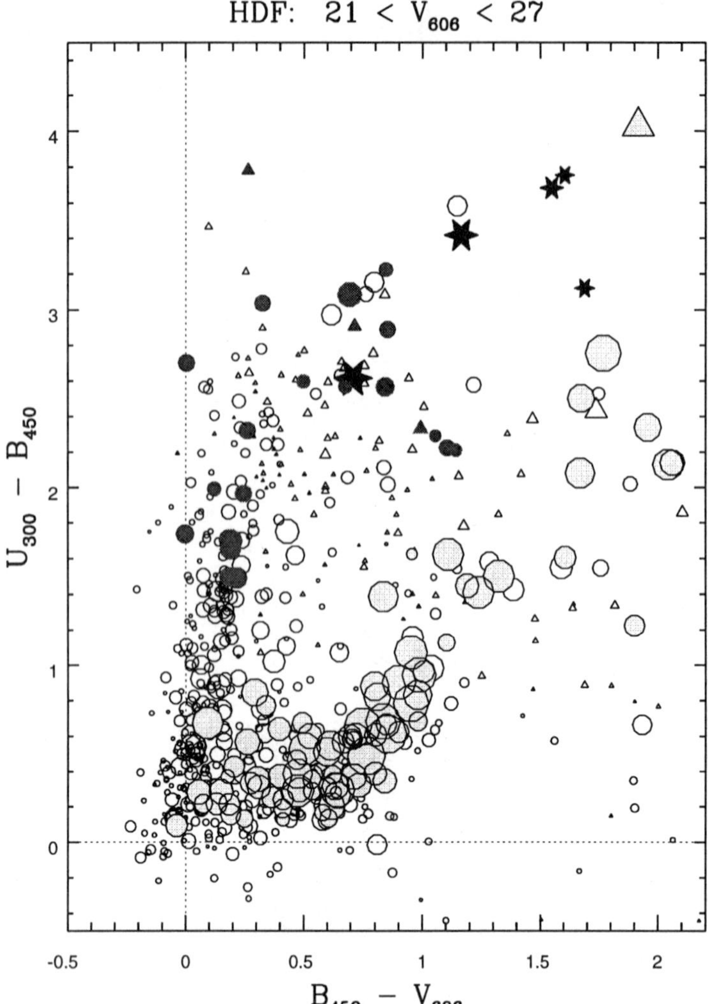

FIGURE 2. Color-color diagram of faint galaxies in the Hubble Deep Field, illustrating the "plume" of Lyman break objects rising from $U_{300} - B_{450} = B_{450} - V_{606} = 0$. These are nearly all galaxies at $z > 2$. Spectroscopically confirmed objects in this redshift range are shown as darker filled symbols; galaxies with measured redshifts $z < 2$ are shown as light filled circles, and stars are indicated by star-shaped points. Triangles mark lower limits (1σ to the $U_{300} - B_{450}$ color for objects undetected in U_{300}. Symbol size scales inversely with apparent V_{606} magnitude.

hence the rather small comoving volume which it samples. This limits its utility for statistical studies (e.g., of luminosity functions, redshift distributions, etc.), as small number statistics, galaxy clustering, and field-to-field variations may introduce significant uncertainties.

2.1. *Color selection of Lyman break galaxies in the HDF*

My own favorite "view" of the Hubble Deep Field is that shown in figure 2, a color-color diagram of galaxies in the HDF. One of the most prominent features of this diagram is the dramatic "plume" of galaxies rising nearly vertically from the zero color point (i.e., flat spectrum galaxies) up toward very red $U_{300}-B_{450}$ colors. These are the high redshift, star forming galaxies—objects whose 912 Å Lyman discontinuities are entering into and passing through the F300W bandpass, shifting them into a portion of color-color space which is unpopulated by low redshift objects.

Figure 3 illustrates the redshift dependent effect of the Lyman limit and Lyman α spectral "breaks" on the colors of galaxies with spectroscopically confirmed redshifts $z > 2$. To first order, the galaxies in the high redshift "plume" form a redshift sequence ordered by $U_{300}-B_{450}$ color. In practice, variation in individual galaxy spectra shuffle this sequence somewhat. Dynamic range for the U_{300} photometry also affects this ordering, as it is only possible to set limits on the $U_{300} - B_{450}$ colors of fainter objects, and the numerical values of these limits therefore depend on the apparent magnitude of the galaxies. The lower panel of figure 3 shows the increasing effect of the Lyman α forest (and, at $z \gtrsim 3.5$, of the Lyman limit) on the B_{450}–V_{606} color. This effect was used by Lowenthal et al. (1997) as an additional criterion for identifying high redshift galaxy candidates (cf. also Fruchter, this volume). Using color selection criteria such as those of Steidel et al. (1996a,b; see also below), the application of a color cut such as $B_{450} - V_{606} \leq 1.2$ results in an upper redshift bound $z \lesssim 3.5$. It is this reddening of $B_{450} - V_{606}$ which is responsible for the "tilt" of the plume in the UBV color-color diagram shown in figure 2—the higher redshift U_{300} dropout galaxies "fan out" toward redder B_{450}- V_{606} colors. Figure 4 shows a different 2-color diagram of HDF galaxies, this time using V_{606} - I_{814} colors which are relatively unaffected by Lyα absorption at $z < 4$. Here, the "plume" remains more or less vertical.

The Lyman break color technique is a simple form of photometric redshift selection. Here we do not attempt to accurately estimate the redshifts of individual objects, but simply to use color criteria to select galaxies in a particular redshift interval and exclude foreground (and background) objects. The redshift selection function of the method depends on the particular color criteria adopted, on the intrinsic dispersion in the ultraviolet spectral properties of star forming, high redshift galaxies, on cosmic variance in the intergalactic transmission along different lines of sight, and on the distribution of photometric measurement errors. This redshift selection function can be estimated using spectral models along with realistic simulations of photometric errors, or can be measured directly by obtaining enough spectroscopic redshifts to define it empirically. We now know that the Lyman break galaxy population exhibits strong clustering in redshift space (Steidel et al. 1998; cf. also Giavalisco, this volume, and figure 7 below). Therefore in order to determine the redshift selection function empirically one must study many independent sightlines in order to average over the effects of large scale structure. We have done this for our ground-based survey, and thus feel that we understand the redshift selection function quite well. For the HDF, we have only the one "realization" of the redshift distribution, and thus the empirical selection function cannot properly be determined—here we must rely to some degree on models, although these models can be informed by the information on galaxy spectral properties which has been gained from the ground-based samples.

The F300W filter in WFPC2 is substantially bluer than U bandpasses used at terrestrial observatories. As such, U_{300} photometry is sensitive to the passage of the Lyman break at significantly lower redshifts than is the case for, e.g., the $U_nG\mathcal{R}$ system used by Steidel et al. For example, roughly 90% of galaxies in our ground-based $U_nG\mathcal{R}$ survey lie in the redshift range $2.6 < z < 3.4$. By $z = 2$, however, the Lyman limit is already well into the WFPC2 F300W bandpass and produces a recognizably large color signature. Followup spectroscopy has already identified U_{300}-dropout galaxies down to $z = 2.01$, and others which did not successfully yield redshifts quite likely lie at redshifts slightly below 2, where the absence of strong spectral features in the wavelength range presently

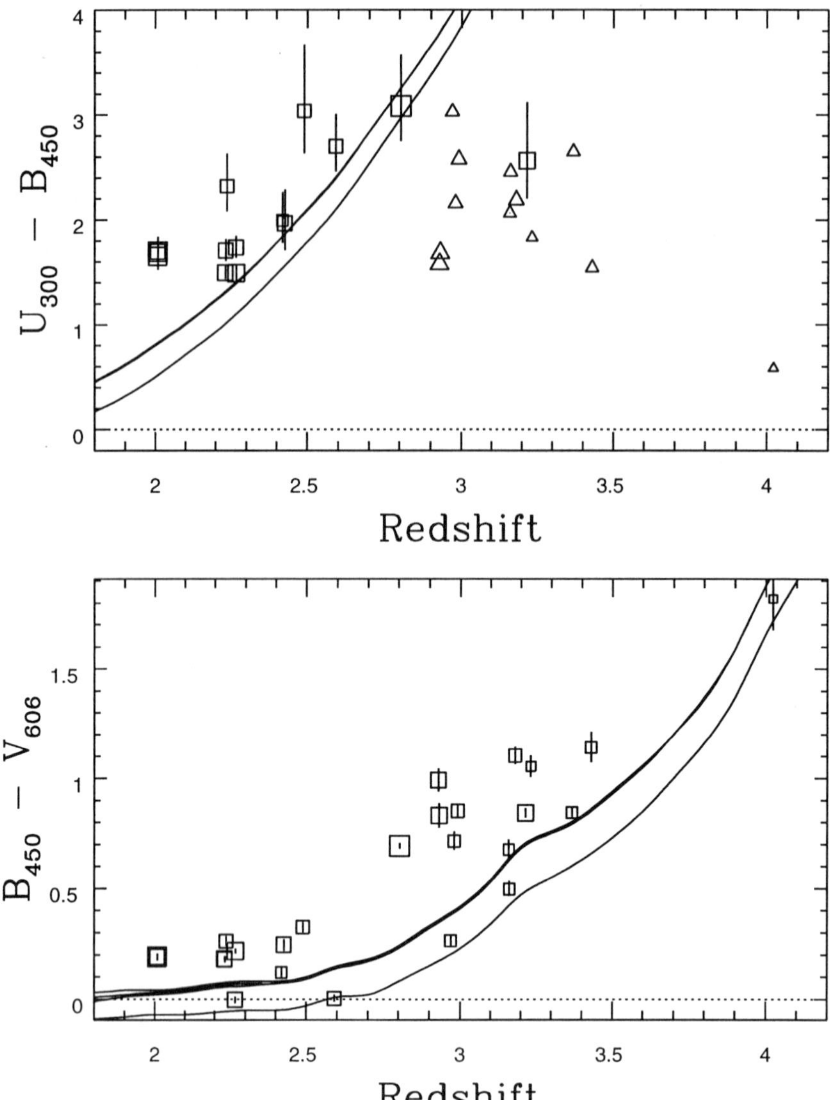

FIGURE 3. U_{300}–B_{450} and B_{450}–V_{606} versus redshift for spectroscopically confirmed $z > 2$ galaxies in the HDF. These plots illustrate the effect of Lyman limit and Lyman α absorption on the colors of high redshift galaxies. The solid lines show predicted colors of actively star-forming galaxies of various sorts using the Madau (1995) prescription for mean intergalactic transmission. In the top panel, triangles mark lower color limits for galaxies with $S/N < 2$ in the U_{300} band. All galaxies with $z < 2.9$ are detected in U_{300}, while all but one galaxy with $z > 2.9$ have $S/N < 2$. (The exception, 4-858.0, is formally detected with $S/N \approx 2.5$; this may partially result from the red leak in the U_{300} filter or from systematic measurement error, but could also indicate photometric contamination from a foreground object.) The lower panel shows the progressive reddening of the B_{450}–V_{606} color with redshift, primarily due to the effect of the Lyman α forest. (Compare with figure 10 for galaxies from the ground-based sample.)

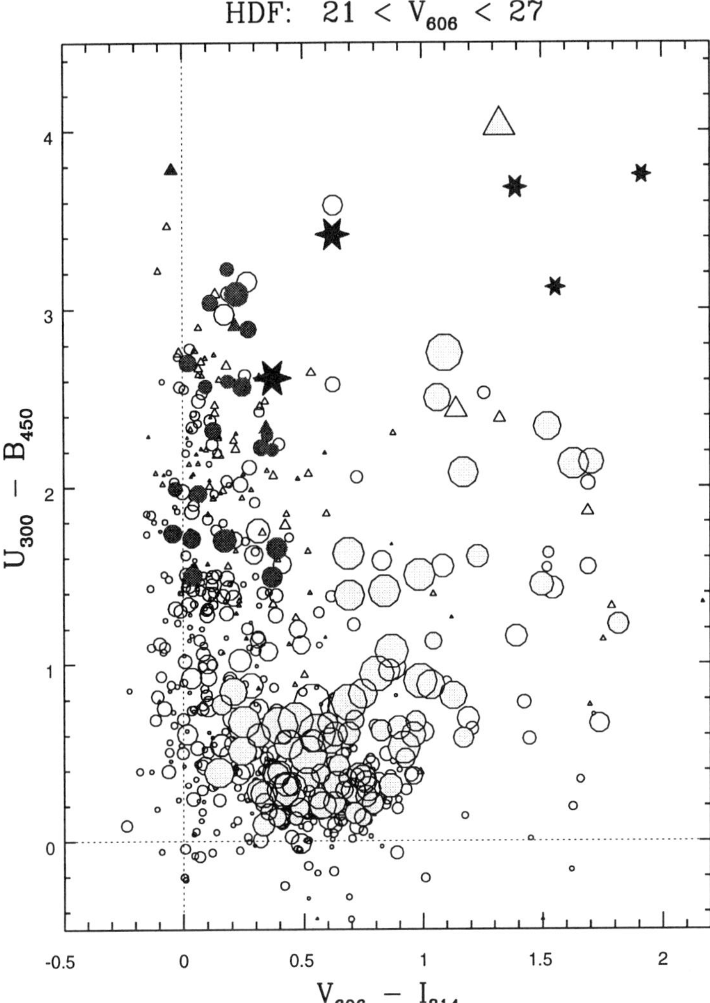

FIGURE 4. Another multi-color view of the HDF. Here, the horizontal axis plots the $V_{606} - I_{814}$ color, which is relatively insensitive to intergalactic Lyman α absorption for redshifts $z < 4$. In this particular multi-color space the "plume" of high redshift galaxies stands nearly vertical, without the tilt seen in figure 2. Symbols are the same as in figure 2.

accessible to LRIS at Keck makes secure redshift measurement difficult. The commissioning of a new blue channel on LRIS will soon make it possible to measure redshifts for many more U_{300} dropouts in the HDF.

The sensitivity of the U_{300} filter to Lyman break galaxies at lower redshifts has benefits and disadvantages. It opens a larger volume of redshift space to Lyman break selection, thus increasing the number of objects which may be identified this way. At the same time, as already noted, it makes many of them more difficult to confirm with optical spectra. However, the ease of selecting galaxies at $2 < z < 2.5$ has one spectroscopic advantage: this is the range of "magic redshifts" for near-infrared spectroscopy, where the [OII], [OIII], Hβ and Hα lines are all accessible in the J, H and K bands. This may make the HDF Lyman break galaxies particularly useful for studying nebular emission line diagnostics; already Elston et al. have observed Hα from several HDF Lyman break galaxies at $z \approx 2$ (see below).

Various groups have created photometrically defined samples of high redshift galaxy candidates in the HDF. The color selection criteria differ, and no "definitive" method has yet been established. The number of Lyman break candidates therefore varies from sample to sample depending on the criteria which are adopted. For example, Madau et al. (1996) defined conservative selection criteria based on models of galaxy color distributions in order to select $z > 2$ galaxies while avoiding significant risk of contamination from objects at lower redshift. These criteria, however, miss some of the galaxies which are now spectroscopically confirmed to have $z > 2$. Using the wealth of redshift information now available in the HDF we can revisit this question and refine the selection criteria. As it happens, applying the same criteria used by Steidel et al. for selecting Lyman break galaxies in the ground-based $U_n G\mathcal{R}$ color system does an excellent job of isolating $z > 2$ galaxies in the HDF. Specifically, we may use the so-called "marginal" criteria of Steidel et al., namely

$$U_{300} - B_{450} \geq B_{450} - V_{606} + 1.0, \qquad B_{450} - V_{606} \leq 1.2.$$

This successfully recovers all 24 spectroscopically confirmed Lyman break galaxies in the HDF at $2 < z < 3.5$. As is the case for the ground-based sample, the only substantial contaminants are galactic stars, some of which also satisfy these criteria, but these are easily recognized in the WFPC2 images and excluded. (All obvious stars with these colors have also already been observed spectroscopically, as it turns out.) The only known low-redshift galaxy which satisfies these criteria is, of all things, the brightest galaxy in the HDF: an elliptical in WF2 at $z = 0.089$. This is not surprising, as ellipticals have colors similar to those of K-stars, and are thus "interlopers" for the same reason until their redshifts become large enough ($z \approx 0.1$) for k-correction effects to move them out of the color selection region. There should be very few $z < 0.1$ early-type galaxies in the small volume of the HDF which populate this part of color–color space.†

Excluding stars and the few obvious $z < 0.1$ interlopers, there are approximately 187 galaxies in the HDF with $V_{606} < 27.0$ which satisfy these color criteria and thus probably fall in the redshift range $2 \lesssim z \lesssim 3.5$. The exact number depends somewhat on the choice of what constitutes a single galaxy with multiple clumps versus separate objects with small angular separations. E.g. should the "quad galaxy," 4-858.0 at $z = 3.22$, be considered a single galaxy with four pieces or four separate objects? Such distinctions may be merely semantic or may be very important depending on the question being asked (cf. Colley et al. (1996)).

2.2. *HDF high–z roundup*

To date, 26 galaxies in the HDF have spectroscopic redshifts $z > 2$. 23 of these were pre-selected as Lyman break objects for spectroscopic observation. One galaxy, 4-445.0, was observed by Cohen et al. (1996) as part of a magnitude limited spectroscopic sample, but qualifies photometrically as a Lyman break galaxy nevertheless. One other was was observed serendipitously (2-585.1—see below), but also satisfies the Lyman break color criteria defined above. Finally, one galaxy (3-577.0) was observed as a gravitational lens candidate; it is too faint for a robust U_{300} Lyman break measurement and so does not qualify (see below). Table 1 lists these objects, including several galaxies from our own observations which are previously unpublished. There are many more Lyman break candidates, primarily U_{300} dropouts, which have spectroscopically accessible

† For those particularly worried about foreground contamination, an additional color cut, $V_{606} - I_{814} \leq 0.5$, should successfully exclude most interlopers at $z < 0.1$.

ID	RA	Dec	z	V_{606}	Reference
2-449.0	12:36:48.332	62:14:16.67	2.008	23.68	S
2-585.1	12:36:49.808	62:14:15.18	2.008:	23.85	E
3-118.1	12:36:54.724	62:13:14.74	2.232	24.41	*
2-903.0	12:36:55.071	62:13:47.05	2.233	24.60	L
2-525.0	12:36:50.120	62:14:01.04	2.237	24.80	*
2-82.1	12:36:44.077	62:14:09.98	2.267	24.54	L
4-445.0	12:36:44.637	62:12:27.34	2.268	24.08	C
2-824.0	12:36:54.617	62:13:41.28	2.419	25.23	L
2-239.0	12:36:45.883	62:14:12.10	2.427	24.54	*
2-591.2	12:36:53.175	62:13:22.76	2.489	24.91	*
4-639.1	12:36:41.715	62:12:38.84	2.591	24.74	S
4-555.1	12:36:45.344	62:11:52.67	2.803	23.41	S
1-54.0	12:36:44.094	62:13:10.75	2.929	24.44	*
4-52.0	12:36:47.687	62:12:55.98	2.931	24.37	L
4-289.0	12:36:46.944	62:12:26.09	2.969	25.17	*
4-363.0	12:36:48.297	62:11:45.88	2.980	25.05	L
2-643.0	12:36:53.427	62:13:29.38	2.991	24.87	L
2-76.11	12:36:45.357	62:13:46.98	3.160	25.32	L
2-565.0	12:36:51.186	62:13:48.79	3.162	25.17	*
2-901.0	12:36:53.607	62:14:10.25	3.181	24.87	L
4-858.0	12:36:41.233	62:12:03.00	3.216	24.28	S,L,Z
3-243.0	12:36:49.817	62:12:48.88	3.233	25.60	L
3-577.0	12:36:52.247	62:12:27.18	3.360	27.25	Z
2-637.0	12:36:52.747	62:13:39.08	3.368	25.27	L
2-604.0	12:36:52.407	62:13:37.68	3.430	25.25	L
3-512.0	12:36:56.117	62:12:44.69	4.022	25.85	*

References:
S: Steidel et al. 1996b; L: Lowenthal et al. 1997; C: Cohen et al. 1996;
Z: Zepf et al. 1997; E: Elston et al. private communication;
*: this paper (Steidel et al. observations, unpublished)

TABLE 1. HDF Galaxies with spectroscopic redshifts $z > 2$

magnitudes, and with considerable effort this list could eventually more than double in length. Few of the B_{450} dropout objects ($z \sim 4$ candidates) are bright enough to tempt the spectroscopist, but if they have strong emission lines a few may still yield redshifts.

Here are a few notes on individual Lyman break galaxies, including corrigenda to some errors in our first paper on HDF spectroscopy (Steidel et al. 1996b).

2-449.0: This galaxy was erroneously identified as having $z = 2.845$ in Steidel et al. 1996b (object C2-05 in that paper). The broad band spectral energy distribution of the galaxy, both in the UV and near-IR, appears to be inconsistent with that redshift (e.g., the Lyman break amplitude is too small for the galaxy to be at $z = 2.8$). We reanalyzed the spectrum and found that $z = 2.008$ is more likely to be the correct redshift. The galaxy was then observed by Elston et al. in the infrared to search for Hα emission, both with narrow band imaging at the IRTF and with the Cryogenic Spectrograph (CRSP) at KPNO. Both observations detected strong line emission, confirming the new redshift.

2-585.1: This galaxy, a U_{300} Lyman break galaxy with a particularly dramatic morphology (see figure 15 below), lies several arcseconds away from 2-449.1 (see above). In their CRSP observations of 2-449.1, Elston et al. also detected line emission from 2-585.1 which fortuitously fell on their spectrograph slit. The line was confirmed with a subsequent observation targeting 2-585.1 itself. Presuming the line to be Hα, the redshift is approximately the same as that of 2-449.1. Both galaxies have very similar $U_{300} - B_{450}$ colors, again suggesting similar redshifts.

3-550.0: This galaxy, object C3-02 from Steidel et al. 1996b, is not included in the table above. As with 2-449.0 (above), the colors are in most respects not consistent with the published redshift ($z = 2.775$); the relatively small U_{300}–B_{450} color and near-IR SED both suggest a lower redshift. The original spectrum has very poor signal-to-noise, and we now feel that the published redshift is probably incorrect. Reobservations of this galaxy have thus far been inconclusive. We suspect that the galaxy has $z \lesssim 2$, making spectroscopic redshift confirmation difficult with LRIS at the present time.

3-577.0: This is the faintest object (with $V_{606} > 27$!) in the HDF with a reported redshift. Hogg et al. (1996) proposed that this is part of a gravitational lens system consisting of a red foreground elliptical, a large blue arc, and 3-577.0 as a suggested counterimage of the arc. Zepf et al. (1997) observed the system and detected a weak emission line from 3-577.0 which they interpret as Lyman α. Their spectra did not yield redshifts for the other components of the system. Although the emission line is very weak and the proposed redshift should perhaps be regarded as tentative, it is plausibly correct. 3-577.0 is too faint to be robustly measured as a "dropout," as the U_{300} photometry is not deep enough to set a strong enough constraint on the $U_{300} - B_{450}$ color. However, the colors are not *inconsistent* with it being at $z \sim 3$. Moreover, photometrically the "arc" (3-593.0) does qualify as a Lyman break galaxy, and probably has $z > 2$. Thus it remains plausible that the two objects are one and the same, gravitationally lensed. Arguing against this is the absence of any emission line in Zepf et al.'s spectrum of the arc, and the fact that the two objects have apparently different $B_{450} - V_{606}$ colors (although they are virtually identical at $V_{606} - I_{814}$). To explain this would require that the long slit observation, which crossed the arc perpendicularly, must simply have missed the line emitting region of the magnified galaxy. The $B_{450} - V_{606}$ colors, if the galaxies are at $z = 3.36$, could conceivably be affected by different amounts of foreground Lyman α forest absorption, while the $V_{606} - I_{814}$ colors are free of this effect. Further observations of this system are warranted.

3-512.0: Of the candidate $z \approx 4$ "B_{450}-dropouts" tabulated in Madau et al. 1996, only three objects have $I_{814} < 26$. 3-512.0 is the brightest of these at V_{606}. The galaxy is in fact detected in the B_{450} image, and does not drop out completely. If this is indeed a high redshift object, then the Lyman limit has only partially passed through the filter bandpass, suggesting $z \approx 4$. We observed this galaxy with LRIS at the Keck Observatory in May 1996 and again in March 1997. Both observations detected a single emission line at λ6105 Å. If the emission line is identified with Lyman α, this would imply a redshift $z = 4.02$. No other significant lines were detected. In particular, if the observed line were [OII] λ3727 Å or [OIII] λ5007 Å from a low-redshift object our spectrum would cover the wavelength ranges where other strong optical lines are expected, and none are seen. The coincidence between color-selection as a "weak" B-dropout and the redshift derived on the assumption that the line is Lyman α is suggestive. The spectrum (see figure 6) shows a flux discontinuity across the emission line, as would be expected for a $z \approx 4$ galaxy due to intervening absorption from the Lyman α forest. Moreover, the emission line profile is asymmetric, with a sharp blue side and extended red wing. This characteristic profile is often seen in high redshift quasars and galaxies (cf. the $z = 4.9$ galaxy of Franx et al. (1997)), where it is due to absorption of the blue side of Lyman α by intervening neutral hydrogen. Although a single-line redshift cannot be regarded as 100% secure, the additional circumstantial evidence suggests that this is indeed a galaxy at $z = 4.02$.

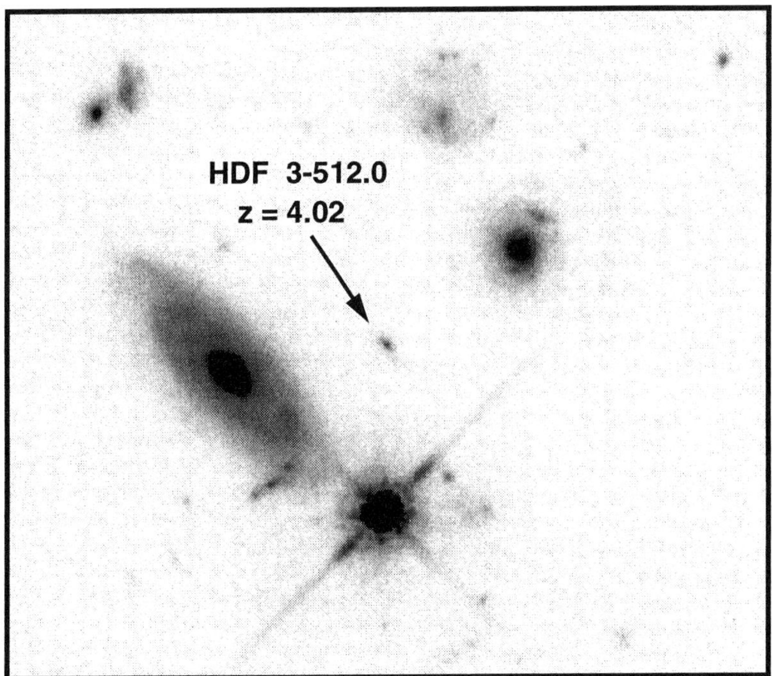

FIGURE 5. Image of 3-512.0, one of the brightest B_{450}-dropout galaxies in the HDF. A single emission line detected in its spectrum suggests that 3-512.0 has $z = 4.02$.

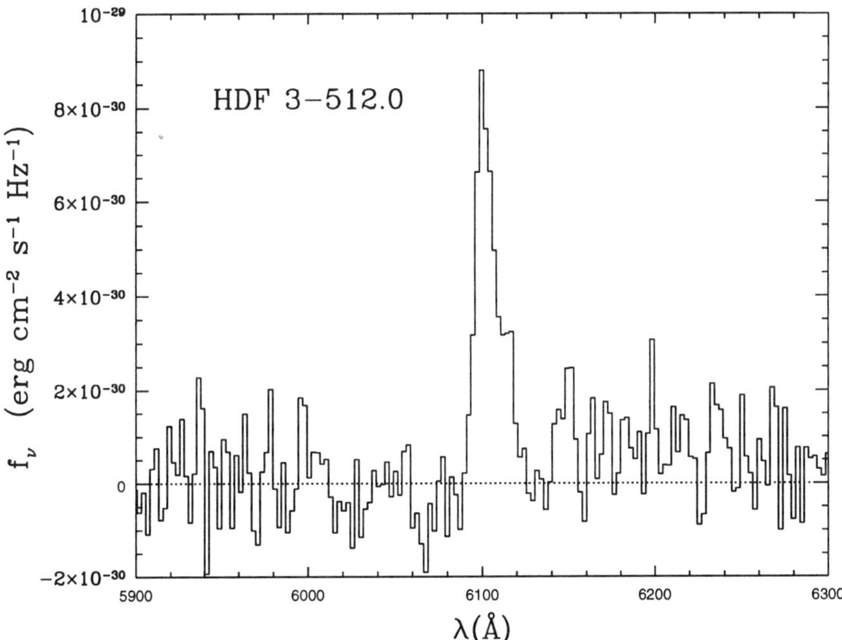

FIGURE 6. Spectrum of 3-512.0, showing the emission line detected at λ6105 Å. The asymmetric line profile and the continuum discontinuity across the line are both consistent with this being Lyman α at $z = 4.02$.

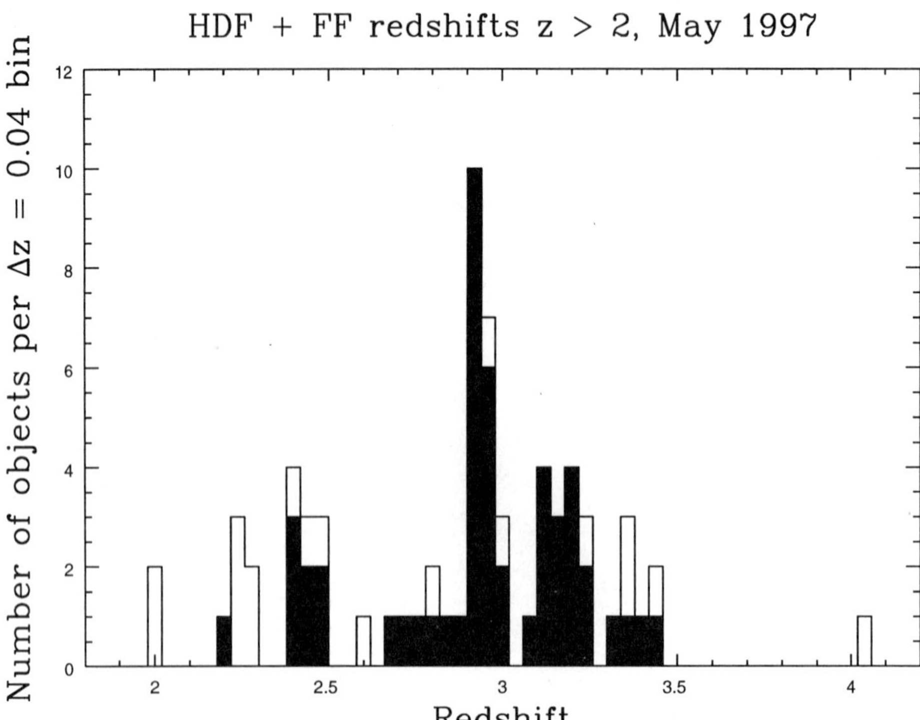

FIGURE 7. The overall redshift distribution of galaxies with $z > 2$ in the HDF and its Flanking Fields. The filled portion of the histogram are galaxies selected from our ground-based $U_n G\mathcal{R}$ imaging, primarily outside the central HDF, while the open histogram includes additional Lyman break galaxies in the central HDF.

As part of our general survey for Lyman break galaxies we have used the Palomar 200-inch telescope to image a wider region (roughly $9' \times 9'$) which includes the HDF and its flanking fields, selecting candidates using our standard $U_n G\mathcal{R}$ color criteria. The overall redshift distribution of $z > 2$ galaxies now known from the central HDF and the surrounding region is shown in figure 7. As has been often noted from surveys at $z < 1$, the redshift distribution in this narrow pencil beam is highly non-random, with prominent "spikes" where the galaxy density is large. The existence and significance of strong clustering in the Lyman break galaxy population is discussed in several papers now in press: Steidel et al. (1998) report on a strong redshift "spike" seen in another survey field, while Giavalisco et al. (1998; see also this volume) have measured the angular correlation function of Lyman break galaxies. Figure 7, as well as data we have collected in other fields, shows that these highly overdense redshift-space structures are ubiquitous at $z \approx 3$—the case studied in Steidel et al. (1998) is not unique, and we find similar spikes in essentially all of our survey fields once sufficiently large numbers of redshifts have been collected. The HDF is no exception.

3. Statistics of Lyman break galaxies

What fraction of the faint galaxy population lies at these large redshifts? This was the question addressed by Guhathakurta et al. (1990), who were among the first to apply the Lyman break technique to faint galaxy samples. At the magnitude limits of their sample

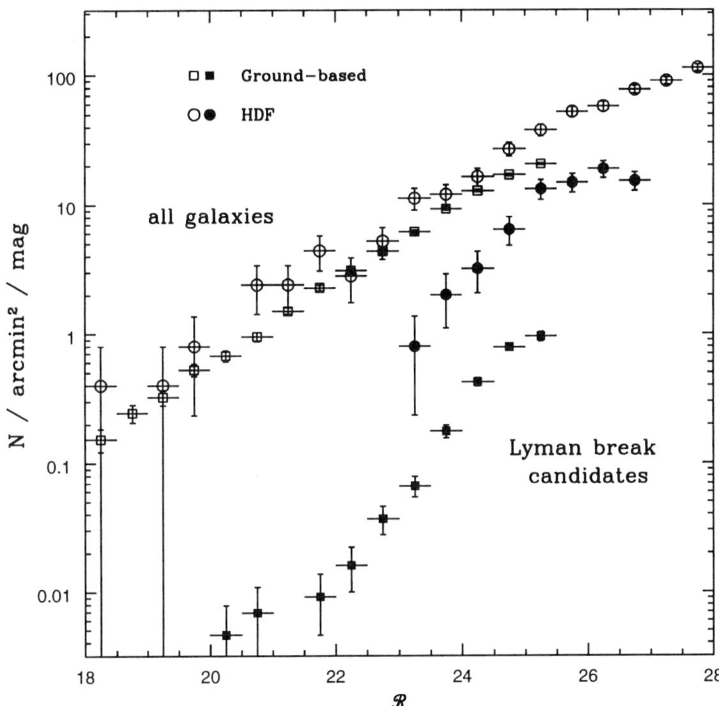

FIGURE 8. Number counts of Lyman break galaxies from ground-based and HDF samples compared to the overall counts of faint field galaxies. R magnitudes are approximated for HDF galaxies as $(V_{606} + I_{814})/2$.

they found few U-band dropout objects and concluded that only a small fraction of faint galaxies lie at $z \approx 3$. However this fraction rises as we go fainter; the exact proportion depends on the color selection (different criteria probe different volumes at high redshift) and the photometric depth of the sample. In figure 8 I show the overall number counts of galaxies in our ground-based survey and in the HDF, compared to the number of Lyman break candidates in each. Brighter than $\mathcal{R} = 22.9$, all of the color-selected objects in the ground-based sample have turned out to be galactic stars (plus a small number of high redshift QSOs)—we have yet to find a Lyman break galaxy with $\mathcal{R} < 22.9$ in the 0.24 square degrees which we have surveyed. Fainter than this, the number of Lyman break objects rises rapidly, reaching 5% of the galaxy population at $\mathcal{R} = 25.0$. In the HDF the counts of Lyman break galaxies are higher. This is partially due to the larger volume probed by the color selection technique in the HDF because of the bluer F300W filter. It may also reflect genuine redshift evolution in the galaxy population, however, since the HDF U_{300}-dropout galaxies have lower mean redshifts than do the objects in the ground-based sample, and the total UV luminosity density in galaxies is evidently rising from $z \approx 4$ to 2 (Madau et al. 1996). At $R \approx 26.5$, nearly 1 in 4 galaxies in the HDF is probably at $z \gtrsim 2$.

With a robust technique for identifying large numbers of galaxies in a particular redshift range we may quantify various statistical properties of the population even without complete spectroscopic redshift information. Here I will consider luminosity and color distributions of Lyman break galaxies. Another example is the angular correlation function, which is addressed in the contribution of Giavalisco to this volume.

3.1. *Luminosity functions*

The redshift selection function of our ground-based $U_n G \mathcal{R}$ survey is now well categorized, with more than 400 spectroscopic redshifts measured in many independent survey fields. For a particular color-defined subset of our sample we find that approximately 90% of the galaxies lie at $2.6 < z < 3.4$. The front-to-back "depth" of this redshift range is small, photometrically speaking (i.e., in terms of distance modulus), and thus the counts of Lyman break galaxies, even without confirming redshifts, primarily reflect their intrinsic luminosity distribution.

The physical significance of the "luminosity function" of Lyman break galaxies is different from that of the more familiar optical luminosity functions that have been determined locally and out to $z \approx 1$ from the CFRS, AUTOFIB, CNOC and Hawaii Surveys. By observing $z \approx 3$ galaxies through optical bandpasses we are measuring their luminosities at rest-frame ultraviolet wavelengths of approximately 1500 Å. For young galaxies, ultraviolet continuum emission arises mainly from hot, massive stars, modulated by the absorbing effects of dust. In the absence of extinction, the ultraviolet luminosity thus primarily reflects an *instantaneous* property of a galaxy: its star formation rate. The UV luminosity declines rapidly after the cessation of star formation as the O and B stars which produce it burn off the main sequence. In more local galaxy samples, the rest-frame optical light used to define luminosity functions manifests some integral over the past star formation history of a galaxy, and thus better describes its total stellar content. It is therefore most straightforward to interpret the UV luminosity function (UVLF) of Lyman break galaxies as a distribution of star formation rates in the population. The complication, however, is that the effects of extinction on UV emission can potentially be large, and are at present mostly unknown for $z \approx 3$ galaxies. I return to this issue below, considering only the "raw" luminosities here.

Figure 9 presents a composite luminosity distribution for Lyman break galaxies derived from the ground-based and HDF samples. Here I briefly describe the procedures used in constructing this diagram in order to point out some of the inherent uncertainties. A detailed discussion and analysis will be presented in a forthcoming paper (Dickinson et al. 1998).

The overall luminosity distribution has been normalized using data from our ground-based sample. Here the statistics are good thanks to the large number of Lyman break objects and the use of many survey fields to average over local fluctuations, and the redshift selection function is well characterized through extensive spectroscopy. At brighter magnitudes ($\mathcal{R} < 24.5$) there is significant contamination from galactic stars (plus a few QSOs). At those magnitudes, therefore, only spectroscopically confirmed galaxies have been used. From $24.5 < \mathcal{R} < 25.5$, the number counts from Figure 9 are used with a small, modeled correction for incompleteness. Spectroscopy has confirmed that the stellar and QSO contamination is very small at these magnitudes. The measured redshift selection function of the ground-based sample is used to normalize the survey volume. The true normalization is probably higher, as I have assumed that the color criteria are 100% efficient at the peak of the selection function ($z = 3.0$). Comparison of our ground–based Lyman break sample in the HDF and the WFPC2 images demonstrates that we do miss some fraction of candidates, even at relatively bright magnitudes, due to photometric confusion with foreground galaxies. For example, 4-555.1, a galaxy at $z = 2.803$ which is one of the brightest Lyman break objects in the HDF (and which is illustrated

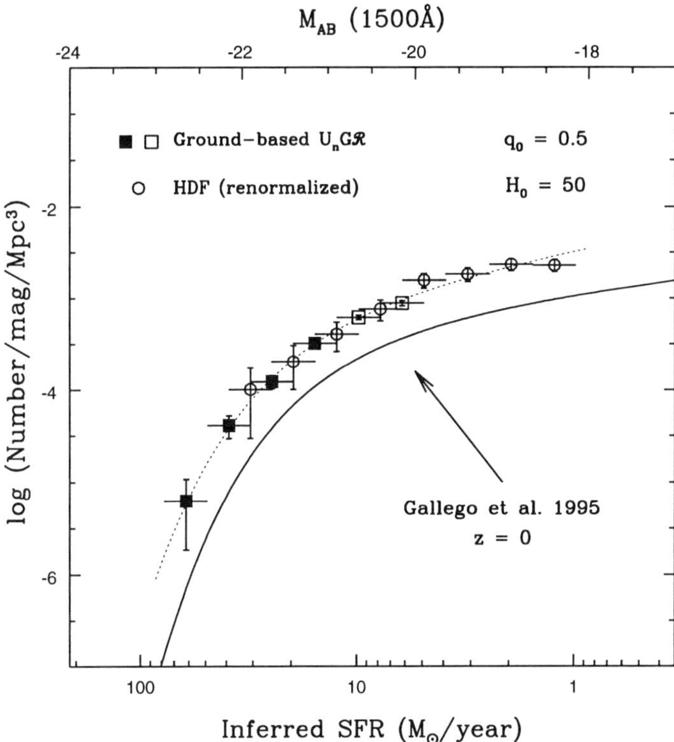

FIGURE 9. The ultraviolet luminosity function of galaxies at $z \approx 3$. Filled and open squares are derived from spectroscopically and photometrically selected galaxies from the ground-based survey, while circles are renormalized counts of objects from the HDF. Absolute magnitudes are computed on the AB system: $M_{AB} = -21$ corresponds to a specific luminosity of 1.1×10^{29} erg s^{-1} Hz^{-1}. The local Hα luminosity function of Gallego et al. (1995) is shown for comparison. Both UV and Hα luminosities are converted to star formation rates (bottom axis) using a Salpeter IMF. See text for discussion.

in figure 1 above) is missed in the ground-based sample due to flux contamination by a foreground elliptical galaxy only $\sim 2''$ away.

The ground-based data, reaching $\mathcal{R} = 25.5$, probes only the relatively bright end of the Lyman break galaxy luminosity function. The HDF offers an opportunity to sample lower luminosities. The difficulty with the HDF sample, however, is that its redshift selection function is quite uncertain; only 24 U_{300}-dropouts have measured redshifts. Moreover, the spectroscopic success rate is probably declining rapidly at $z < 2.3$ due to the lack of strong spectral features in the wavelength range currently accessible to LRIS. Finally, there is only one HDF and it covers a *very* small region of the sky. As noted above, we now know that Lyman break galaxies are strongly clustered, and thus any single, narrow pencil beam survey will encounter a highly non-random galaxy distribution. This complicates the determination of the redshift selection function, and may compromise any attempt to normalize the luminosity function simply because field-to-field variations cannot be averaged away. The Southern HDF, planned for 1998, will offer a second "realization" of the F300W dropout sample for comparison, but still the total volume surveyed will be small.

In addition, as noted above, the redshift range of the HDF Lyman break galaxies is much broader than that of the ground-based sample, and extends to lower redshifts.

HDF U_{300}-dropouts with measured redshifts have $\langle z \rangle \approx 2.7$, but it is likely that the mean redshift of the complete photometrically selected sample is lower still. Madau et al. (1996) suggest that the UV luminosity density of the universe, an integral over the luminosity function, was rising steeply with time at this cosmic epoch, a result derived from the HDF data by comparing numbers of U_{300}- and B_{450}-dropout objects. Therefore the luminosity function itself may have evolved rapidly at these redshifts, providing an additional uncertainty for splicing the ground-based and HDF Lyman break samples together. Also, this broader redshift range means that the photometric "depth" of the survey (in terms of relative distance modulus, front to back) is larger. This complicates the transformation from apparent magnitude distribution to luminosity function, both by blurring the distance modulus conversion and by making k-correction effects somewhat larger.

For the present purposes, I have used the HDF sample mainly to provide an indication of the UVLF slope at faint luminosities. Magnitudes of HDF galaxies are transformed to luminosities assuming $\langle z \rangle = 2.6$. Initially we normalize the survey volume by assuming unit selection efficiency over the range $2 < z < 3.5$. The resulting HDF luminosity function is still higher than the ground-based counts over the range of luminosity overlap. In part this may be due to incompleteness in the ground-based sample, but it is likely that it also manifests to the effect of redshift evolution. Here, the HDF space densities have been scaled downward to match that of the ground-based sample in the luminosity range of overlap. The faintest data points in figure 9 should be regarded with caution, as the sample may suffer as-yet unquantified incompleteness effects at its photometric limits. Thus while the luminosity function appears to be flattening, the apparent slope should be taken as a limit to the true value.

One measure of the distribution of galaxy star formation rates in the local universe is the Hα luminosity function of Gallego et al. (1995). In figure 9, both the $z \approx 3$ UVLF and the Hα measurements at $z \approx 0$ have been converted to star formation rates (SFR) using consistent assumptions of a Salpeter IMF spanning 0.1–125 $M_\odot$ (e.g., Madau et al. 1998; adopting the Kennicutt 1983 conversion for Hα, which assumes a somewhat different IMF, increases the Gallego et al. SFR values by 26%). The SFR distribution at $z = 3$ is strikingly like that measured at $z = 0$, spanning a similar range but with more galaxies per unit volume forming stars at any given rate. The characteristic "L^*" of the $z = 3$ and $z = 0$ SFR distributions are approximately 14 and 10 $M_\odot$ yr^{-1}, respectively, and the faint end slopes are similar.

These similarities may very well be coincidental. No correction has been made for the effects of extinction in the $z = 3$ galaxies (the Gallego et al. Hα data does include an extinction correction). As we will see below, these corrections could be quite significant. Also, the Lyman break luminosity function in figure 9 is plotted for an Einstein-de Sitter cosmology. For an open universe the $z \approx 3$ data translates downward and to the left relative to the local data, i.e., toward higher luminosities/SFRs but lower space densities. Thus both extinction and cosmology could work in the direction of increasing the typical star formation rates inferred for high redshift galaxies.

The physical state of star formation in the distant galaxies may be quite different than that in the "typical" galaxy actively forming stars in the nearby universe. The sizes of Lyman break galaxies are much smaller than those of ordinary galaxies with similar star formation rates nearby (Giavalisco et al. 1996b, Lowenthal et al. 1997), and their ultraviolet surface brightnesses are much higher, comparable to those of powerful starburst galaxies today (Meurer et al. 1997, Giavalisco et al. 1996a). Again, this suggests that the similarity of the SFR distributions at $z \approx 0$ and 3 may be, to a certain extent, a coincidence.

The integral over the best fit to the UVLF shown in figure 9 gives a comoving 1500 Å luminosity density at $z \approx 3$ of 2.1×10^{26} erg s^{-1} Hz^{-1} Mpc^{-3}, corresponding to a star formation density (using the same Salpeter conversion) of 0.026 $M_{\odot}$ yr^{-1} Mpc^{-3}, again neglecting any correction for extinction. This value should be regarded as a lower limit because of the possibility of incompleteness in the ground-based sample, although comparison between WFPC2 and ground-based Lyman limit samples suggests such incompleteness is not likely to exceed 30%. Varying the procedure used to construct the data set changes the form of the luminosity function somewhat, and the integrated luminosity density varies from 1.6 to 3.5×10^{26} erg s^{-1} Hz^{-1} Mpc^{-3} because of these systematic changes—this is the dominant uncertainty, significantly exceeding shot noise in the data. As noted above, the integrated luminosity density of the HDF Lyman break galaxies appears to be somewhat larger than that implied from the $z \approx 3$ UVLF which is normalized by the ground-based sample, but uncertainties about the redshift selection function of the HDF sample make it hard to estimate by how much. If real, the excess could be the consequence either of clustering in the small HDF volume or of redshift evolution of the galaxy population, with a larger luminosity density present at the lower redshifts probed by the HDF Lyman break color selection.

3.2. *Colors of high redshift galaxies and measures of extinction*

The preceding discussion of the UVLF of Lyman break galaxies and their star formation explicitly neglected the effect of extinction, which could be strong at ultraviolet wavelengths even if the dust content of these galaxies is relatively small. Indeed we are reasonably sure that these objects do contain some amount of dust, as their Lyman α emission lines are generally much weaker than expected from their UV-derived star formation rates under the assumption of Case B recombination (Steidel et al. 1996a). Lyman α, however, is a resonance line and is easily extinguished with small amounts of dust, so there is little constraint on the dust content from this spectral feature.

A dust-free, star forming galaxy should have a very blue UV continuum—flat in f_ν units if star formation has been proceeding for $\gtrsim 10^8$ years, or even bluer for very young starburst populations. If we examine the actual UV spectral slopes of Lyman break galaxies, however, we find that they are mostly redder than flat spectrum in f_ν. Figure 10 plots the $G - \mathcal{R}$ color of more than 400 galaxies from our ground-based sample versus redshift. For this color combination knowing the redshift of each galaxy is important because Lyman α forest opacity can effect the flux measured in the G-band, making colors redder at larger redshifts independent of extinction internal to the galaxy. In figure 10, the predicted colors of star-forming galaxies are plotted versus redshift for various amounts of internal extinction. The unreddened models define the blue envelope of the color distribution, with nearly all galaxies being redder than the colors expected from a naked star forming galaxy. This effect is also seen in local starburst galaxy samples (e.g., Calzetti et al. 1994), and has been studied in Lyman break galaxies by Meurer et al. (1997). Although for some objects this reddening of the UV spectral slope may be due to ageing of a starburst with a declining star formation rate, it is likely that some or most of the effect is indeed due to extinction.

The difficulty in interpreting UV spectral slopes as a measure of extinction is that the inferred luminosity corrections are tremendously sensitive to the form of the reddening law at UV wavelengths. For local starburst galaxies, an effective attenuation law has been derived which relates UV slope to total extinction (cf. Calzetti et al. 1994; Calzetti 1997). The wavelength dependence of this attenuation law in the near-UV is very "grey," such that a small change in the UV color requires a large change in the total extinction. Varying the UV slope of the extinction law, as well as its normalization,

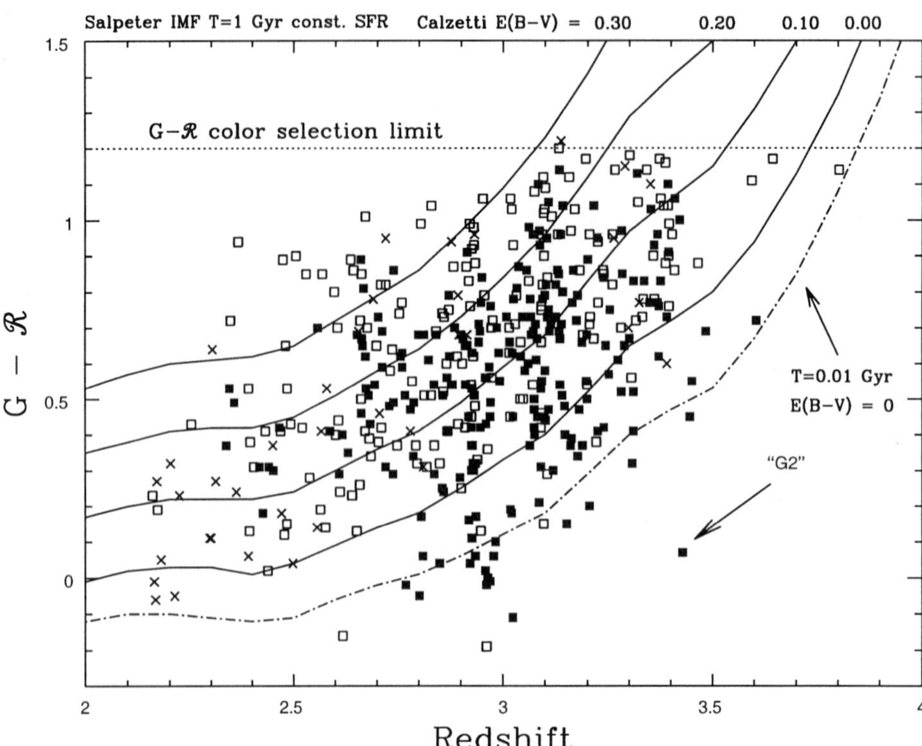

FIGURE 10. $G - R$ color of Lyman break galaxies versus redshift. The various symbol types mark different color-selected subsamples and are not important here. The solid lines show the predicted colors of actively star forming galaxies with various amounts of reddening, using the Calzetti et al. (1994) starburst extinction law. The bluest model track shows an unreddened starburst model with very young (10^7 year) age, i.e., O-star dominated and intrinsically bluer than flat spectrum. The unreddened models define the blue envelope of the color distribution. Some objects which are bluer than the unreddened models have colors affected by strong Lyman α emission lines in the G-band, e.g., the galaxy labeled G2 which is an AGN with very strong line emission.

can dramatically change the derived total suppression of UV galaxy luminosities. This is illustrated in figure 11, where I have used the observed colors of Lyman break galaxies in our spectroscopic sample to infer the UV extinction at 1500 Å under the assumption of two dust attenuation laws: the SMC extinction curve and the local starburst extinction prescription of Calzetti (1997). Without correction, the star formation rates derived for the brightest galaxies in our sample are $\sim$ 50 $M_{\odot}$/year. With SMC extinction, they slightly exceed 100 $M_{\odot}$/year, and the net correction to the overall star formation rate of the Lyman break population is a factor of two. For the starburst attenuation law, the most luminous galaxies approach SFRs of 2000 $M_{\odot}$/year, and the net correction to the *observed* population of galaxies is a factor of 7.1; the actual effect of dust on the global star formation rate would be larger because some intrinsically luminous but reddened galaxies would disappear from of a flux-limited sample. Meurer et al. (1997), using an extinction relation calibrated for local starbursts using correlations between UV spectral slope and far-infrared emission, derive even larger correction factors of $\sim$ 15$\times$ for the Lyman break population, in part due to different assumptions about the spectral slope of the underlying, unreddened continuum.

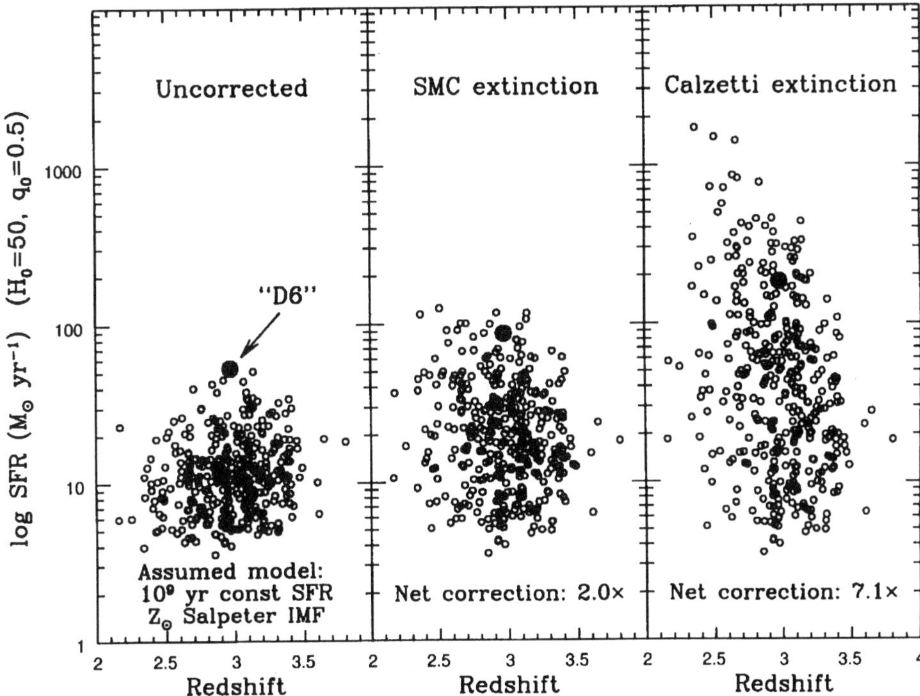

FIGURE 11. Estimates of the effect of extinction on the UV luminosities and derived star formation rates of Lyman break galaxies. The left panel shows computed star formation rates for objects with redshifts in our survey. The most luminous galaxy in the sample, labelled D6, is shown with a large symbol for identification in each panel. In the center and right panels, the $G - R$ color of each object has been used to compute reddening using SMC and Calzetti extinction laws and the appropriate correction has been applied to the star formation rates. At bottom, the net correction to the *observed* population of objects is tabulated.

At present, we have little direct information about the true effects of extinction on the UV luminosities of Lyman break galaxies. While it is plausible that the attenuation law for local starburst galaxies may apply to their high redshift ancestors, the inferred effect on the global UV luminosity is so sensitive to small variations in the extinction law that one wishes for independent data at other wavelengths which could be used to verify star formation rates. In the future, far-infrared measurements of thermally reradiated dust emission in Lyman break galaxies may be possible from SIRTF or WIRE, or with sub-millimeter observations with the instruments like SCUBA. In the meanwhile, we have begun a program of near-infrared spectroscopy at UKIRT to measure Balmer line emission from Lyman break galaxies, and thus provide an independent measure of their star formation rates which should be less sensitive to extinction. This work is painfully slow compared to optical multislit spectroscopy for measuring redshifts, requiring night-long exposures on one galaxy at a time. Preliminary results are reported by Pettini et al. 1997, and suggest that the star formation rates derived from the Balmer emission may be a few times larger than those inferred from the UV continuum. A much larger sample is needed before we can make secure statements, but at least the problem is addressable by observation. Moreover, the same is now true at low redshift. Recently, Tresse & Maddox (1997) have derived an Hα LF for CFRS galaxies at $z \approx 0.2$, while Treyer et al.

(1997) presented a near-UV LF for galaxies at similar redshifts. Comparison of these two LFs, converted to star formation rates, suggests that the UV luminosity density is underestimated of this local sample is suppressed by ~ 1 magnitude, presumably due to the effects of dust.

4. Infrared properties

One difficulty with the HDF WFPC2 data for studying galaxies at $z > 1$ is that their optical rest-frame light is redshifted beyond the WFPC2 filter bandpasses. The WFPC2 images thus provide only an ultraviolet view of the $z > 1$ universe. While this has advantages for detecting and studying active star formation in very distant galaxies, it makes it difficult to compare their properties to those of objects in the nearby universe at familiar rest-frame wavelengths. We may address this problem by collecting data on the nearby universe in the ultraviolet, or by studying HDF galaxies in the near-infrared.

The HDF has been imaged in the infrared by several groups. Len Cowie and colleagues used the CFH and UH 2.2m telescopes to obtain images of the central deep field in J and an "H+K" notched filter, and have imaged a much wider surrounding region ($9' \times 9'$) to shallower depths. Hogg et al. (1997) used the NIRC camera on Keck to obtain deep images of two small ($\approx 40'' \times 40''$) fields within the HDF, and the Caltech group has also covered a wider surrounding region with the Palomar 60-inch. My collaborators and I imaged the central HDF using IRIM on the KPNO Mayall 4m telescope over the course of ten nights. The field of view of IRIM is well matched to that of the WFPC2, providing easy coverage of the complete HDF. Data was collected in the J, H and K bands, and reach formal 5σ limiting magnitudes in a $2''$ diameter aperture of $J = 23.5$, $H = 22.3$ and $K = 21.9$. These images are available to the community, and have been used by several groups presenting results at this symposium. For further information and access to the data, please see `http://www.stsci.edu/ftp/science/hdf/clearinghouse/irim/irim_hdf.html`.

A complete discussion of the infrared properties of galaxies in the HDF is beyond the scope of this presentation. Here I restrict my attention to a few simple color properties of $z > 2$ Lyman break selected galaxies. Although WFPC2 optical photometry in this paper is presented in AB units, I will use standard (Vega-normalized) infrared magnitudes in the discussion that follows. For reference, the approximate conversion to AB units for the K_S bandpass used for the IRIM observations is $K_{AB} = K_{\rm Vega} + 1.86$.

4.1. *Colors and luminosities*

Figure 12 plots $V_{606} - K$ colors for galaxies in the HDF with spectroscopically confirmed redshifts (plus a few stars). At $0 < z \lesssim 1$, galaxy colors are nicely bounded by the range expected for star forming "irregulars" to old ellipticals. The "red envelope" of galaxies at $z \approx 1$ is only slightly bluer than are giant elliptical galaxies in the local universe. (Indeed, perhaps the most luminous galaxy in the HDF at rest-frame optical wavelengths is a giant elliptical at $z = 1.012$.) Figure 13 shows the colors of nearly all optically selected U_{300}-dropout galaxies in the HDF brighter than $V_{606} = 27$. (A few have been excluded due to photometric confusion with other objects in the near-IR data). Most Lyman break galaxies with $V_{606} < 25.5$ are detected in the infrared images. Beyond that magnitude there is increasing incompleteness due to the limited depth of the IR data.

Lyman break galaxies mostly have colors in the range $2.5 < V_{606} - K < 4.5$ (although we cannot rule out that some of the fainter objects are intrinsically bluer). As can be seen from the model tracks in figure 12 these colors (shifted to the UV-optical rest frame) would be normal for actively star forming spirals (Sb–Sc) in the local universe. They are

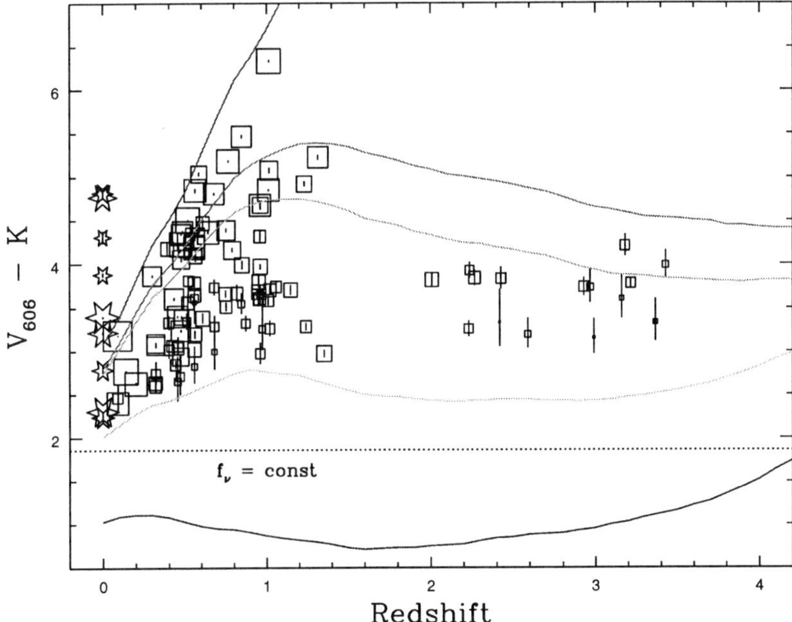

FIGURE 12. Color versus redshift for galaxies with spectroscopic identifications in the HDF. The symbol sizes scale inversely with apparent K magnitude; star-shaped symbols at $z = 0$ are known galactic stars. The solid lines are *unevolving* tracks of k-corrected color vs. redshift for various model galaxy spectral energy distributions designed to span the range (reddest to bluest) from old ellipticals to actively star forming irregulars. The bluest model track, with $V_{606} - K \approx 1$ at most redshifts, is an extremely young (age $= 10^7$ year) unreddened starburst. There are no spectroscopically observed galaxies in the HDF which approach this color.

all much redder (as indeed are all HDF galaxies) than the expected colors of an extremely young, unreddened starburst spectrum—indeed, there are few faint galaxies anywhere in the universe which are as blue or bluer than a flat (f_ν) spectrum over a long (optical-IR) wavelength baseline. The observed color does not, of course, tell us *why* the Lyman break galaxies have these colors; the effects of dust extinction on the rest-frame UV emission could have a significant impact on the observed colors (see previous section). The typical ("L^*" from figure 9) Lyman break galaxy at $z \approx 3$ has $K = 21.5 \pm 1$, corresponding to rest-frame $M_V = -22 \pm 1$ for $H_0 = 50$, $q_0 = 0.5$, i.e., similar to L_V^* locally.

Figure 14 shows a color-magnitude diagram for a K-selected sample of objects in the HDF. Different symbol types code the objects as U_{300}-dropout Lyman break objects (with or without spectroscopic confirmation), galaxies with spectroscopically measured $z < 2$, stars, or unobserved objects (but which presumably mostly have $z < 2$, since they do not qualify as Lyman break candidates). The optical-IR colors of the Lyman break galaxies are quite typical for the general population of faint galaxies in the HDF.

Two galaxies in figures 13 and 14 have significantly redder optical-IR colors ($V_{606} - K$) than the majority of Lyman break objects. Although they both satisfy the color selection criteria defined in §2.1, it is not certain that these are also at $z > 2$. In addition to their relatively bright infrared magnitudes, they are also much redder in $V_{606} - I_{814}$ than the majority of Lyman break galaxies ($V_{606} - I_{814} \approx 0.7$ to 1.0—compare with figure 4), and their overall spectral energy distributions are quite different than those of the known $z > 2$ galaxies in the HDF. At present, neither object has a spectroscopic redshift; future

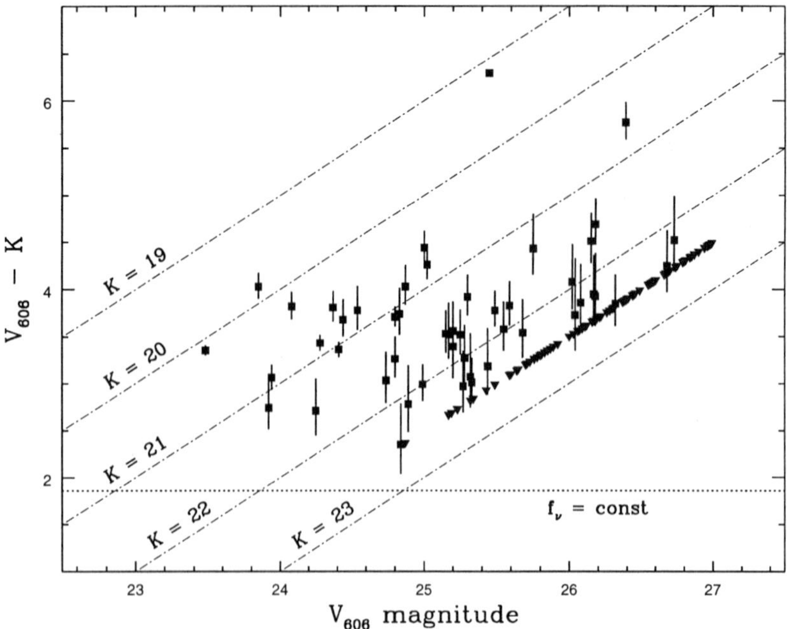

FIGURE 13. $V_{606} - K$ colors of U_{300} Lyman break selected galaxies in the HDF. Note that only a small subset of these galaxies have spectroscopic redshifts. Objects with $K > 22.5$ (i.e., beyond the robust detection limit of our IR data) are plotted as lower limits (triangles) at that magnitude.

observations should establish whether these are unusually red and luminous members of the Lyman break population or interlopers into the color selected sample from lower redshifts.

4.2. *Morphologies*

Existing ground-based near-IR images of Lyman break galaxies in the HDF and elsewhere are mostly inadequate for answering questions about the rest-frame optical morphologies of these objects. Our KPNO IRIM images of the HDF, for example, have a resolution of $\approx 1''$, much larger than the typical scale lengths associated with the rest-frame UV light seen in WFPC2 images (Giavalisco et al. 1996b). However, a few simple conclusions can be drawn from the existing data. In general, the centroid position of the infrared emission is spatially coincident with that of the optical images, suggesting that in most cases there is no substantial "displacement" between the dominant sources of rest-frame UV and optical luminosity in these objects. In the HDF, however, there is one spectacular exception found in galaxy 2-585.1, the $z = 2.01$ object whose redshift was serendipitously measured by Elston et al. (see §2.2 above). The WFPC2 images of this galaxy show a chain-like assembly of blobs $\sim 3''$ long, making it rather unusual among the Lyman break objects, which are mostly quite compact. The near-IR emission is spatially extended, with its peak/centroid located just beyond one end of the "chain," coincident with diffuse optical emission seen in the WFPC2 images. In this object it appears that spatially segregated stellar populations and/or extinction are affecting the observed morphology—the high surface brightness, star forming "blobs" of the chain structure protrude radially outward from a larger, red host galaxy.

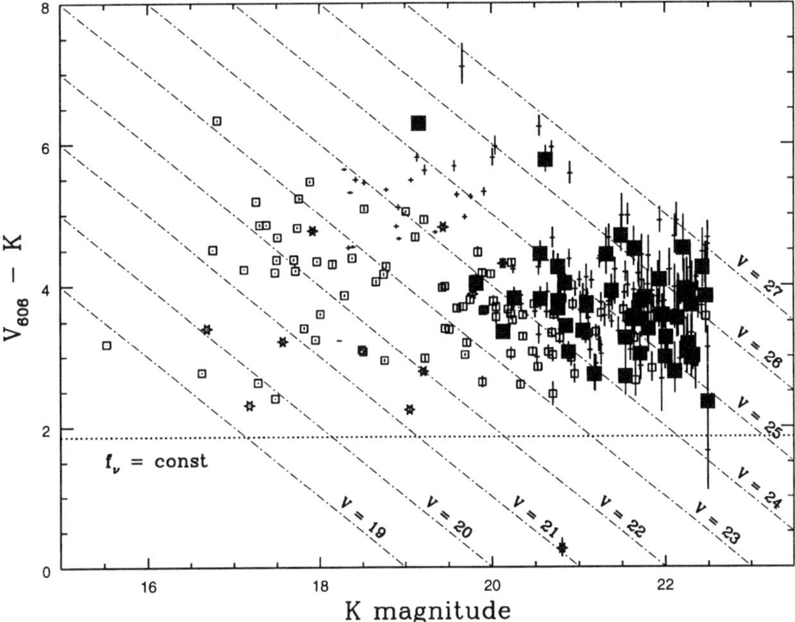

FIGURE 14. $V_{606} - K$ color versus K magnitude for objects with $K < 22.5$ and $V_{606} < 27.0$ in the HDF. Lyman break selected galaxies are plotted as filled boxes. Galaxies with spectroscopic redshifts $0 < z < 2$ are plotted as unfilled boxes; galactic stars are shown as star symbols. Objects without spectroscopic observations are plotted as small crosses.

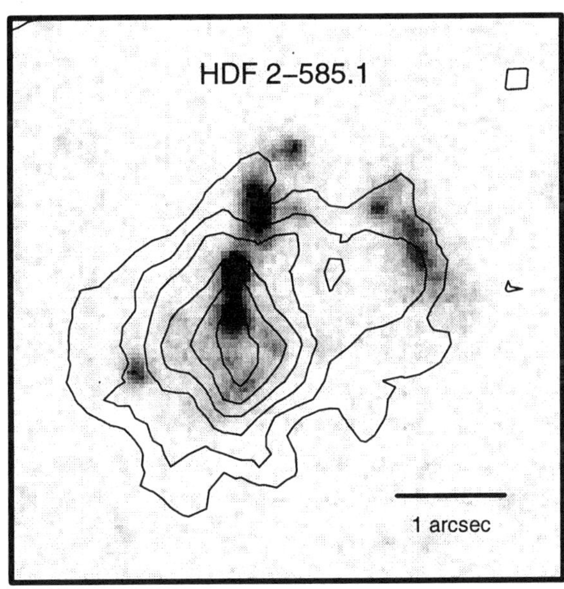

FIGURE 15. V_{606} WFPC2 image of the $z = 2.01$ Lyman break galaxy 2-585.1 (greyscale) overlaid with contours from the ground-based infrared image. Although the $\approx 1''$ resolution of the ground-based image does not allow morphological details to be discerned, there is a clear spatial offset between the peak of the IR emission and the brightest structures visible in the rest-frame UV. The object toward the upper right is not a Lyman break candidate and thus is unlikely to be related to 2-585.1.

5. Summary

In this presentation I have tried to use HDF data to illustrate the ways in which multi-color photometry can be used first to select galaxies at high redshift and then to learn something about their intrinsic properties. The great depth of the HDF and the precision of its photometry makes it ideal for illustrating and applying the Lyman break color selection technique: cf. the remarkable prominence of the high redshift "plume" in figures 2 and 4. By $V_{606} = 26.5$, nearly 1 in 4 HDF galaxies is a Lyman break candidate and thus is likely to be at $z > 2$. Spectroscopy of HDF galaxies at all redshifts has proceeded at a remarkable pace. After only two observing seasons, the central HDF is almost certainly the piece of celestial real estate with the highest surface density of measured galaxy redshifts (now $\approx 24/\text{arcmin}^2$, of which $\sim$20% are at $z > 2$).

The distribution of ultraviolet luminosities of $z \approx 3$ galaxies, converted to star formation rates using a simple prescription, spans a similar range to that of galaxies in the local universe. Schechter function fits give a characteristic 1500 Å specific luminosity of $M_{AB} \approx -21$ (for $H_0 = 50$, $q_0 = 0.5$), corresponding to a star formation rate of $\sim 14\ M_\odot$/year. Very few Lyman break galaxies have "raw" SFRs $> 50\ M_\odot$/year. If these measurements are taken at face value, these objects cannot produce the total stellar mass of $\sim L^*$ galaxies in short timescales as traditional "monolithic" formation scenarios for elliptical galaxies would require. This, then, would suggest that massive galaxy formation takes place either at still higher redshifts where we have yet to look, or proceeds by the hierarchical assembly of smaller objects as expected in theories such as CDM. However, the possible extinction corrections to the UV luminosities of Lyman break galaxies are highly uncertain and could be quite large. Their UV colors are redder than expected from spectral models of "naked" star forming galaxies, a fact which could easily be explained by the presence of dust. The derived extinction corrections based on these colors, however, are extremely sensitive to the form of the extinction law in the ultraviolet, and can range from factors of 2 to > 7. New observations in the infrared, and eventually at far-IR and sub-millimeter wavelengths, will provide an independent measure of star formation rates which can be useful for addressing this question.

Characteristic rest-frame optical luminosities of Lyman break galaxies, as measured from infrared photometry, are $M_V = -22$. Their UV-optical rest-frame colors galaxies span a range which would be typical for normal spirals in the nearby universe. Detailed morphological study of $z \approx 3$ galaxies at rest-frame optical wavelengths must await observations with NICMOS, which will take place in 1997–1998.

Although Lyman break color selection is a simple technique compared to more sophisticated photometric redshift methods, it has the virtue of being relatively model independent and easy to apply and understand. Like all such methods, however, its utility depends strongly on the degree to which it is tested and calibrated by follow-up spectroscopy, which requires substantial effort on large ground-based telescopes. With hundreds of Lyman break redshifts in hand, we can begin to carry out fairly sophisticated analyses of the luminosity function, clustering, and other properties of galaxies at $z \approx 3$. For the HDF, despite the impressive observing efforts to date, we are unlikely ever to succeed in collecting hundreds of redshifts for $z > 2$ galaxies with existing telescopes and instrumentation. However, we may take advantage of the insights gained from studying the Lyman break galaxy population in non–HDF data sets to aid interpretation of the HDF objects, and thus to use the HDF to push the method to different flux and redshift limits.

Regardless of how much we learn about high redshift galaxies in the HDF, we must remind ourselves of what a small volume the HDF probes. The entire comoving volume over which U_{300}-dropouts have been found in the HDF, $2 < z < 3.5$, is only $18000h_{50}^{-3}$ Mpc^3 for $q_0 = 0.5$. Locally, that would correspond to a sphere with radius $16.2h_{50}^{-1}$ Mpc—not even reaching the Virgo cluster! We must therefore be cautious about how representative HDF galaxies are in any statistical sense, particularly in light of recent evidence for strong clustering at $z > 2$. The pencil-beam geometry of the HDF volume ensures that it will traverse a wider range of large scale structures than would the corresponding 14.2 Mpc radius sphere locally, but such a geometry brings its own complications for some applications. For example, clustering may introduce large fluctuations in $N(z)$ which can seriously complicate analyses of the redshift evolution of galaxy properties, global luminosity density, etc. It is therefore risky to extrapolate too far from the HDF to the properties of galaxies in the high redshift universe as a whole. Ultimately, however, the insights gathered from the HDF, when calibrated with data from ground-based, large-volume surveys, should provide a powerful means of understanding the early stages in the evolution of normal galaxies.

I would like to extend special thanks to my collaborators Chuck Steidel, Mauro Giavalisco, Max Pettini, Kurt Adelberger, and Mindy Kellogg, for endlessly interesting discussions, much hard work, and for allowing me to present materials in advance of publication. The same thanks also go to Adam Stanford, Peter Eisenhardt, Richard Elston, and Matt Bershady for their collaboration on the KPNO infrared imaging program. Finally, I would also like to thank my colleagues at ST ScI from the HDF Team, and the editors of this volume for their considerable patience.

REFERENCES

Calzetti, D., Kinney, A. L., & Storchi-Bergmann, T. 1994, *ApJ* **429**, 582.

Calzetti, D. 1997 in *The Ultraviolet Universe at Low and High Redshift* (ed. W. Waller). AIP Press, in press (astro-ph/9706121).

Cohen, J. G., Cowie, L. L., Hogg, D. W., Songaila, A., Blandford, R., Hu, E. M., & Shopbell, P. 1996, *ApJ* **471**, L5.

Colley, W. M., Rhoads, J. E., Ostriker, J. P., & Spergel, D. N. 1996, *ApJ* **473**, L63.

Gallego, J., Zamorano, J., Aragón-Salamanca, A., & Rego, M. 1995, *ApJ* **455**, L1.

Franx, M., Illingworth, G., Kelson, D. D., van Dokkum, P. G., & Tran, K. 1997, *ApJ* **486**, L75.

Giavalisco, M., Livio, M., Bohlin, R. C., Macchetto, F. D., & Stecher, T. P. 1996a, *AJ* **112**, 369.

Giavalisco, M., Steidel, C. C., & Macchetto, F. D. 1996b, *ApJ* **470**, 189.

Giavalisco, M., Steidel, C. C., Adelberger, K. L., Dickinson, M., Pettini, M., & Kellogg, M. 1998, *ApJ*, submitted.

Guhathakurta, P., Tyson, J. A., & Majewski, S. R. 1990, *ApJ* **357**, L9.

Hogg, D. W., Blandford, R., Kundic, T., Fassnacht, C. D., & Malhotra, S. 1996, *ApJ* **467**, L73.

Hogg, D. W., Neugebauer, G., Armus, L., Matthews, K., Pahre, M. A., Soifer, B. T., & Weinberger, A. J. 1997, *AJ* **113**, 2338.

Kennicutt, R. C. 1983, *ApJ* **272**, 54.

Lowenthal, J. D., Koo, D. C., Guzman, R., Gallego, J., Phillips, A. C., Faber, S. M., Vogt, N. P., Illingworth, G. D., & Gronwall, C. 1997, *ApJ* **481**, 673.

Madau, P. 1995, *ApJ* **441**, 18.

Madau, P., Ferguson, H. C., Dickinson, M., Giavalisco, M., Steidel, C. C., & Fruchter, A. 1996, *MNRAS* **283**, 1388.

MADAU, P., POZZETTI, L, & DICKINSON, M. 1998, *ApJ*, in press (astro-ph/9708220).

MEURER, G. R., HECKMAN, T. M., LEHNERT, M. D., LEITHERER, C., & LOWENTHAL J., 1997, *AJ* **114**, 54.

PETTINI, M., STEIDEL, C. C., DICKINSON, M., KELLOGG, M., GIAVALISCO, M., & ADELBERGER, K. L. 1997, in *The Ultraviolet Universe at Low and High Redshift* (ed. W. Waller). AIP Press, in press (astro-ph/9707200).

SONGAILA, A., COWIE, L. L., & LILLY, S. J. 1990, *ApJ* **348**, 371.

STEIDEL, C. C., & HAMILTON, D. 1992, *AJ* **104**, 941.

STEIDEL, C. C., PETTINI, M., & HAMILTON, D. *AJ* **110**, 2519.

STEIDEL, C. C., GIAVALISCO, M., PETTINI, M., DICKINSON, M., & ADELBERGER, K. L. 1996a, *ApJ* **462**, L17.

STEIDEL, C. C., GIAVALISCO, M., DICKINSON, M., & ADELBERGER, K. L. 1996b, *AJ* **112**, 352.

STEIDEL, C. C., ADELBERGER, K. L., DICKINSON, M., GIAVALISCO, M., PETTINI, M., & KELLOGG, M. 1998, *ApJ* **492**, 428.

TRESSE, L., & MADDOX, S. J. 1997, *ApJ*, in press (astro-ph/9709240).

TREYER, M. A., ELLIS, R. S., MILLIARD, B., & DONAS, J. 1997, in *The Ultraviolet Universe at Low and High Redshift* (ed. W. Waller). AIP Press, in press (astro-ph/9706223).

WARREN, S. J., HEWITT, P. C., IRWIN, M. J., MCMAHON, R. G., BRIDGELAND, M. T., BUNCLARK, P. S., & KIBBLEWHITE, E. J. 1987, *Nature* **325**, 131.

WILLIAMS, R. E., ET AL. 1996, *AJ* **112**, 1335.

ZEPF, S. E., MOUSTAKAS, L. A., & DAVIS, M. 1997, *ApJ* **474**, L1.

Gravitational lensing in the Hubble Deep Field

By ROGER D. BLANDFORD

Theoretical Astrophysics, 130-33 Caltech, Pasadena, CA 91125

The present state of gravitational lensing studies on the Hubble Deep Field (HDF) is summarized. To date, no secure cases of strong (i.e., multiple imaging) lensing have been identified, although a few candidates are still undergoing spectroscopic study. A search of the few galaxies that dominate the strong lensing cross section has also failed to turn up convincing examples. This may be consistent with a model in which the strong lensing probability is dominated by compact groups of galaxies and most lines of sight having a lensing cross section an order of magnitude smaller than average. Nearly 2000 images have been scrutinized in search of evidence for week gravitational lensing. As expected, there is no significant signal attributable to the effects of large scale structure; the field is too small. However, a statistically significant detection of galaxy-galaxy lensing appears to be present. The signal is strongest if the foreground deflector sample is restricted to those brighter galaxies that have measured redshifts both on the HDF and the surrounding flanking fields. The magnitude of the measured tangential elongation of the images of the background galaxies is consistent with a model in which the foreground galaxies have masses and sizes somewhat smaller than their local counterparts and the background galaxies (which are too faint for spectroscopy) are distributed widely in redshift. Future prospects are briefly summarized.

1. Introduction

The unprecedented depth and image quality of the Hubble Deep Field (HDF) allows new studies of gravitational lensing. As there are good images of over two thousand galaxies of which a large fraction are believed to be located at large redshift, we expect that they should exhibit strong gravitational lensing with the same frequency as distant radio sources and quasars. This implies, naively, ~ 10 cases of multiple imaging should be observable on the HDF. In addition all of the images of high redshift sources should be distorted at the level of several percent as a result of weak lensing by a combination of large scale structure and individual galaxy perturbations. This distortion, cannot be measured in an individual galaxy, because we do not know its intrinsic shape. However, it can be measured statistically by seeking large scale polarization of the images over a patch of sky and a negative correlation of the observed background galaxy images with the directions on the sky of the closest deflector galaxies. The HDF provides an ideal testbed for seeking these effects because the image quality is far higher than will be routinely obtained in much larger surveys using ground based telescopes or shorter exposures with HST. It can therefore be used to calibrate imaging and to quantify the degradation caused by atmospheric distortion, sky noise and telescope optics. We discuss these issues in turn.

2. Strong lensing

Theoretical and observational studies have converged upon a rough rule of thumb that the incidence of observable multiple imaging of cosmologically distant sources is about 0.002 (e.g., Schneider, Ehlers & Falco 1992, Myers 1996). This deceptively simple rule should be applied with caution. In particular, it is a serious underestimate behind the cores of rich clusters. In addition, it appears that although individual galaxies should have high enough surface densities to multiple image a sufficiently distant background source sufficiently close to the optic axis and that these are responsible for most secure examples, in practice they are assisted by neighboring galaxies, often in the form of a compact group (Kundić et al. 1997, in preparation). Furthermore, a more accurate estimate must depend upon the observed flux, because "there is no multiplication without magnification" and so the strongly lensed sources will be intrinsically fainter than the single sources with which they are being compared by a factor which, in turn, depends upon the slope of the faint source counts at the flux limit of the sample under consideration. The strong lensing frequency also depends upon the redshift distribution of both the source galaxies and the deflector galaxies as well (rather weakly) the world model. It turns out that, in practice, the strong lensing frequency is dominated by deflector galaxies with redshifts $z_d \sim 0.5$ and source galaxies with redshifts $z_s > 1.5$. Allowing for these effects, we estimate that there could be as many as 4000 observable galaxies on the HDF and roughly eight of these might be strongly lensed.

Armed with this realization, Hogg et al. 1996 scrutinized the HDF image, soon after it was released, looking for the most promising candidates on purely morphological grounds. Specifically, they sought faint image doubles and quads with similar colors arranged with respect to a brighter galaxy in a manner consistent with standard lensing geometries. Despite experimentation with semi-automatic procedures, like those used in the CLASS survey, it turned out that, in this instance, visual inspection was the most efficient way to identify candidates for multiple imaging. A ranked list of 12 candidate lenses was produced and these were proposed as suitable candidates for spectroscopic follow up.

2.1. *J123652+621227*

This source was the most intriguing candidate of Hogg et al. 1996. The proposed lens is a $I = 24$ red, elliptical galaxy. It is surrounded by 16 faint galaxies within $5''$ that can be partitioned into diametrically opposed pairs of similar colors, plus a tangential arc. However, the separation of the images was much larger than would be expected for a normal elliptical galaxy at the redshift $z \sim 1$, indicated by the deflector galaxy color Coleman, Wu & Weedman 1980). Furthermore, the $V = 26.1$ arc can be associated with a $V = 27.4$ counter-arc. Zepf, Moustakis & Davis 1997 have attempted spectroscopy using a slit that covers both the arc and the counter-arc. Although it is not possible to measure a redshift for either object, Zepf et al. claim that the spectra are sufficiently dissimilar that they images are unlikely to have a common source. Of course, this does not rule out the possibility that the arc is "naked" (which occurs when the cusp associated with the tangential caustic protudes beyond the radial caustic) so that no counter-image is formed. Direct spectroscopy of the elliptical galaxy (Cohen, private communication) produced features consistent with $z = 0.41$, but the signal to noise is too low for this to be treated with any confidence.

2.2. *J123656+621221*

In this case, the proposed lens is a $I = 24$ red, elliptical galaxy and the image is a single, blue tangential arc located only $0.8''$ from the deflector. In this case the impact parameter is so low that, if it can be demonstrated that the arc has a significantly higher redshift than the deflector, then multiple imaging is almost inevitable. Cohen has measured a tentative redshift $z_d = 0.483$ for the elliptical. More integration is needed before this value can be treated as secure. Even less confidence can be attached at this stage to the best guess redshift for the arc, $z_s = 1.02$ (Zepf, Moustakis & Davis 1997).

2.3. *J123649+621322*

In this case, we have a $R = 24$ blue galaxy accompanied by a single red arc. Spectroscopy by Cohen has produced two candidate redshifts for the blue galaxy. Unfortunately, even if one of these can be confirmed, it will still be necessary to obtain an arc redshift which will be an even more challenging observation.

2.4. *J123641+621204*

Here, there are four bright images arranged in a quad. No lensing galaxy is visible, but this is not uncommon in *bona fide* lenses. Steidel et al. 1996 have measured a redshift $z_s = 3.226$ for this object, but find differences between the components and doubt that this is a multiple-imaged source.

3. Weak lensing by large scale structure

As light propagates through a gravitational potential, its path will be deflected. In terms of the local Newtonian potential, ϕ, a ray will be deflected through an angle

$$\alpha = 2\nabla_\perp \int ds\phi/c^2 \tag{3.1}$$

where ds is an element of distance measured along the ray and the derivative is in the plane normal to the unperturbed ray. This implies that the angular displacement of a cosmologically distant source due to large scale structure on scale $\sim k^{-1}$ is $\sim \Omega(k_m)(ck_m/H_0)^{3/2} \sim 1'$, where $\Omega(k)$ is the mass fraction of an Einstein-De Sitter Universe clustered on scale k and $k_m \sim (10\ \mathrm{Mpc})^{-1}$ is the scale on which the power spectrum of relative density fluctuations peaks. It is amusing that the whole HDF is probably translated laterally by more than its size by large scale structure.

However, none of this is directly observable. It is only differential displacements that we can hope to measure. The simplest way to measure large scale structure using weak gravitational lensing is to use the images of individual galaxies and measure their distortion. In the case of the HDF, we prepared a catalog of over 2500 sources based on the second release of HDF data. In this catalog, sources were identified using a recursive algorithm that listed groups of five or more contiguous pixels with signals in excess of $2.5\sigma_{\mathrm{sky}}$ in the F606W frame. This defines the source area and fluxes are then measured over each of the four HDF bands. Centroids and intensity-weighted moments are next defined using

$$\bar{x} = \frac{\sum_i x_i I_i^W}{\sum_i I_i^W} \tag{3.2}$$

where x_i is the $x-$ coordinate of the i'th pixel. Exponents $W = 0, 0.5, 1$ were tried and the results were not very sensitive to the choice. Second moments were calculated for each image and from these, the galaxy ellipticities, ϵ, and position angles, ϕ, were computed. (Subsequent to the preparation of this catalog, Williams et al. 1996 published the HDF catalog which produced similar results but which was superior in its handling of blends.) Using the "orientations," ($\chi = \epsilon e^{2i\phi}$ to linear order), of each individual galaxy image we can obtain an unbiased estimate of the gravitationally-induced "polarization," ($p =< \chi >$), over some area of sky, under the assumption that the individual position angles are uniformly distributed in $[0, \pi]$. (Note that, as long as our measurement of galaxy shapes is unbiased, it is not necessary to determine the position angles of the galaxies with high accuracy.)

Any weak lensing study is a battle against systematic error. For example, trailing of the telescope or imperfect image stacking can lead to a spurious signal. The best way to counteract such effects is to work directly with the data as processed—to regard the effects of the post-processing as essentially similar to the effects of the telescope and (for ground observations) of the atmosphere. We can then calibrate using the stellar images. In the case of the HDF, there are too few stars to map aberration across the field of view. However, they do provide a confident upper limit on the rms distortion, which can be used to limit systematic effects and in simulations to measure the "polarizability" of the images as a function of the number of pixels. This is necessary to decide how much to weight images of a given size in estimating the polarization. The point spread function is also important for quantifying the effects of seeing on weak lensing studies using ground-based telescopes.

Unfortunately, the proper size of the HDF field is ~ 1 Mpc and, consequently, it is about ten times too small to probe the scales where the maximum signal is expected to be found Blandford et al. 1991, Miralda-Escudé 1991. Calculations of the rms size of the predicted mean polarization on a individual chip range from $\sim 0.01 - 0.02$, depending upon the source redshift distribution and assumptions about the density fluctuation power spectrum. The upper limit on the polarization on the individual chips is $\sim 0.02 \pm 0.04$, not inconsistent with expectation, but not particularly interesting either.

4. Galaxy-Galaxy lensing

As first discussed by Webster 1985 and Tyson et al. 1984, individual deflector galaxies at modest redshifts ($z_d \sim 0.5 - 1$) are expected to produce a weak tangential stretching of the images of background source galaxies of greater redshift ($z_s \sim 1 - 3$). If we make a very simple model of the potential of the deflectors, presumably dominated by dark matter), as truncated isothermal spheres, then, in round numbers, the Einstein ring subtends an angular radius $\theta_c \sim 1''$, and the halos are expected to continue out to angular radii $\theta_H \sim 10''$. The cross section for weak lensing by these galaxy halos is therefore roughly a hundred times that for strong lensing by the galactic nuclei. If we measure the halo surface density by the roughly constant circular velocity V_c, then the predicted polarization of background galaxy displaced by an angle θ from the galaxy center is

$$
\begin{aligned}
p &\sim \frac{2\pi V_c^2}{\theta c^2}; \qquad \theta < \theta_H \\
&\sim \frac{2\pi\theta_H V_c^2}{\theta^2 c^2}.
\end{aligned} \tag{4.3}
$$

If we concentrate on sources with impact parameters close to the halo radii of intervening galaxies, we find that the expected tangential polarization signal will be $p \sim 0.1$ and this will be contributed by all, sufficiently distant, sources within $\theta \sim \theta_H$ and, to a pretty good approximation, the polarizations due to different galaxies will add linearly Schneider & Rix 1997. The "noise" in this measurement is the individual galaxy ellipticity which has a value ~ 0.3. Now, the number of lens galaxies within an angle θ of a given source increase as $N \propto \theta^2$, whereas the signal decreases $\propto \theta^{-1}$, for $\theta < \theta_H$. Therefore the signal to noise of the measurement is maximized if we include all lenses within a few θ_H of the source and ignore those with larger impact parameters.

Galaxy-galaxy lensing is not subject to systematic error to the same degree as large scale structure because the most serious errors are diminished by the angular averaging. However, there are two additional potential sources of error. The first is that when the source is too close to the deflector, the isophotes can overlap so that galaxies appear to be elongated radially. The second is that some designated source galaxies will actually be companions of the deflector and will be subject to Newtonian tidal forces that may stretch them out in tail along their orbits.In practice, both of these effects can be avoided by excluding impact parameters $\theta < 5''$.

A positive measurement of this effect has previously been reported by Brainerd, Blandford & Smail 1996 who analyzed a $9.6' \times 9.6'$ red image taken under conditions with average seeing $\sim 0.8''$ with the COSMIC imager on the 5m Hale telescope. The detected galaxies were partitioned into $\sim$ 500 sources with $23 < r < 24$ and a similar number of deflectors with $r < 23$. Source-deflector separations in the range $5'' < \theta < 34''$ were included. A polarization signal $p \sim 0.01$ was measured, which is consistent with a simple model of the deflector galaxies in which the circular velocity of an L^* galaxy is $\sim$ 220 km s^{-1} and the physical radius of the halo of an L^* galaxy is $\sim 50h^{-1}$ kpc. (This last quantity is only poorly constrained by the analysis.) In addition, Griffiths et al. 1996 reported detecting this effect using HST images drawn from the MDS survey. They were able to distinguish ellipticals from spirals and found that the effect was stronger for the former. Turning to the HDF itself, Dell'Antonio & Tyson 1996 reported a detection of galaxy-galaxy lensing for pairs of galaxies with $\theta < 5''$, consistent with an L^* circular velocity of $\sim$ 260 km s^{-1}.

Each of these studies considered sources and deflectors in a blind, pairwise manner. In practice, this dilutes the signal by weighting all pairs equally, including those where $z_z < z_d$. Clearly, if we know the redshift distribution of the deflector galaxies, then we should be able to measure a much larger signal to noise ratio from a given set of imaging data. These studies assumed the source redshift distribution and focused on describing the mass distribution in the deflectors. In the present analysis of the HDF, we have taken the opposite tack and restrict our deflector sample to those galaxies with known redshifts (cf. Cohen these proceedings), and, consequently, well-determined luminosities. We then address the scientifically more pressing question of the redshift distribution of the sources, which are mostly far too faint for spectroscopic study. The deflector field is formed by over 400 galaxies of which about 100 are on the HDF while the remainder are in the flanking fields. This ensures that the average distribution of the deflector galaxies about the source galaxies is isotropic on the sky a neutralizes a potentially serious source of systematic error. A subset of these galaxies have nine wavelength photometry and the remainder have six wavelength photometry. We can therefore interpolate their rest frame B magnitudes and so estimate L/L^*. We then adopt the deflector galaxy model of Brainerd, Blandford & Smail 1996 in which the circular velocity and halo radius scale according to

$$V_c = 220(L/L^*)^{1/4} \text{ kms}^{-1}$$
$$r_H = 50h^{-1}(L/L^*)^{1/2} \text{ kpc.} \tag{4.4}$$

We also adopt a specific world model—a FRW open universe with $\Omega_0 = 0.3$—although our results turn out to be pretty insensitive to this choice. Using this model of the deflector galaxies, it is possible to predict the deflector field, $p(X, Y, z_s)$, where X, Y are angular coordinates on the sky.

Our total source sample comprises $\sim$ 1000 galaxies each covering > 15 $0.04''$ pixels. This includes some galaxies as faint as $V = 28$. The typical ellipticity is $\epsilon = 0.3$. In one simple approach to analyzing the data, we assume that all the sources are at high z_s and form the complex correlation coefficient $C = < p\chi^* >$ averaging over the whole source sample. The expectation of this statistic should be real; the imaginary part, measuring the correlation expected if each source were to be rotated through 90°. The value obtained is $C = 0.0008 + 0.0000i$. The expected error in C is ~ 0.0003 and we have a $\sim 3\sigma$ detection of galaxy-galaxy lensing. The expected value on the basis of our deflector galaxy model is ~ 0.003. After correcting for the image polarizibility, we find that the signal is roughly a third of that expected on our simple model.

A more detailed analysis of galaxy-galaxy lensing will be presented shortly. Results consistent with those described above have been presented here by Hudson et al. Their analysis uses color redshifts for both the lens and the source galaxies and is directed toward measuring the properties of the masses of the halos of the deflector galaxies.

5. Discussion

To summarize this report, we still have no good examples of strong lensing on the HDF. There remain some suggestive possibilities but nothing is secure. How do we reconcile this with our original expectation based upon the radio gravitational lens surveys? We suspect that there are two important effects at work. The first is that, unlike the compact radio sources, the faint optical source population is probably not all at high redshift; a significant fraction of these galaxies may belong to a population of local dwarf galaxies. The second possible contributing factor is a bit more subtle and has larger ramifications, at least for gravitational lensing. It is being found that a large fraction of the secure galaxy gravitational lenses are in compact groups of galaxies with similar redshifts. This may contribute to the explanation of two puzzles involving the existing lens sample—the preponderance of quads over doubles and the existence of spiral galaxy lenses. This will be discussed in greater detail elsewhere (Kundić et al. 1997 in preparation).

On this basis, it is reasonable to hypothesize that the field galaxy lensing rate really is low. We can test this on the HDF by taking a different approach to gravitational lensing. Strong gravitational lensing invariably involves the luminous regions in the cores of deflector galaxies. The incidence is relatively insensitive to reasonable assumptions about the distribution of dark matter. We can therefore measure the surface brightness of the luminous matter and use the locally determined mass to light ratio ($\sim 5h$ for spirals, $\sim 10h$ for ellipticals Kochanek, C. S. 1996 and references cited therein), to compute the surface density distribution. (Lens mass to light ratios are about twice this value Kochanek, C. S. 1996.) It is then straightforward to compute the deflection field and the cross section for multiple imaging as a function of the source redshift. Carrying out this exercise on the HDF, we find that the integrated cross section (for $z_s \sim 3$) is dominated by elliptical galaxies with $z_d \sim 1$ and totals ~ 5 arcsec^2 (cf. Im, Griffiths & Ratnatunga 1997). Even with the faintest source galaxies, the probability of forming just one strong lens on the HDF is only ~ 0.5. It might be thought hard to detect a faint source galaxy against a bright deflector, especially as the irregular blue arcs one is seeking could easily be confused with spiral arms. However, as most of the potential lens galaxies are actually red ellipticals, it is possible to identify strong lenses in this manner. Simulations using images from the HDF confirm this. A search of the deflector images reveals only one possible example of strong lens consistent with our expectations. Nevertheless, given a much larger area imaged to HDF depth and an associated program to determine redshifts either photometrically or spectroscopically, it should be possible to quantify the incidence of strong lensing in this manner. A fuller description of this topic is in preparation.

Similar remarks apply to the analysis of galaxy-galaxy lensing. The simplest rationalization of our results is that the deflector galaxies are somewhat less massive than local galaxies relative to their luminosities consistent with passive evolution and that the source galaxies are not all as distant as we have presumed. The prospects for developing this approach using a larger field are excellent. Instead of assuming a source redshift distribution, it is possible to invert the observed incidence of lensing to infer the redshift distribution of the faintest galaxies and partition them directly between the first protogalactic fragments destined to merge to form the contemporary galaxy distribution, smaller galaxies at intermediate redshift that have faded to local obscurity and a hitherto undetected population of low luminosity dwarf galaxies. This will also be discussed at greater length.

I am most indebted to my colleagues, Tereasa Brainerd, Judith Cohen, Chris Fassnacht, David Hogg, Tomislav Kundić and Sangeeta Malhotra for collaboration on the above. Thanks are also due to the HST team led by Bob Williams for executing such an imaginative project so well. Support under NASA grant AR-06337.IS-94A and NSF grant 95-29170 is gratefully acknowledged.

REFERENCES

BLANDFORD, R. D., SAUST, A.-B., BRAINERD, T. G. & VILLUMSEN, J. V. 1991, *MNRAS* **251**, 600.

BRAINERD, T. G., BLANDFORD, R. D. & SMAIL, I. R. 1996, *ApJ* **466**, 623.

COLEMAN, G. D., WU, C. C. & WEEDMAN, D. W. 1980, *ApJS* **43**, 393.

DELL'ANTONIO, I. P. & TYSON, J. A. 1996, *ApJ* **473**, L17.

GRIFFITHS, R. E., CASERTANO, S., IM, M. & RATNATUNGA, K. U. 1996, *MNRAS* **282**, 1159.

HOGG, D. W., BLANDFORD, R.ED., KUNDIĆ, T., FASSNACHT, C. D.& MALHOTRA, R. 1996, *ApJ* **467** L73.

IM, M. S., GRIFFITHS, R. E. & RATNATUNGA, K. U. 1997, *ApJ* **475**, 457.

KOCHANEK, C. S. 1996, *ApJ*, **466**, 638

MIRALDA-ESCUDÉ, J. 1991, *ApJ* **380**, 1.

MYERS, S. G. 1996 in *Astrophysical Applications of Gravitational Lensing IAU 173* (ed. C. S. Kochanek & J. N. Hewitt). p. 317. Kluwer.

SCHNEIDER, P., EHLERS, J. & FALCO, E. E. 1992, in *Gravitational Lenses.* Springer.

Schneider, P. & Rix, H.-W. 1997, *ApJ* **474**, 25.

STEIDEL, C. C., GIAVALISCO, M., DICKINSON, M. & ADELBERGER, K. L. 1996, *AJ* **112**, 352.

TYSON, J. A., VALDES, F., JARVIS, J. F. & MILLS, JR., A. P. 1984, *ApJ* **349**, L1.

WEBSTER, R. L. 1985, *MNRAS*, **213**, 871.

WILLIAMS, R. E. ET AL. 1996, *AJ* **112**, 1335.

ZEPF, S. E., MOUSTAKIS, L. A. & DAVIS, M. 1997, *ApJ* **474**, L1.

White dwarf stars and the Hubble Deep Field

By STEVEN D. KAWALER

Department of Physics and Astronomy, Iowa State University, Ames, IA 50011 USA

1. Introduction

The value of the *Hubble Deep Field* to study of the remote regions of the observable Universe is difficult to overstate, as the many exciting results of this workshop are ample testament. However, an added bonus of the HDF is that, although it a very narrow angle survey, the depth of the HDF results in its sampling a significant volume of the halo of our galaxy. Thus it is useful for the purposes of detecting (or placing upper limits on the distribution of) intrinsically faint stars, such as white dwarfs. In this capacity, the HDF can say some important things about the possible constituents of the baryonic dark matter in the halo of the Milky Way, as well as constrain the physics of white dwarf formation and cooling.

Such an investigation is timely since white dwarfs could provide a significant fraction of the total mass of the halo of the Milky Way. The MACHO observations of microlensing events in the direction of the LMC suggests that up to 50% of the dark matter halo of the Milky Way could be comprised of faint Population II white dwarf stars (Alcock et al. 1997). Thus, constraints on the population of halo white dwarfs from the HDF can directly address this possible partial explanation of the nature of the dark halo of the Milky Way.

In this review, I hope to illustrate how the HDF can be used to constrain the luminosity function of halo white dwarfs. I begin with a brief summary of the observed white dwarf luminosity function (WDLF) of the galactic disk, and show how the HDF serves as a probe of the WDLF for the halo. I then review the theoretical background used in interpreting the WDLF in terms of the theory of white dwarf evolution and cooling, and the history of star formation in the galaxy. We are then in a position to explore the theoretical WDLF, beginning with the WDLF of the disk. We then move to a theoretical examination of the WDLF of the halo population. The results of searches for white dwarfs on the HDF can then be examined in terms of the halo white dwarf population.

1.1. *White dwarf stars*

White dwarf stars represent the final stage of evolution of stars like our sun. They represent the ultimate fate of all stars with masses less than about 8 $M_{\odot}$, and are the natural consequence of the finite fuel supply of these stars. Upon exhaustion of their nuclear fuels of hydrogen and helium, these stars lack sufficient mass to take advantage of the limited return of carbon fusion, and are doomed to gravitational collapse. Upon reaching a radius comparable to that of the Earth, the inner cores of these stars support themselves almost entirely by electron degeneracy pressure. Their final transformation is through the gradual release of heat that has been stored within them during the prior stages of nuclear burning.

There is a simple and compelling reason to care about white dwarf stars. It can be shown that *98% of all stars are or will be white dwarfs!* In terms of stellar mass, 94% of all matter in stars is either already locked-up in white dwarfs, or is in stars that will eventually become white dwarfs; a mere 6% of matter is destined to be, or already incorporated in neutron stars, and an insignificant amount is on the black hole path.

The basic theory of white dwarf stars was stimulated by their identification, early in this century, as a class of stars with observable luminosity but very small radius. The solution to this mystery, by Chandrasekhar and Fowler, neatly integrated the then-new sciences of quantum mechanics and relativity with astrophysics. Their work was essentially a complete description of the inner workings of cool white dwarf stars. While Chandrasekhar's theory is indeed elegant, in the later half of this century we have come to recognize that these stars represent a rich storehouse of information on the evolution of all stars. This information is available through examination of the luminosity function of white dwarf stars.

The relatively recent discovery of white dwarfs is evidence that they are relatively hard to find. Almost all white dwarfs that we know of are in the immediate solar neighborhood; 50% of known white dwarfs lie within 24 parsecs of the Sun. White dwarf discoveries have come from surveys of stars with large proper motions, searches for faint blue objects, and examination of faint members of common proper motion pairs. All three produce candidate objects that require spectroscopic follow-up observations for confirmation. Spectroscopic follow-up has identified a large number of white dwarf stars. However, the selection effects involved in such surveys are thorny to account for, and make statistical studies suspect.

The most successful recent surveys for producing new white dwarf identifications have been surveys for faint but excessively blue stars. A prime example is the Palomar-Green (PG) survey (Green et al. 1986) for blue objects out of the galactic plane. While the PG survey (and others like it) are designed for discovery of quasars, they are almost optimally designed to pick out white dwarf stars. Once again, spectroscopic follow-up is required, but such surveys have an advantage in that they produce magnitude-limited samples, and therefore are without the peculiar selection effects associated with proper-motion identifications. With the success of the PG survey, others are underway, such as the Montreal-Cambridge survey (Demers et al. 1986) and the Edinburgh-Cape survey (Stobie et al. 1987, Kilkenny et al. 1991).

Because of the relatively narrow range in masses of white dwarfs (and their confinement to a line of constant radius in the H-R diagram) the Greenstein colors allow a fair determination of the absolute visual magnitude M_v. In this system, the (U-V) color provides a useful indicator for the temperature of hot white dwarfs, while the (G-R) color index works best for cooler white dwarfs. Conversion of M_v to bolometric magnitude M_{bol} requires model atmosphere bolometric corrections, such as given in Greenstein (1976) for cool hydrogen-atmosphere white dwarfs. Of course, when parallax measurements are available, more accurate estimates of individual values of M_v and M_{bol} are possible.

1.2. *The observed white dwarf luminosity function and the age of the Galactic disk*

Using estimates of M_v, one can construct a luminosity function for white dwarfs. The luminosity function is a representation of the relative space densities of white dwarfs with given absolute magnitudes, usually plotted in terms of number of stars per cubic parsec per unit bolometric magnitude, versus luminosity. This luminosity function for low luminosities by Liebert et al. (1988) used data on stars in the Luyten (1979) survey. An example of the observed luminosity function is shown in Figure 1, with data from (1988). The luminosity function increases steadily with decreasing luminosity. As could be expected, the cooler a white dwarf is, the more slowly it cools and fades, and so the number increases with lower temperature and luminosity.

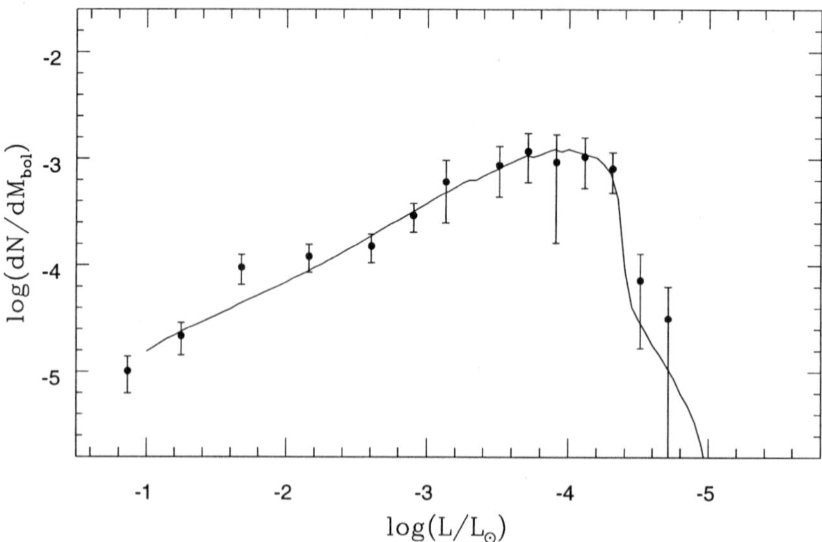

FIGURE 1. The luminosity function of DA white dwarfs. Data points are from Fleming et al. (1986) and Liebert et al. (1988). The line is a representative theoretical luminosity function using the pure carbon white dwarf models of Winget et al. (1987) and a disk age of 8 Gyr.

However, there is a steep turn-down in the luminosity function below $10^{-4.4}$ $L_{\odot}$ that requires explanation. This turn-down is the result of the finite time that white dwarfs in the solar neighborhood, and by extension in our galaxy, have had to cool. The time it takes for white dwarfs to fade to below this luminosity must be longer than the age of the galaxy itself. Thus the luminosity of this cut-off is a direct measure of the age of our galaxy. For Figure 1, I show a sample theoretical luminosity function calculated with simple models of pure carbon white dwarfs published by Winget et al. (1987). Matt Wood's results (1992) using the best available white dwarf models and observed luminosity functions, indicate that the age of the galactic disk in the vicinity of the Sun is 9.3 ± 1.5 Gyr. From the shape of the luminosity function, it is apparent that the star formation rate has been roughly constant over the history of the disk. Had there been bursts of star formation at some times, these bursts would have produced bumps along the observed luminosity function that are not seen in the data (Wood 1992, Iben & Laughlan 1989).

1.3. *Other ways of finding white dwarfs: the HDF and MACHO searches*

The techniques described above are adequate for studying relatively bright white dwarfs that lie reasonably close to the Sun. As such, they are targeted for white dwarfs from the disk of the galaxy. Halo white dwarfs, on the other hand, could escape detection in these surveys by several means. If they are not close to the Sun, such halo white dwarfs could easily be fainter than the magnitude limits of these surveys. With the high velocities expected for halo stars, halo white dwarfs brighter than the limiting magnitude of current surveys could have enormous proper motions; thus they would have escaped detection by having proper motions larger than the upper limit of proper motion studies.

The HDF represents one approach to finding halo white dwarfs: extremely deep surveys over a small range of the sky that can find very faint stars. The discussion below (from Kawaler 1996) shows that the HDF samples a reasonable volume of the halo. Another novel approach is to search for microlensing events, where lensing of distant stars occurs through the action of an intervening halo white dwarf.

1.3.1. *The halo as probed by the HDF*

For a small survey area such as the HDF, the volume V of space sampled out to a distance d (in pc) given an area of A square arc minutes is simply

$$V = 2.82 \times 10^{-8} A d^3 \ \text{pc}^3 \, . \tag{1.1}$$

The HDF covered an area of approximately 4 square arc minutes; therefore since it looked out of the plane, taking $d = 500$ pc suggests that it samples a disk volume of only 14 pc^3. With such a small volume sample, it is not surprising that no disk white dwarfs are seen in the HDF; the approximate space density of white dwarf stars is 3×10^{-3} per cubic parsec (Liebert et al. 1988).

For a putative population of white dwarfs in the halo, the HDF has sampled a much larger volume. Assuming a limiting magnitude of m_l, and a white dwarf with an absolute magnitude M_f (in the same band as the limiting magnitude), then the effective volume contained within can be written in terms of the limiting magnitude of the survey and the absolute magnitude of the faintest white dwarf:

$$\log(V) = -4.550 + \log(A) + 0.6(m_l - M_f) \tag{1.2}$$

where V is in pc^3 and A is in square arc minutes. Using some representative numbers for the HDF, $m_l \approx 27.5$ and $M_f \approx 16.5$ yields a search volume of 450 pc^3.

Now, assume the density distribution of the halo is spherically symmetric about the center of the galaxy, and follows a standard "softened potential" distribution such as described in Binney & Tremaine (1987, p. 601, with $\gamma = 2$). With this form for the halo mass distribution, and parameter for the halo as in Bahcall & Soneira (1980), the mass of the halo sampled by the HDF field (in solar masses) can be computed (for the details, see Kawaler (1996). The halo mass sampled by the HDF ranges from 1.05 $M_\odot$ to 7.8 $M_\odot$ for a reasonable range of M_f and m_l. Thus the HDF in principle samples several solar masses of halo material.

1.3.2. *MACHO searches*

The MACHO collaboration has published an analysis of 8 microlensing events in the direction of the LMC, which they attribute to halo objects in the Milky Way (Alcock et al. 1997). The duration of these events suggests that the lensing objects have masses between 0.1 and 1.0 $M_\odot$. The lensing objects must be of extremely low luminosity; therefore ordinary dwarf stars are ruled out. Lensing at this rate allows the MACHO collaboration to estimate that up to 50% of the dark matter halo of the Milky Way could therefore be comprised of white dwarf stars. Considering the large mass of the galactic halo, this implies that halo white dwarfs must be extremely abundant. Direct observational constraints on the halo white dwarf luminosity function from observations (Liebert et al. 1988) are not necessarily in conflict with the MACHO result; however the luminosity function of white dwarfs in the disk must then contain a fair fraction of halo white dwarfs.

Can the lenses responsible for the MACHO results be white dwarfs, and still be consisted with the white dwarf space density found by Liebert et al. (1988)? Given the age of the halo as determined from, for example, globular cluster studies, the observed downturn in the white dwarf luminosity function along with the absence of any white dwarfs with $M_{\text{bol}} > 16.2$ constrains the star formation history for the generation of halo stars that might have produced such a halo white dwarf population (Tamanaha et al. 1990,

Adams & Laughlin 1996). Under conventional assumptions about star formation early in the history of the galaxy, Adams & Laughlin (1996) conclude that the observations of Liebert et al. (1988) already limit the fraction of the dark matter halo that can be attributed to white dwarfs to less than 25%. On the other hand, Tamanaha et al. (1990) show that extreme conditions, such as an enormous burst of star formation early in the history of the galaxy, could produce a massive number of white dwarfs in the halo that could escape detection by traditional ground-based studies (see below).

2. The basics of white dwarf cooling

To understand the WDLF requires combining the star formation rate as a function of time and mass, the processes by which main sequence stars become white dwarfs, and the rate at which white dwarf stars, once formed, fade and cool. Therefore, as Don Winget is fond of saying, the history of star formation in the disk of our galaxy is written in the coolest white dwarf stars. In this section, I concentrate on one element of the theoretical white dwarf luminosity function: the theory of white dwarf cooling. For more details see, for example, Kawaler (1997) and Van Horn (1971) and references in between. With an understanding of white dwarf cooling, the observed luminosity function can constrain the other inputs. In the next section, we will combine white dwarf cooling with the remaining inputs needed to model the complete WDLF.

2.1. *Simple "Mestel" cooling theory*

In a remarkable 1952 paper, Leon Mestel laid the groundwork for the study of white dwarf cooling (Mestel 1952). Following the discovery of the peculiar nature of white dwarf stars, the question arose as to what their power source might be. Nuclear burning, identified as the energy source in ordinary stars, was an early suspect, but Ledoux & Sauvenier-Goffen (1950) showed that if white dwarfs were powered by nuclear burning, then they would be vibrationally unstable. Since at that time white dwarfs were known to be non-variable, their conclusion was that another mechanism was needed. Mestel (1952) identified the luminosity source as release of stored thermal energy.

The reservoir of thermal energy stored in the core of a white dwarf star is slowly depleted by leakage into space. The cooling process is not unlike that undergone by a heated brick placed in a cool environment. Such a brick would cool rapidly when exposed to the cool outside air unless surrounded by some sort of insulating blanket. With insulation, the cooling of the brick is slowed as the temperature gradient between it and its surroundings is reduced by the poor thermal transport properties of the blanket. In a white dwarf star, heat is transported through the degenerate core by the efficient process of electron conduction, while in the nondegenerate envelope energy is transported by the (much less efficient) form of photon diffusion. Thus in a white dwarf star, the degenerate core corresponds to the hot brick, and the nondegenerate outer layers play the role of an insulating blanket.

Mestel (1952) began with the equations of stellar evolution, and then made several simplifying assumptions for the case of white dwarf stars. Nuclear energy generation and gravitational contraction are assumed to play no role. The specific heat in the electron-degenerate, isothermal (temperature T_c) core is set by the (nondegenerate) ions. Combining these assumptions, and integrating through the degenerate core, the luminosity of a white dwarf star can be expressed in terms of the total stellar mass, core composition, and rate of change of the core temperature.

Atop the degenerate core is the nondegenerate envelope, which contains (by assumption) only a small fraction of the mass of the star. If the envelope is radiative, then the rate of energy transfer is governed by the radiative opacity of the material. Adopting a Kramers opacity law and also assume a zero boundary condition leads to the so-called radiative zero solution for T as a function of P (see for example Hansen & Kawaler 1994). Matching this solution to the degenerate core yields an expression relating the stellar luminosity to T_c and the core composition.

Combining these two expressions for the luminosity and integrating with respect to time yields the time needed to cool (actually, fade) to a given luminosity:

$$t_{\rm cool} = 9.41 \times 10^6 \text{ yr} \left(\frac{A}{12}\right)^{-1} \left(\frac{\mu_e}{2}\right)^{4/3} \mu^{-2/7} \left(\frac{M}{M_\odot}\right)^{5/7} \left(\frac{L}{L_\odot}\right)^{-5/7} \tag{2.3}$$

Some features to note about this remarkable result include the dependence of $t_{\rm cool}$ on A and M. Cores with larger A (cores composed of heavier elements) cool faster than "lighter" cores. If we assume representative values for mass ($0.60\ M_\odot$), atomic weight ($A = 14$, a 50/50 mix of carbon and oxygen), $\mu = 1.4$, $\mu_e = 2$, and a luminosity corresponding to the faintest white dwarfs ($L = 10^{-4.5}\ L_\odot$), then this simplified analysis results in a cooling time of 7×10^9 years.

2.2. *Complications at the cool end: crystallization*

This estimate of the cooling time is remarkably close (within 30%) of the most modern computation of the cooling age of these coolest white dwarf stars. This despite the fact that it leaves out several effects that are now known to be very important: neutrino cooling, prior evolutionary history, crystallization effects, etc. Van Horn (1971) reviews the Mestel theory and evaluates the precision of the assumptions; Iben & Tutukov (1984) discuss the Mestel cooling law in light of their more detailed evolutionary models.

The Mestel law assumes that the white dwarf interior is an ideal gas, in which there are no electromagnetic interactions between the nuclei. But this is, after all, material in the core of a white dwarf star, where densities are enormous by terrestrial standards. Above some density (and at finite temperature) the ions are indeed sufficiently crowded that their electrostatic interaction can affect the equation of state. As white dwarfs cool, the effects of Coulomb interactions become more important. Also, since the interiors of white dwarfs are roughly isothermal, the Coulomb effects are largest at the center. More details relevant to white dwarf interiors are available from numerous sources; see for example Shapiro & Teukolsky (1983) and references therein.

When the density becomes large enough, Coulomb effects overwhelm those of thermal agitation, long-range forces organize the small-scale structure of the material, and the gas settles down into a crystal. For conditions relevant to white dwarf stars, with central densities of order 10^6 [g/cm^3], oxygen crystallizes at about $3.4 \times 10^6\ K$, while carbon crystallizes at $2.1 \times 10^6\ K$. These central temperatures are reached when the white dwarf has been cooling for about 10^9 years. Thus crystallization can be an important process in the evolution of white dwarf stars. Note that the crystallization condition assumes a one component plasma, while material within most white dwarfs is probably a mixture of at least carbon and oxygen. The actual process of crystallization must be extremely complex.

2.2.1. *Effects of crystallization on the cooling rate*

As a white dwarf cools, the specific heat for the degenerate interior changes in ways most familiar to condensed matter physicists. Here, I only sketchily summarize the behavior of the specific heat; further details (in language appropriate for astrophysicists) are clearly described by Shapiro & Teukolsky (1983). In an ideal perfect gas, such as the ions within a white dwarf, the specific heat at constant density is independent of temperature, and proportional only to the mean atomic weight μ. In the limit of complete degeneracy (such as electrons in a white dwarf core), the internal energy is independent of the temperature. For such a system, the specific heat is identically zero. Therefore, as white dwarf material becomes degenerate, the specific heat of the electrons becomes very small. Since the ions are nondegenerate, however, the specific heat of the ions remains unchanged. Therefore the specific heat of electron-degenerate material is determined solely by the specific heat of the ions which is a constant for a given composition.

When crystallization begins, the latent heat release associated with the growing long-range ordering provides an additional luminosity source. Lattice vibrations (in three dimensions) ultimately result in a doubling of the specific heat as the material cools towards crystallization. With a higher specific heat, the rate of change of the luminosity decreases with time as the latent heat of crystallization provides an additional source of energy. In average white dwarfs, this occurs at $\log L/L_{\odot} \approx -3.6$ to -4.2. In this luminosity range, a white dwarf at a given luminosity is probably older than the Mestel law would indicate.

As the core cools further, the internal energy of the lattice is determined by quantum mechanical effects. Upon cooling below the Debye temperature, the material reaches the Debye cooling phase, with a specific heat that drops in proportion to the temperature. The rate of change of luminosity increases dramatically; a low specific heat means that the star cannot hold in thermal energy easily. Thus once a white dwarf cools below the Debye temperature, its luminosity plummets faster than the Mestel law alone would predict. In this Debye cooling phase, a white dwarf at a given luminosity is probably a bit younger than the Mestel law would suggest.

Figure 2 shows a representative white dwarf cooling curve, kindly provided by Matt Wood, as a solid line. The dashed line represents Mestel cooling. At the low luminosity end, the bump corresponding to the release of latent heat of crystallization is clearly evident. The lowest luminosity portion shows the effects of Debye cooling

2.2.2. *Fractionation during crystallization: another energy source?*

Along with the latent heat of crystallization, another possible source of energy in a cooling white dwarf is rooted in the fact that the material undergoing crystallization is of mixed composition. Most white dwarf stars have degenerate cores composed of a mixture of carbon and oxygen, along with at least trace abundances of other heavier elements.

The temperature at which material may crystallize depends on the atomic weight and charge of the material. As indicated earlier, the temperature at which oxygen can crystallize is approximately 50% higher than the temperature at which carbon does so. Thus considered separately, oxygen will begin to solidify within the still (degenerate) gaseous carbon environment. What will happen to the growing oxygen crystalline material? Will it condense out forming precipitating grains that sink towards the stellar center? Will the material remain mixed but "slushy" until it cools below the crystallization temperature for carbon?

The question of differential crystallization and possible fractionation is a very important one. If it occurs in a way that causes element separation, then the gravitational potential energy released during the settling acts as an energy source that can slow the cooling of the star as a whole. If plotted in Figure 2, the effects would be to stretch out the solid curve to the right (longer times) at low luminosities. Currently a very active area of research, the situation is explored in some detail by Segretain et al. (1994) and others (see for example Isern et al. 1997), who conclude that such a process can slow the evolution of carbon/oxygen white dwarf stars to the lowest observed luminosities by up to two billion years or more.

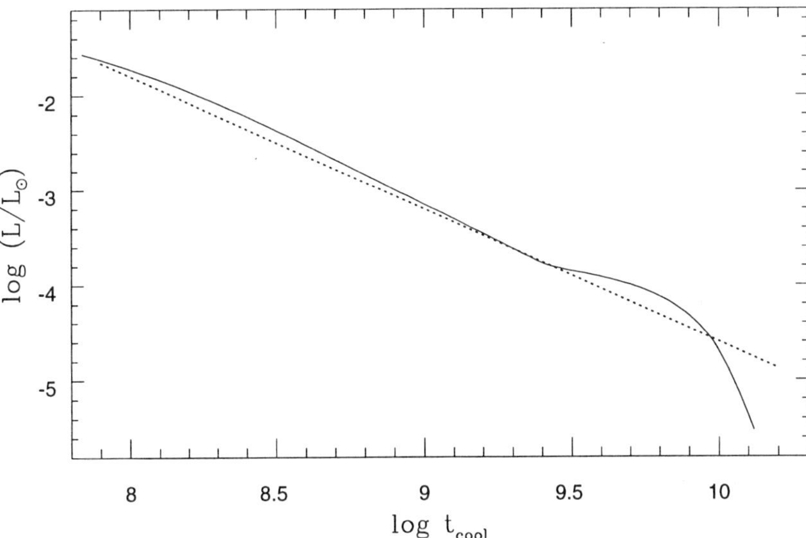

FIGURE 2. A representative cooling curve for a 0.60 $M_\odot$ white dwarf model (solid line) with a carbon core, and stratified outer layer of helium (10^{-2} $M_\odot$ and hydrogen (10^{-4} $M_\odot$), kindly provided by Matt Wood. This model includes the effects of crystallization, but not fractionation (since it is pure carbon in the core). The dashed line is the comparable cooling curve based on the Mestel law. The differences illustrate how the Mestel cooling law is modified by inclusion of crystallization effects.

2.3. *Realistic calculations*

In constructing realistic cooling curves, modelers of white dwarf stars try to include all of the known physics within the model. The physical properties enter the cooling curve as the constitutive relations that provide the coefficients of the equations of stellar structure and evolution. These models use realistic conductive and radiative opacities, sophisticated equations-of-state, including the effects of degeneracy, Coulomb interactions, and crystallization.

Three representative sequences from the calculations of Wood (as quoted by Winget et al. 1987) are shown in Figure 3. This figure shows several of the basic scalings of the Mestel law hold for these complete models. First note that at a given luminosity the more massive models are older, as expected from the above equation. All of the tracks parallel the Mestel law; that is, they show a general power-law slope of -1.4. Note that the luminosity of the 0.80 $M_\odot$ model shown begins to drop quickly at a luminosity below 10^{-4} $L_\odot$. This is the effect of crystallization; these models have already largely crystallized, and they are showing the effects of Debye cooling discussed in a previous section. The 0.60 $M_\odot$ model begins Debye cooling at a lower luminosity and later time; this implies that more massive models cool more slowly, but that they crystallize earlier.

Before the downturn, the cooling curve flattens compared to the Mestel law. This is consistent with the expectations, described above, that as crystallization begins, the specific heat rises with the release of latent heat of crystallization. This "extra" luminosity source slows the evolution. Once largely crystallized, the models enter the Debye cooling stage resulting in the acceleration of the luminosity drop of the white dwarf models. Because of this, even though they cool more slowly initially, we expect that the faintest

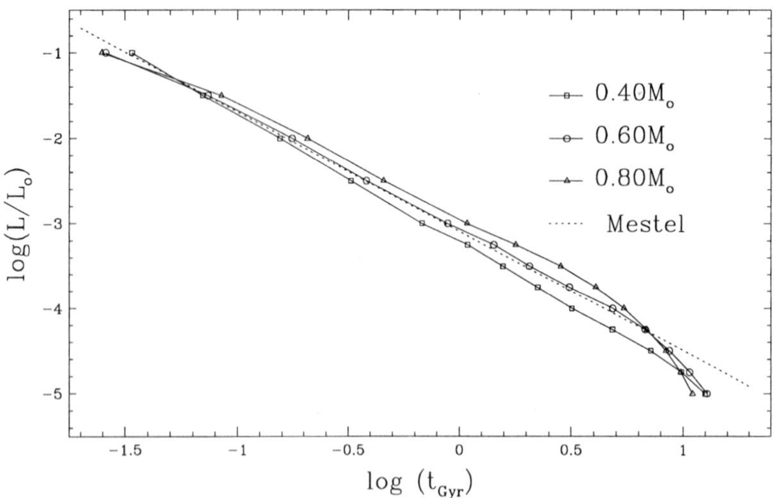

FIGURE 3. Cooling curves for representative white dwarf evolutionary sequences. Data from Winget et al. 1987; figure from Hansen & Kawaler 1994

white dwarfs in the galaxy have, on average, higher mass than slightly more luminous (and therefore younger still) white dwarfs.

The most complete models of cooling white dwarfs, and the white dwarf luminosity function, is the work of Wood (1992), who computed many sequences of evolving white dwarf models with many different combinations of parameters. Since white dwarf stars at the cool end of the luminosity function have largely forgotten their initial conditions, Wood used a homologous set of starting models. Convergence of the evolutionary tracks was essentially complete by the time the models reached a luminosity of roughly 0.1 $L_\odot$.

3. The WD luminosity function of the disk

3.1. *Constructing theoretical WDLFs*

With theoretical cooling curves, we need a way to see if the theoretical inputs to the models are realistic. The principal observational contact of the theory of white dwarf cooling is the observed *luminosity function* of white dwarf stars: the number of white dwarfs per cubic parsec per unit luminosity (or bolometric magnitude), usually denoted as Φ.

The cooling rate of white dwarfs is one input into the construction of a theoretical luminosity function. As described by Wood (1992), a theoretical determination of the luminosity function requires evaluation of the expression

$$\Phi(L) = \int_{M_{\rm low}}^{M_{\rm hi}} \psi(t)\, \phi(M)\, \frac{dt_{\rm cool}}{d\log(L/L_\odot)}\, dM \tag{3.4}$$

at a given value for the population's age. In this expression, $\psi(t)$ is the star formation rate at the time of the birth of the white dwarf progenitor, and $\phi(M)$ is the initial mass function for stars with initial mass M. The mass limits of the integration cover the mass range of main sequence stars that produce white dwarfs. The upper mass limit is simply the largest main-sequence mass that produces white dwarfs (about 8–10 $M_\odot$). The lower mass limit is approximately the main-sequence turn-off age for the population age; simply put, stars lower than this mass have not yet produced white dwarfs. One may obtain this mass limit by estimating the main sequence lifetime with a relation such as

$$\log t_{ms} = 9.921 - 3.6648(\log M) + 1.9697(\log M)^2 - 0.9369(\log M)^3 \qquad (3.5)$$

from Iben & Laughlin (1989. The initial mass function (IMF) of stars in the galaxy, $\phi_s(M)dM$ is the number of stars formed per year per cubic parsec within an interval of masses between M and $M + dM$. Salpeter (1955) found that

$$\phi\, dM = 2 \times 10^{-12} M^{-2.35}\, dM \;\; \mathrm{stars/yr/pc^3}, \qquad (3.6)$$

which is the famous "Salpeter mass function." For the disk, one begins by assuming a constant star formation rate ψ_o with a value adjusted to match the observed white dwarf space density.

The remaining quantity is the inverse of the rate of change of the white dwarf luminosity; it is a time scale for luminosity change that we'll denote as $\tau_{\rm cool}$. This number is provided by the cooling theory for white dwarfs given the white dwarf luminosity and mass. In the form above, however, $\tau_{\rm cool}$ needs to be specified at a given white dwarf luminosity and progenitor mass M. To determine $\tau_{\rm cool}$, one needs to know the mass of the white dwarf that a star with a main-sequence mass M produces, the time it takes to cool to luminosity L, and the fading rate at that luminosity. The white dwarf mass corresponding to an initial mass M comes from empirical studies of the $M_i - M_f$ relation, which has been explored extensively for the past two decades. A useful parametric form is presented by Wood (1992) as

$$M_f = 0.49 \exp(0.095 M_i)\ . \qquad (3.7)$$

Given this mass for the white dwarf, and the luminosity L, white dwarf cooling models can be consulted to yield the cooling age t_{cool}, its derivative, and therefore $\tau_{\rm cool}$.

Iben & Laughlin (1989) show how the luminosity function appears for several simplified relations for the star formation rate, etc.; this paper is an excellent starting point for those who wish to work with the white dwarf luminosity function for various purposes. Figure 4 shows an example of a simplified luminosity function and the effect of the population age on it. Quite simply, the older the population the lower the luminosity of the cutoff. The stars populating the lowest luminosity bin are the oldest white dwarfs in the sample; with increasing age, they reach lower luminosity and pile up because of the lengthening cooling time scale.

When crystallization begins, the cooling rate temporarily slows, and the luminosity function will rise above the Mestel curve. Debye cooling will cause a drop in Φ with the associated accelerated cooling. Of course, realistic calculations of the luminosity function are needed for detailed comparison with the observed luminosity function.

Figure 5 shows a sample of realistic luminosity functions constructed using realistic white dwarf models computed (and kindly provided) by Matt Wood (described in Oswalt et al. 1996). Note that the general slope of this cooling curve is nearly $-5/7$, but with a depression at high luminosities and a bump near the final cutoff. Also shown in Figure 5 is the observed luminosity function of Liebert et al. (1988). Agreement between the theoretical and observed luminosity functions above the drop-off is very good; those portions that disagree may result from changes in the star formation rate and/or the IMF over time. Such possibilities provide an intriguing use of the white dwarf luminosity function to explore the history of star formation in our galaxy (or, in the near future, in other stellar populations such as globular clusters).

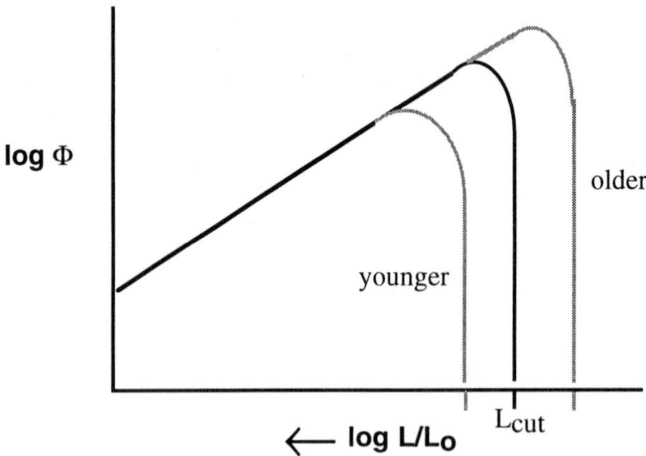

FIGURE 4. Schematic luminosity functions for populations of three different ages, from Kawaler (1997). This illustrates the case of a constant star formation rate with time, a Mestel cooling law ($t_{\rm cool} \propto L^{-5/7}$), and three disk ages.

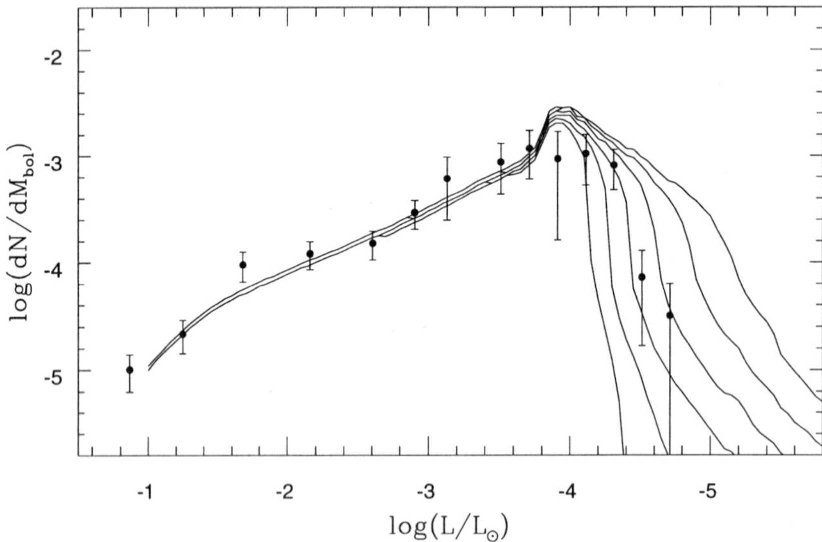

FIGURE 5. Theoretical white dwarf luminosity functions (constructed with DA white dwarf evolutionary sequences computed by Matt Wood) for several input disk ages. The drop-off in luminosity corresponds to disk ages from 7 Gyr (at the highest luminosity) to 12 Gyr. Also shown is the observed white dwarf luminosity function.

Figure 5 clearly shows the effect of increasing age on the white dwarf luminosity function; older populations have fainter white dwarf stars. The low-luminosity cutoff decreases in luminosity with increasing population age; the figure shows luminosity functions for ages from 7 Gyr to 12 Gyr. Inspection of this figure reveals that the observed luminosity function is consistent with an age for the disk of the Milky Way of between 8 and 11 Gyr, with a best value of about 9 Gyr.

Input physics ("X")	$\frac{dt_{\rm disk}}{d\log X}$	real range
$m_{\rm helium}$	–0.7 Gyr	$\leq 1.4^*$ Gyr
$A_{\rm core}$ (C? O?)	–16 Gyr	$\approx 2^*$ Gyr
conductive opacity	+7.6 Gyr	≤ 0.3 Gyr
radiative opacity	+1.4 Gyr	≤ 0.06 Gyr
$Z_{\rm env}$	+1.4 Gyr	≈ 0.2 Gyr
fractionation	—	< 2 Gyr
other stuff	?	< 1 Gyr?

* measurable with pulsation observations

TABLE 1. Dependence on $t_{\rm disk}$ on input physics

3.2. *Sensitivity to the input physics*

The derived age of the galactic disk depends on the input physics used in the computation of the white dwarf cooling rates. Given an observed, or otherwise fixed, luminosity function for comparison, the uncertainties in the derived ages follow from uncertainties in various properties of the white dwarf models. These properties, the sensitivity of the age to them, and the real range of possible ages, are summarized in Table 1. This table uses information from Winget & Van Horn (1987) and Wood (1992). It lists how the time for a white dwarf with a mass of 0.60 $M_\odot$ to drop to a luminosity of $10^{-4.4}$ $L_\odot$ changes with various changes in the input physics.

For example, if the mass of the surface helium layer is increased by one decade, the cooling time for a 0.60 $M_\odot$ white dwarf will decrease by 0.7 Gyr. In this tabulation, the values of $m_{\rm helium}$, $A_{\rm core}$, and $Z_{\rm env}$ are uncertain because of unknowns in the prior evolution of the white dwarf stars—they depend on how the star became a white dwarf star, on the $^{12}C(\alpha,\gamma)^{16}O$ nuclear reaction cross section, and on the trace metal content in white dwarf envelopes. The "real range" represents the current uncertainty in the cooling time given current uncertainties in the listed parameters. The dominant uncertainties arise from the thickness of the surface helium layer and on the core composition. The combination of these uncertainties yields a current best-estimate of the age of the galaxy of about 9.3 ± 1.5 Gyr.

Fortunately, through observations of the pulsating white dwarf stars, there is a way to measure these quantities that is independent of approaches that use the observed luminosity function (see Kawaler 1997 for details). These observed pulsations allow us to measure the depth of subsurface transition zones and, therefore, the thickness of the surface helium layer in the pulsating DO and DB white dwarfs. Other pulsators place constraints on the core composition through the rate of period change.

4. The white dwarf luminosity function of the halo

As described earlier, the white dwarfs that are likely to appear on the HDF are halo white dwarfs. This section discusses the issues that need to be addressed to examine such a halo population.

4.1. *Differences from the disk component*

Nearly all of the work in the field so far has been on the white dwarf luminosity function of the Galactic disk, simply because that is where the observed white dwarfs live. When considering the white dwarf component of the galactic halo, however, many of the inputs into constructing theoretical WDLFs must be changed, based on what we know (or think we know) about stellar properties and star formation during the initial collapse of our galaxy.

4.1.1. *Star formation rate: a burst*

Whereas star formation has continued in the disk from the early history of the galaxy to the present, the stars of the halo formed over a very brief period very early on. Therefore, in modeling the WDLF for halo stars, a nearly universal assumption is that halo stars formed in a single burst at some time in the past (Tamanaha et al. 1990). The time of this burst is the age of the oldest stellar component of the Milky Way.

With this assumption, the star formation rate $\psi(t_h)$ becomes simply a constant, with a value equal to the total number of stars (of all masses) produced in the burst. Associated with this is the time t_h of the burst; adjusting t_h allows examination of how the WDLF depends on the halo age.

This simplifies the numerical calculation of the WDLF for the halo; if we consider the star formation rate such a δ function, then

$$-\frac{dN}{dM_{\rm b}} = \frac{\log(10)}{2.5}\,\psi(t_h)\,\phi(M)\,\tau_{\rm cool}\left[\frac{dt_{\rm ms}}{dM_{\rm i}} + \frac{dt_{\rm cool}}{dM_{\rm wd}}\frac{dM_{\rm wd}}{dM_{\rm i}}\right]^{-1} \tag{4.8}$$

4.1.2. *The initial mass function*

Several lines of evidence point to an IMF for the halo that was quite different than the IMF of the disk. The observed scarcity of low-mass halo stars requires an IMF for halo stars (i.e., Bahcall et al. 1994) that does not continue to rise at low mass (as does the Salpeter IMF for the disk). At the high-mass end, the disk IMF applied to the halo would have produced a higher metallicity (through supernovae) in halo stars than is observed (Ryu et al. 1990). Thus the IMF for the halo is a function that is peaked at an intermediate mass.

Additional theoretical evidence leads to an IMF that is parameterized conveniently, following Adams & Laughlin (1996), as

$$\ln\left[\frac{dN}{dM}(\ln M)\right] = A - \frac{1}{2\sigma^2}\left[\ln\left(\frac{M}{m_c}\right)\right]^2 \tag{4.9}$$

where the parameter A is a normalization parameter, σ characterizes the width of the distribution, and m_c is related to the mass at the peak of the distribution. Adams & Laughlin (1996) choose base values for these parameters are $m_c = 2.3$ and $\sigma = 0.44$, though allowable values range from m_c of 2.0 to 4.0 and σ ranging from 0.10 to 0.40 for a halo population. They also show that for a disk population, the Population I IMF has values for these parameters closest to $\sigma = 1.57$ and $m_c = 0.15$.

4.1.3. *The initial-final mass relation*

The initial-final mass relation was derived empirically using Population I stars, typically in moderate-aged open clusters. Theoretical initial-final mass relations have reproduced the observed relation by imposing mass loss on the Asymptotic Giant Branch using a variety of uncertain mass-loss prescriptions. Because of the dependency of the initial-final mass loss relation on such mass loss, it is likely that the relationship will be different for the metal-poor progenitors of the halo white dwarfs.

Such low-metallicity stars will have envelope opacities that may be significantly smaller than their Pop I counterparts. The most likely mechanism driving mass loss on the AGB is radial pulsation (such as in Mira variables; see Bowen and Willson 1991). If so, mass loss rates on the AGB for Pop II stars may have been smaller... leading to white dwarfs

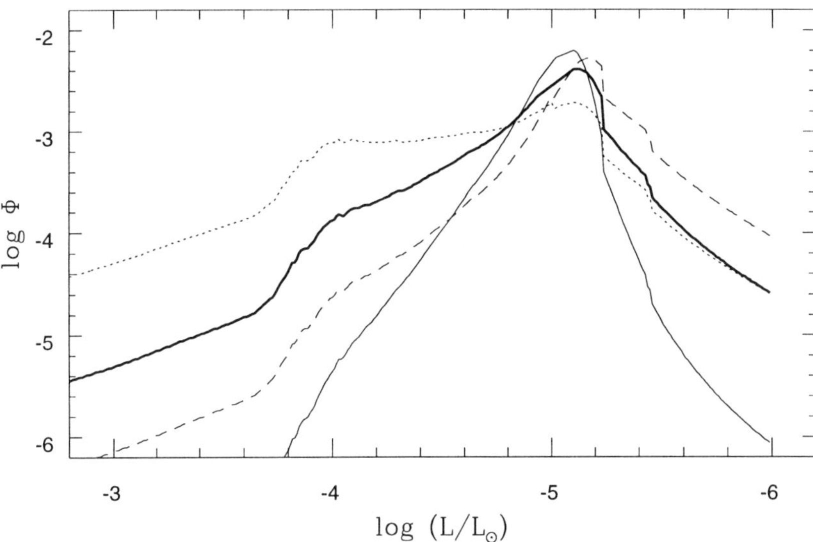

FIGURE 6. Theoretical WDLFs for halo models at 12 Gyr. The heavy solid line corresponds to an IMF with parameters $m_c = 2.3$ and $\sigma = 0.44$. A narrower IMF ($\sigma = 0.24$) is shown as a thin line, and an IMF centered at higher mass is shown by a dashed line ($m_c = 3.3$). The dotted line shows the halo WDLF assuming a disk IMF, for comparison.

growing to much larger masses inside of them than Pop I stars. This would steepen the initial-final mass relation, and bring the lower-mass limit for Type II supernovae well below the 8 to 10 $M_\odot$ that it is for Population I.

4.2. *Models of the halo WDLF*

We now have all the ingredients we need to explore the shape of the WDLF for halo white dwarf stars. For more details about some of the results described in this section, see Tamanaha et al. (1990), Adams & Laughlin (1996), Chabrier et al. (1996), and Graff et al. (1998). Here we assume a total number density of halo white dwarfs of 4×10^{-3} pc^{-3}, which corresponds to a mass density of approximately 25% of the dark halo.

Considering first the halo WDLF in isolation, the influence of the IMF on the WDLF is shown in Figure 6. For this figure, I used Wood's DA white dwarf models to compute the halo WDLF using the equation shown above. All three curves correspond to a halo age of 12 Gyr, but with different values for the parameters governing the IMF. The heavy line shows the WDLF corresponding to the reference values of $m_c = 2.3$ and $\sigma = 0.44$. Narrowing the IMF by reducing σ results in a narrower WDLF (compare the heavy and light lines in Figure 6), while increasing the value of m_c shifts the WDLF to lower luminosities. For reference, the dotted line shows the WDLF using the IMF for the disk.

The effect of the halo age on the WDLF of halo white dwarfs is similar to that of increasing m_c. Older halos produce WDLFs that reach lower luminosities with broader peaks. This is clearly a result of the fact that the white dwarf population has had more time to cool to lower luminosities. The entire distribution moves towards lower luminosity. In Figure 7, the evolution of the WDLF for a halo white dwarf population is clearly evident. Under these circumstances, the older the halo, the greater the chance that halo white dwarfs may have escaped detection by current surveys. In addition, if

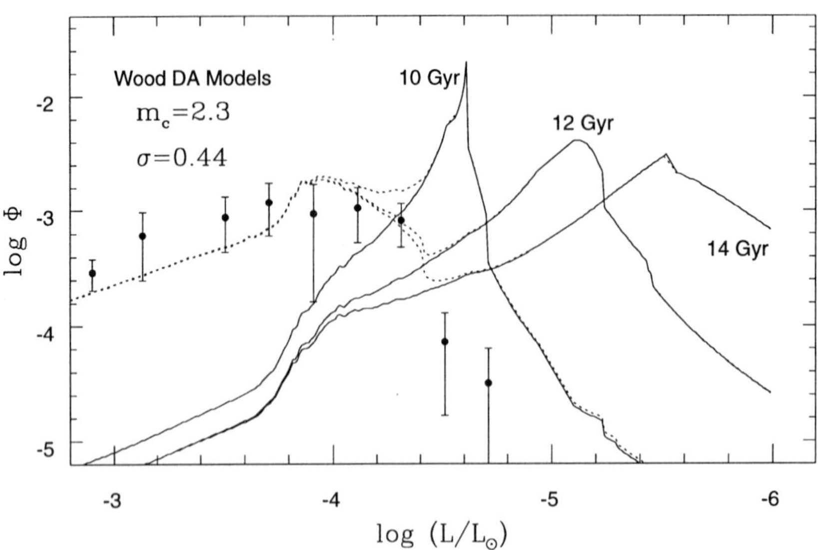

FIGURE 7. Theoretical WDLFs for halo models with fixed IMF parameters, and the same total space density, at ages of 10, 12, and 14 Gyr. The dotted lines are the complete WDLF including disk stars. The Liebert, et al. (1988) observed disk WDLF is also shown.

the IMF for the halo is skewed towards higher mass stars at the expense of low mass stars (i.e., a larger m_c) then the halo portion of the composite WDLF will be forced to lower luminosity.

More complete discussions of the form of the halo WDLF can be found in the papers cited at the beginning of this subsection; Figures 6 and 7 should serve to illustrate the main points about the dependence of the halo WDLF on age and the IMF. For example, Adams & Laughlin (1996) explore a range of values of the parameters of the IMF (using pure carbon white dwarf models)—and the consequences for other observables including the residual gas left over after the mass loss that halo stars must have undergone in the production of halo white dwarfs. Graff et al. (1998) consider the effects of fractionation during crystallization on the halo WDLF, as well as the many selection effects associated with surveys, and the issue of the bolometric correction for such cool objects.

5. Observational constraints on halo white dwarfs

The MACHO results have already been discussed, and are provocative indeed. There is a firm upper limit on the number of halo white dwarfs: the total mass of halo white dwarfs must not exceed the mass of the galactic halo! At or near the position of the Sun, then, the halo white dwarf density must be less than about 1.7×10^{-2} pc^{-3}. This is a very large number that exceeds (by nearly an order of magnitude) the density of white dwarfs currently known. Interestingly, if halo white dwarfs are to account for a significant fraction of the halo mass, then the number density cannot be significantly lower than this. Thus if the MACHO interpretation of their data is correct, such white dwarfs must have escaped detection by current surveys.

5.1. *Presence (or absence) of halo white dwarfs in existing white dwarf surveys*

How do studies of the white dwarf luminosity function that were restricted to the solar neighborhood constrain the *halo* WDLF? This issue has been addressed by Adams & Laughlin (1996) who show how the luminosity function of white dwarfs in the solar neighborhood as reported by Liebert et al. (1988) limits the number density of halo white dwarfs. The last data point in the disk WDLF is an upper limit of approximately 6×10^{-5} pc$^{-3}M_{\rm bol}^{-1}$ at $\log L = -4.7$. For luminosity functions that derive from standard models of cooling white dwarfs, Adams & Laughlin (1996) show that their luminosity function must quickly rise to several $\times 10^{-4}$ pc$^{-3}M_{\rm bol}^{-1}$ and remain high down to very low luminosities if the age of this population is not excessively larger than the age of the oldest globular clusters.

This is apparent from examination of Figure 7, which compares the WDLF to a "standard" computation of the WDLF including a halo population. At $\log L = -4.7$, the theoretical WDLF lies above the observed upper limit for all halo ages, using the most likely parameters of the IMF. Recall that these theoretical WDLF calculations assume that the total mass of halo white dwarfs is 25% of the total halo mass. To be consistent with the last WDLF point from Liebert et al. (1988), the contribution of white dwarfs to the mass of the halo must be much less than 25%.

The fraction of the halo mass contributed by white dwarfs can be greater if the width of the IMF is much narrower than $\sigma = 0.44$, as shown in Figure 6. If $\sigma = 0.24$ then a halo age of 14 Gyr can have a white dwarf contribution of 25% and still be consistent with the lowest luminosity WDLF point. Thus the IMF width must be extremely narrow for the halo white dwarfs to contribute significantly to the mass of the halo and to have largely escaped detection in traditional searches for white dwarfs. This one of the essential points made by Adams & Laughlin (1996) and Graff et al. (1998).

With this constraint alone, the interpretation of lensing events in the MACHO project as white dwarfs requires a halo age of 14 Gyr or more along with a rapid rise in the luminosity function at lower luminosities. Can this prediction be addressed by the HDF? Read on...

5.2. *White dwarfs (or lack thereof) on the HDF*

If, as suggested by the MACHO results, the halo dark matter of the Milky Way is up to 50% (by mass) halo white dwarfs, then up to half of the halo mass sampled in the HDF can be white dwarf stars. The mass sampled is dependent on the minimum absolute magnitudes of these halo white dwarfs and the magnitude limit of the images. Kawaler (1996) computes the expected number of white dwarfs on the HDF at any magnitude with minimal assumptions about the WDLF and the magnitude distribution of halo white dwarfs. Following Adams & Laughlin (1996), this section will describe more precisely the way that the WDLF of the halo is affected at different luminosities by the result of searches for white dwarfs on the HDF.

Despite heroic efforts at finding them, there are few obvious stars in the HDF images (apart from a few "bright" 20th magnitude stars); the task of discriminating between stellar and nonstellar objects at very faint magnitudes requires extreme care (see the review by John Bahcall in this volume). For example, Flynn et al. (1996) report that no white dwarfs exist down to $V = 26.3$ for objects with $2.5 > V - I > 1.8$. They do find bluer stellar objects ($V - I$ between 0 and 1.8) that are consistent in number as well as color with low–mass main sequence stars. Similar results have been reported by Mendez et al. (1996). Thus it appears that there are no white dwarfs on the HDF down to $V < 28$ or so. As discussed by Flynn et al. (1996) and by Bahcall (these proceedings), confident discrimination of true stellar objects from compact faint galaxies is not possible at higher magnitudes on the HDF.

While disappointing, this upper limit still important when trying to constrain the white dwarf population of the halo. We can use it to further constrain the shape of the WDLF at very low luminosities. Here we use equation (1.2) from the Introduction; this equation shows the volume sampled by the HDF given these figures, and allows calculation of an upper limit to the space density of white dwarf stars at each luminosity bin. For simplicity, we consider as an upper limits to the space density $1/V$, with V determined with equation 1.2, and zero bolometric correction (but see Graff et al. 1998). This then gives the upper limit for white dwarfs based on the HDF null result as

$$\log \Phi_{\rm up} = -6.80 - 0.6 m_{V,\rm lim} - 1.5 \log\left(\frac{L}{L_\odot}\right) . \qquad (5.10)$$

For each magnitude fainter that searches for white dwarfs on the HDF can go, the WDLF upper limit drops by 0.6 dex at a given luminosity.

Figure 8 shows the HDF limit on the halo contribution (as a dotted line, assuming a limiting magnitude of $V = 28$) to the WDLF along with several models of the WDLF and the data for disk white dwarfs. In Figure 8, as earlier, the models are for a halo white dwarf population with a total mass equal to 25% of the mass of the dark halo. Each panel represents model WDLFs with the same disk age (9 Gyr) and halo ages of 12 Gyr and 14 Gyr, with the 14 Gyr luminosity function lying to the right in all three panels. As can be surmised by examining Figure 7, a halo age of 10 Gyr is clearly inconsistent with the data in all cases.

The top panel of Figure 8 shows the model WDLF using the fiducial parameters for the IMF. While the WDLF falls below the HDF limit for both ages, the lowest luminosity disk WDLF points are clearly inconsistent with the theory. Shifting the IMF to higher masses, as illustrated by the middle panel of Figure 8, helps drop the theoretical WDLF near the disk cutoff, but is still inconsistent with the observations. The bottom panel illustrates that the only way for a halo WDLF to satisfy the constraints of the disk observations and the HDF is to narrow the IMF. Bringing the width parameter down to $\sigma = 0.24$ satisfies both constraints for a halo age of 14 Gyr and older. As the trend in Figure 8 shows, a younger halo with a narrower distribution might satisfy the disk constraints, but would then rise above the HDF upper limit at slightly lower luminosities than the disk cutoff. Another point to consider is that the models used to construct the WDLFs in Figure 8 do not include the effects of fractionation of the crystallizing white dwarf cores. Graff et al. (1998) point out that the such an effect increases the age constraints by about 2 Gyr.

There are a variety of other ways to compute the theoretical halo WDLF to address the observations. Again, the reader is urged to consult the papers by Tamanaha et al. (1990), Adams & Laughlin (1996) and Graff et al. (1998) for more details of the influence of various parameters on the halo white dwarf luminosity function.

The results presented in Figure 8 are intended to be representative and schematic. They show the essential results of the lack of finding white dwarfs on the HDF. First, the lack of white dwarfs on the HDF is not a surprise given what we have learned from studies of the white dwarf luminosity function of the galactic disk. Second, the contribution of white dwarfs to the Milky Way's massive dark halo can be limited based on this lack of detection. For reasonable parameters for the halo white dwarf population, it is difficult for this null result on the HDF (and the disk WDLF) to be consistent with a halo comprised of more than 25% white dwarfs by mass. Third, if the limits of the HDF (or other very deep surveys) can be pushed fainter, the MACHO results require that some

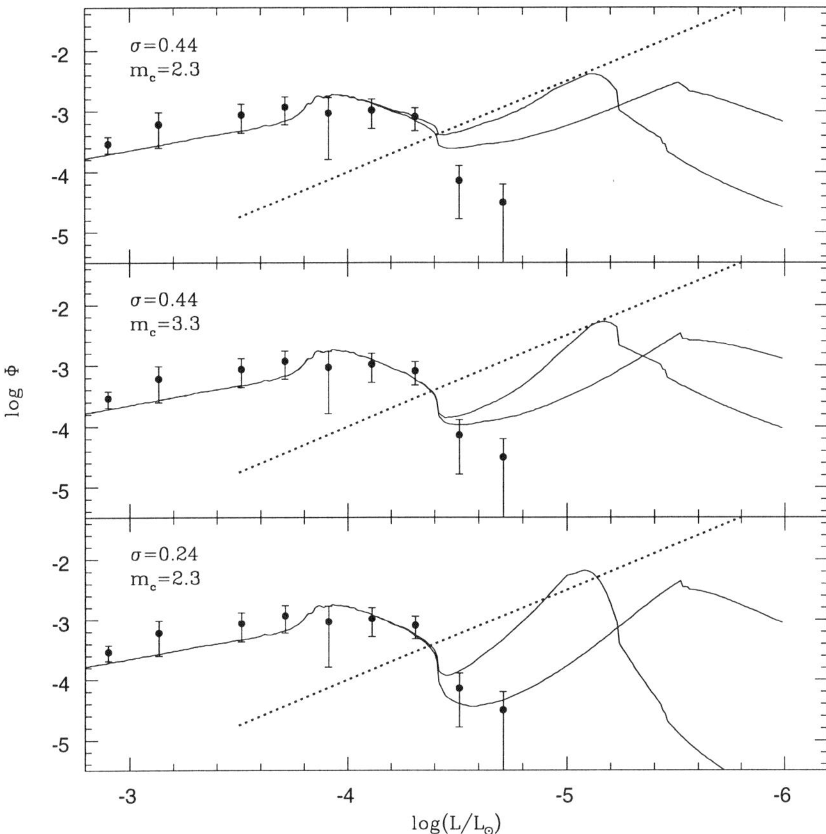

FIGURE 8. The WDLF for the disk + halo. The data points are the observed disk WDLF from Liebert et al. (1988); the dotted line represents an upper limit based on the lack of white dwarfs on the HDF (assuming a magnitude limit of $V = 28$. Solid lines represent theoretical WDLFs for DA white dwarfs with a disk age of 9 Gyr, and halo ages of 12 Gyr and 14 Gyr. Older models reach lower luminosities. The three panels represent different parameters for the IMF. In comparison with the top panel, the middle panel shows results for an IMF with a higher central mass, while the bottom panel represents an IMF with a much narrower width. All models assume a halo white dwarf mass equal to 25% of the total mass of the dark halo.

white dwarfs should be found (probably with an $M_{\rm bol}$ of about 17.5 to 19). Finally, we can hope to learn a great deal about halo white dwarfs by their discovery in abundance, even if it is not a chore for which HST is ideally suited.

5.3. *Postscript: white dwarfs on the HDF—sort of*

In closing, the results of the WDLF studies of the disk of our galaxy, along with the newer work on white dwarfs in the halo, allow us to estimate the total number of white dwarfs in the Milky Way. Assuming a uniform distribution of white dwarf stars in the galactic disk, the local density implies that there are approximately 3×10^9 white dwarfs in the disk. We can then use the luminosity function of the disk to estimate the total luminosity from all of these white dwarf stars as approximately $3 \times 10^6\ L_\odot$. Thus, to ballpark accuracy, white dwarfs contribute a fraction of about 3×10^{-5} of the total photon luminosity of our galaxy.

If we take the Milky Way as representative of the galaxies seen on the HDF, we can ask how many photons that were counted by the HDF originated on a white dwarf in a galaxy. Given the number of galaxies on the HDF and the total photon count, approximately 1 photon per galaxy came from a white dwarf. Therefore, white dwarfs have indeed been detected by the HDF; the trick remains identifying which of those photons came from which white dwarf!

It is a pleasure to thank the organizers of this workshop for their help at all stages of this review. Matt Wood graciously provided many white dwarf evolutionary tracks for use in illustrating the properties of the white dwarf luminosity function, and Russ Lavery assisted in the estimate of the white dwarf photon flux on the HDF. Partial support for this work also came through the NASA Astrophysics Theory Program (Grant NRA-96-04-GSFC-052) and from an NSF Young Investigator award (Grant AST-9257049 to Iowa State University).

REFERENCES

ADAMS, F., & LAUGHLIN, G. 1996 *ApJ* **468**, 586.

ALCOCK, C. ET AL. (THE MACHO COLLABORATION) 1997 *ApJ* **486**, 697.

BAHCALL, J., & SONIERA, R. M. 1980 *ApJS* **44**, 73.

BAHCALL, J., FLYNN, C., GOULD, A., & HIRHAKOS, S. 1994 *ApJ* **435**, L51.

BENNETT, D. ET AL. (THE MACHO COLLABORATION) 1995 *BAAS* **28**, 47.07.

BINNEY, J. & TREMAINE, S. 1987 *Galactic Dynamics.* Princeton University Press.

BOWEN, G. & WILLSON, L. A. 1991 *ApJ* **375**, L53.

CHABRIER, G., SEGRETAIN, L., & MERA, D. 1996 *ApJ* **468**, L21.

CHANDRASEKHAR, S. 1939 *An Introduction to the Study of Stellar Structure.* University of Chicago Press.

DEMERS, S., KIBBLEWHITE, E., IRWIN, M., NITHAKORN, D. S., BELAND, S., FONTAINE, G., & WESEMAEL, F. 1986 *AJ* **92**, 878.

FLEMING, T. A., LIEBERT, J. & GREEN, R. F. 1986 *ApJ* **308**, 176.

FLYNN, C., GOULD, A., & BAHCALL, J. 1996 *ApJ* **466**, L55.

GICLAS, H. L., BURNHAM, R., JR., & THOMAS, N. G. 1971 *Lowell Proper Motion Survey, The G-Numbered Stars.*

GICLAS, H. L., BURNHAM, R., JR., & THOMAS, N. G. 1978 *Lowell Obs.Bull.*, No. 163.

GRAFF, D., LAUGHLIN, G., FREESE, K. 1998 *ApJ*, in press.

GREEN, R., SCHMIDT, M., & LIEBERT, J. 1986 *ApJS* **61**, 305.

GREENSTEIN, J. L. 1976 *ApJ* **227**, 224.

HANSEN, C. J. & KAWALER, S. D. 1994 *Stellar Interiors: Physical Principles, Structure, and Evolution.* Springer-Verlag.

IBEN, I. JR. & TUTUKOV, A. 1984 *ApJ* **282**, 615.

IBEN, I. JR. & MACDONALD, J. 1985 *ApJ* **296**, 540.

IBEN, I. JR. & LAUGHLIN, G. 1989 *ApJ* **341**, 312.

ISERN, J., MOCHKOVITCH, R., GARCIA-BERRO, E., & HERNANZ, M. 1997 *ApJ* **485**, 308.

KAWALER, S. D. 1996 *ApJ* **467**, L61.

KAWALER, S. D. 1997 in *Stellar Remnants: Saas Fee Advanced Course 25* (ed. G. Meynet & D. Schaerer. p. 1. Springer.

KILKENNY, D., O'DONOGHUE, D., & STOBIE, R. S. 1991 *MNRAS* **248**, 664.

LEDOUX, P. J., & SAUVENIER-GOFFIN, E. 1950 *ApJ* **111**, 611.

LIEBERT, J., DAHN, C., MONET, D. 1988 *ApJ* **332**, 891.

LUYTEN, W. J. 1969 *Proper Motion Survey with the Forty-eight Inch Schmidt Telescope. XVIII. Binaries with White Dwarf Components.* University of Minnesota.

LUYTEN, W. J. 1979 *NLTT Catalogue*, University of Minnesota.

MENDEZ, R., MINNITI, D., DE MARCHI, G., BAKER, A., & COUCH, W. A. 1996 *MNRAS* **283**, 666.

MESTEL, L. 1952 *MNRAS* **112**, 583.

OSWALT, T. D., SMITH, J. A., WOOD, M. A., & HINTZEN, P. 1996 *Nature*, **382**, 692.

PARESCE, F., DE MARCHI, G. & ROMANIELLO, M. 1995 *ApJ* **440**, 216.

RYU, D., OLIVE, K., & SILK, J. 1990 *ApJ* **353**, 81.

SALPETER, E. E. 1955 *ApJ* **121**, 161.

SEGRETAIN, L., CHABRIER, G., HERNANZ, M., GARCIA-BERRO, E., ISERN, J., & MOCHKOVITCH, R. 1994 *ApJ* **434**, 641.

SHAPIRO, S. & TEUKOLSKY, S. 1983 *Black Holes, White Dwarfs, and Neutron Stars: The Physics of Compact Objects*, Wiley.

STOBIE, R. S., MORGAN, D. H., BHATIA, R. K., KILKENNY, D., & O'DONOGHUE, D. 1987 in *IAU Colloq. 95, The Second Conference on Faint Blue Stars* (ed. A. G. D. Philip, D. S. Hayes, & J. Liebert). p. 493. Davis.

TAMANAHA, C., SILK, J., WOOD, M. A., & WINGET, D. E. 1990 *ApJ* **358**, 164.

VAN HORN, H. M. 1971 in *White Dwarfs* (ed. W. Luyten. p. 97. Reidel.

VON HIPPEL, T., GILMORE, G., & JONES, D. 1995 *A&A*, in press

WINGET, D. E., HANSEN, C. J., LIEBERT, J., VAN HORN, H. M., FONTAINE, G., NATHER, R. E., KEPLER, S. O., & LAMB, D. Q. 1987 *ApJ* **315**, L77.

WINGET, D. E., & VAN HORN, H. M. 1987 in *IAU Colloq. 95, The Second Conference on Faint Blue Stars* (ed. A. G. D. Philip, D. S. Hayes, & J. Liebert. p. 363. Davis.

WOOD, M. 1992 *ApJ* **386**, 539.

Educational uses of the Hubble Deep Field

By MEGAN DONAHUE

Space Telescope Science Institute, 3700 San Martin Drive, Baltimore, MD 21218

The Hubble Deep Field was used as a central theme for an educational activity sponsored by ST ScI. I describe and discuss the HDF lesson plan, poster package, and Web activities designed by teachers and astronomers (including myself) in ST ScI's "Amazing Space" program. I include a interim report of the distribution, usage, and evaluation results and plans. This collaboration is one example of how astronomers can collaborate with teachers to make a useful classroom product. Scientists should not isolate themselves from the on-going discussions about educational standards in science and math nor from the education process itself. Intellectual contact with astronomers and their research through the "astronomer's answers" is one of the most appealing aspects of the HDF activity for teachers and students, revealing the scientific process for the natural human endeavor that it is. I conclude with suggestions for how astronomers and astronomy may become more accessible to educators, students, and the public, through educational Web sites and other projects.

---

1. Introduction

In the summer of 1996, the first year of the ST ScI Amazing Space program, ten teachers spent five weeks at the Space Telescope Science Institute (ST ScI) designing lesson plans and Web activities that were centered around results from the Hubble Space Telescope. Of those teachers, two middle-school teachers, Gina Cash and Kirk Fitch, worked with the guidance of Ray Lucas (ST ScI), and me to construct classroom activities around the theme of the Hubble Deep Field (HDF). Ray Lucas is an HST Program Coordinator who has been active in many other outreach programs; I am a tenure-track astronomer in the Archive Branch whose avocation is public outreach. Gina Cash and Kirk Fitch are both science teachers of middle-school students. Middle school students in the U.S. are children between the ages of 11–13 years, 6th-7th-8th graders. For those you who have visited middle schools or who have children of this age, you know that middle school teachers are by definition saints and miracle workers. They teach children with a wide range of background and maturity. Even if the children's quality of education before this age was remarkably high and uniform, the biological and social maturity spans an incredible range, making instruction a challenge for anyone. On the other hand, middle school students can be enthusiastic practitioners of science. They are still excited by astronomy subject matter, not yet motivated by their peers to be uninterested and "cool."

The teachers also provided me insights into their charges that I may not have extrapolated from my own experiences with college freshmen. While college freshmen have been known to be reduced to trembling and tears at the prospect of not knowing the answer, and must be guided gently and surely to the confidence that perhaps they do know the answer, this lack of self-confidence is not nearly so common in middle school students. Middle school students have their own unique characteristics; they are not simply short versions of 18-year olds.

Ray Lucas and I provided background information and brief tutorials on the context and content of the Hubble Deep Field. I also helped guide the teachers on the design of the Web activities, since I was familiar with the strengths and weaknesses of Web presentations and programs. The teachers, however, were the ones who came up with the final form of the lesson plan, poster, and Web activities. The lesson plan could be done with the poster and handouts independently of Web access, which is important because not every classroom has Web access.

2. Overview of Amazing Space Web Site

The Web site address is *http://oposite.stsci.edu/pubinfo/amazing-space.html.* There is a direct link to the Amazing Space pages from the ST ScI home page at *http://www.stsci.edu.* In my presentation, I gave an on-line tour of the site and the activities designed using the HDF.

The 1996 Amazing Space site contained four lesson plans and a display:

(*a*) Solar System Trading Cards (Web activity and card set).
(*b*) HDF Academy (Web activity, poster, and classroom handouts).
(*c*) Stars: Birth, Life, Death, and Rebirth (Web activity and poster).
(*d*) Student Astronaut Challenge: Schedule the Servicing Mission (Web activity).
(*e*) Galileo to HST (Web display only.)

Within the HDF Academy, the student is presented with 5 modular activities. Each one can be done without particular reference to the other, and most importantly, a single substantive activity can be accomplished well within the time limits of a single class (40–50 minutes.) The modules were designed to be used in context with an astronomy section of a middle-school science course. Before the modules are started, the students should have some idea about what stars are, that the Sun is a star, and that galaxies are large collections of stars. Galaxies are formally introduced and defined mid-way through the activity, but knowing that they are collections of stars is helpful. These ideas are reiterated throughout the modules, but ideally, students should at least have this background presented to them before embarking on these activities.

One important theme in this activity is to connect the student's results with those of the astronomers, inviting the students to (a) be pleased that their answers are similar to those of astronomers, (b) learn that sometimes astronomers disagree about "the answers" amongst themselves, and (c) perhaps even challenge the answers of the astronomers. This theme relates the process of "doing science" with learning scientific techniques, reasoning skills, and concepts, and emphasizes science as a human endeavor. There is a balance to be struck, between the relatively solid knowledge base of mathematics and basic science and the uncertainty of knowledge at the frontiers of science. As we all know, the balance has shifted erratically from declarations of the infallibility of science to the complete irrelevance of scientific knowledge. As scientists, we have rarely played a role in these discussions, but we all know that these concepts now affect the education not only of future scientists, but of all of us. We are not well-served by a public that blindly follows scientific advice, because such a public is just as likely to follow poor or nonsensical science as it is good science, nor are we served by an all-encompassing cynicism of knowledge in which we believe we can "know nothing." What is the balance to be struck here? I don't know the answer offhand, but I do know that scientists should lend their voices to the discussion.

In the next sections, I will describe each of the modules in turn.

2.1. *Introduction: Where is the Hubble Deep Field?*

The student is shown an animation where we zoom in on the field of view of the HDF, starting from the constellation field of the Big Dipper. This animation emphasizes the anonymity of the HDF, or the lack of well-known astronomical features such as bright stars or near-by galaxies. It emphasizes that the HDF is just a picture of a very tiny and not-that-special piece of sky.

The student is requested to construct questions about the HDF that they would like to answer or have answered. They compare those questions with those that astronomers might ask. This activity is intended to show students that the questions that astronomers ask are not necessarily questions that one could only construct after years of training, but they are natural questions generated by human curiosity: How many galaxies are there? What kinds of galaxies are there? How far away are they? What are the weird ones? What can we learn from these galaxies? If the student does not know what the smudges in the picture are, we hope they ask what they are! If they ask why the picture is not square, they can follow a Web link for an explanation. (Yes, we have had Internet accusations of hiding "the secret stuff" and the UFOs in the area of sky not filled by the Planetary Camera.)

2.2. *Stellar statistician*

In the second activity, the student is asked to estimate the number of objects in the HDF by selecting one of the Wide Field (WF) chips, then by selecting one section of that chip (out of 12 pre-defined sections). They count the objects in that one section, enter the number into a calculation form, and derive the number of objects in the WF chip, the number of objects in the HDF chip and finally, the number of such objects in the universe. A more advanced application of this activity would require the students to derive these extrapolations themselves.

Finally, the students are asked to compare their answers to those of astronomers. In my experience, I was able to count galaxies and be remarkably close to the astronomer's "answer" in the end. In a classroom, students can compare their answers to those of their classmates. This comparison activity can lead students to discuss why their answers may differ from each other and from the astronomer's answer. This discussion could range from the various possible definitions of an "object" to the statistics of counting and sampling. This module can be used at several levels of mathematical sophistication, ranging from simply extrapolating the number counts of galaxies from a small area to the entire sky, to a discussion of Poisson statistics and convergence of means.

2.3. *Cosmic classifier*

In the initial stages of many scientific studies, one of the first impulses scientists have is to classify the objects within their sample. In the third activity, the student classifies galaxies and stars by their color and shape. Here again, they compare their answers to the astronomers' "answer."

To get an astronomer's "answer," the teachers polled various astronomers involved in the HDF project throughout the Institute, including Bob Williams (whose answers I think we eventually used, but I cannot swear to that!) The teachers themselves were entertained by the variety of astronomers' replies in this activity. If we astronomers hadn't lost our aura of papal infallibility in the teachers' eyes by then, this project surely accomplished that.

The students then discuss the differences between their answers and the astronomers. Here, students see that there can be differences of opinion about the exact color and shape of any given galaxy. They may notice that there can be a dependence on whether the student used the poster, the class handout, a Web download image, or a display on a computer screen, or even on how an individual defines boundaries between different colors or shapes. Nevertheless, they can notice certain trends of color with shape and begin to ask the next questions: Do galaxies of certain shapes tend to have the same colors? Why are there different kinds of galaxies? Why do they have different colors? What makes a galaxy red or blue?

The premise is that if the student generates the questions themselves, they are more likely to remember at least some of the answers. The student is presented then with content and background describing galaxies, light, stars and star colors. This lesson is an excellent starting point for the discussion of some of the basic concepts of light and the information contained in light.

This lesson is one example of a lesson where a teacher, not grounded in the background, might be intimidated not to teach the lesson because the students may find out that the teachers don't know much more about this subject than they do. I don't know many people comfortable giving a talk for which they are underprepared, and teachers are no different. In order to give the teachers the confidence to teach any astronomy lesson, astronomers need to provide more than enough background for the teachers to handle the questions that might arise as a desired result of the activity. The background should be complete and intelligible to the teachers. Writing successful background texts is not trivial, but it is absolutely essential and can be accomplished with feedback from working teachers.

2.4. *Galactic guide*

The final interactive activity requests the student to estimate the distance order of selected galaxies. The galaxies have a range of sizes and colors. The student orders the selected galaxies as a function of estimated distance. The students then compare their answers to the "right" answer of the astronomer. This is the module that I find the least satisfying, since (a) there is little insight into how the astronomers know the "right" answer in this case, as well as a strong sense that there is no disagreement among astronomers as to the ordering of these galaxies, and (b) there is no way to be "right" except by nearly blind luck. The concept of determining the distance by studying the spectrum is alluded to in the background information, but getting into the expansion of the universe, redshifting of light, and Hubble's Law is a little beyond the scope of the lesson and most middle-school science classes.† The lesson also, in some sense, traps the student with the modest physical intuition that the apparently smaller objects may be the more distant objects into being wrong, since some of the smaller galaxies in the selection are actually some of the closer ones. I have found college freshmen who were unwilling to admit that if a galaxy looked smaller, it just might be farther away than a galaxy that appeared large. They shrug away the prospect of hypothesizing anything about a galaxy by its appearance as merely a random guess on their part. But here again is an example where middle students apparently differ from college freshmen. The middle school teachers told me that showing a student that their physical intuition could be wrong was an important lesson at the age of 12. While this may be true, what happens between the age of 12 and 18 that many students become such fragile intellectual beings once they are in college?

2.5. *The final exam*

In a final "exam," the student is shown a galaxy, one of the more unusual and beautiful galaxies in the HDF, and asked, based on what they have learned about galaxies in the previous lessons, to explain the galaxy's appearance. They are given review lessons and hints to provide the answer. This is the only module that relies on information from the previous modules, but that is obvious from the title.

† In this summer's Amazing Space project (1997), one group of teachers did tackle the distance scale, so the Web site will eventually have lessons on how astronomers measure distance.

3. Educational goals: Process and content

The main goals of the HDF activity were to teach scientific processes as well as scientific content. Both are important and crucial in science education. A student is no more prepared to embark on a scientific inquiry by purely "hands-on" experience than a doctor can make a diagnosis based on the act of examining blood drawn from a vein. The diagnosis must be informed by scientific results derived by others and by the doctor's knowledge of those results. Similarly, it is not enough to teach merely the process of doing science without the context. In something of an oversimplification, the current national standards focus on processes, while the American Association for the Advancement of Science (AAAS) has made recomendations on content as well.

Flavio Mendez, in preparing background information for teachers, has placed the HDF activity into the context of addressing this range of educational goals. He tied those goals to those identified by the National Science Education Standards and the National Council of Teachers of Mathematics,as well as content standards from the Benchmarks for Scientific Literacy from the AAAS. These goals are only preliminary standards, in that they only very generally describe the educational goals of a science program addressed to a broad range of students, and the nature of these goals do not change appreciably from level to level. These goals have the advantage that they are nationally accepted and thus a program that connects with them is more likely to be supported by school boards and subject supervisors. In fact, teachers have used Mendez's lists to construct arguments for the inclusion of Web activities in their classroom.

The scientific processes listed in the national standards for middle school that are addressed by the HDF activity are:

- Classification
- Counting, estimation, mathematics skills, statistics
- Identification of scientific questions
- Construction of logically consistent hypotheses
- Variation of results; comparison of results; judgment
- Communication of results, procedures, and explanations
- Keeping records

The scientific contents of the HDF lessons are (most of which are quoted directly from the Benchmarks for Scientific Literacy from the AAAS):

- Telescopes show more stars and galaxies than are visible to the naked eye.
- The sun is a medium-sized star in a disk-shaped galaxy.
- The universe contains billions of galaxies; a galaxy contains billions of stars.
- Light from the most distant galaxies takes billions of years to reach us; we are seeing them as they were billions of years ago.
- Galaxy colors are affected by stellar content (young=blue, old=red) and dust (dust makes things look redder).

We encourage anyone who puts out educational materials for distribution or on the Web to tie their goals to current national science education standards, and to make these ties explicit in their explanations and background material. The extra effort in doing so provides teachers with even more leverage to use such materials in the classroom.

4. Distribution and usage

The posters and lithograph handouts have been distributed widely. I list the meetings and their distribution numbers in Table 1.

These posters and lithograph sets have proved quite popular at the teacher conven-

NSTA Summit Meeting (San Francisco, Dec. 1996)	2,000 HDF Posters 2,000 HDF Litho Sets
NSTA Annual Meeting (New Orleans, April, 1997)	7,000 HDF Posters and 7,000 Litho Sets 5,000 New Educational Card Sets
NCTM Meeting (Minneapolis, April 1997)	3,000 HDF Posters and 3,000 HDF Litho Sets 3,000 HDF Education Card Sets
NASA Teacher Resource Centers	35,200 HDF sets sent

TABLE 1. Distribution of Posters and Lithographs

Web Page	Number of Visits
Amazing Space Top Level Page	4866
Student Astronaut Challenge	1054
HDF Top Level Page	1259
Stellar Top Level Page	1767
Solar System Trading Cards	1920
Introduction to the HDF	516
1: Stellar Statistician	343
2: Cosmic Classifier	278
3: Galactic Guide	243
4: Final Exam	169

TABLE 2. Web Statistics for April 1997

tions. OPO personnel have provided presentations and demonstrations of the use of these materials at the meetings. I was impressed with these numbers because I usually only generate 50–100 reprints of any given article I have ever written, and the work these teachers accomplished in less than a summer was distributed to thousands of other teachers and thus to tens to hundreds of thousands of students. The potential impact of such an educational activity is inspirational.

Less awe-inspiring, but impressive for a educational release are the Web statistics for the Amazing Space Web site in Table 2. Of course, we all know now the savage potential for a well-developed astronomy Web site with the spectacular success of this summer's Mars mission, with millions of hits (Web visits) at JPL in the brief span of the July 4 weekend. However, the statistics do show some interest in the educational activities being developed at ST ScI. We note that this site has been popular with home schoolers eager for free material for their classrooms as well as traditional classroom teachers.

5. Evaluation: Preliminary results and the formal study

There are 600 teachers in the pilot evaluation study, being conducted by Bonnie Eisenhamer (ST ScI), a educational program evaluator. The scheduled completion is June 1997. I presented results current as of May 1997. 93% of the teachers stated they would use the lessons in their classroom. 96% said they would like to see other activities added (and, as of the summer of 1997, a new batch of activities have been generated by another group of teachers.) The particular aspects of the HDF lesson that the teachers liked are: the modular nature of the lessons, the explicit relevance to the national standards, and the contact with "real live astronomer answers."

Evaluation and testing of the HDF activity was absolutely essential. The initial phase review of the HDF Web activity revealed a serious navigational problem. The structure of the activity was adjusted, to make its modular nature more readily apparent, and the results of this adjustment were spectacular: only 2% of the teachers could navigate and complete the activity prior to the adjustment, while 90% of them could successfully navigate and complete the lesson after the adjustment. Without evaluation, testing, and proper revision, the HDF Web lessons would have been virtually inaccessible to most teachers trying to use it.

In a Web survey of Amazing Space visitors, who identified themselves as teachers, students, or general users, we found the following results:

- 75% teachers said they learned something new about HST
- 78% teachers said they have Internet access at school
- 62% teachers said they have Internet access at home
- 85% students want their teacher to use the lessons in class
- 81% students said they learned something new about HST
- 96% students said they enjoyed viewing the HST images
- 94% general users said they would recommend the site to others
- 89% general users said they would return for more activities

As a Web survey, this one has the usual bias that if the respondent has no Internet access, they can't participate in the survey, thus the questions about access are skewed. Nevertheless, we note an overall positive response of the visitors.

The final phase of evaluation, also being conducted by Bonnie Eisenhamer, is a formal evaluation or an impact/retention outcome study to begin in the summer of 1997. In this study, which will occur on a national scale, internal and external (to ST ScI) evaluators will visit classrooms which are using the HDF and other Amazing Space materials. The evaluators will observe the use and the effects of the lessons plans.

In a broader context, several "First Contact" groups have been established for review of ST ScI educational and outreach activities and materials. These groups are variously composed of educators, the general Web audience, staff of science museums and planetaria, and the press. The Amazing Space content and use will be evaluated by First Contact groups over the next year.

6. What can you do?

Link up! If you have relevant HDF (or other astronomy) Web material, let us know where it is. Send us your URLs! Contact the Office of Public Outreach at the email address *outreach@stsci.edu*.

Prepare for visits! The OPO web site is one of the most heavily visited educational sites on the Internet, so prepare for public and student interest in your site. Here's how:

- Supply context! Teachers are hesitant to use materials if they don't have the background to explain the material or to handle student questions or to exploit the full educational potential of the material.
- Write background material at a level accessible to your readers. Get reader comments from people other than other astronomers.
- Tell users HOW to use these materials in the classroom. Be explicit.
- Ask for feedback from us and from your Web visitors.
- Share your results and send us your suggestions.

Consider producing print materials. Because Web access will come slowly to some parts of the country and the world, many Web activities, if thoughtfully constructed, can be converted successfully to off-line print materials.

7. Why bother?

I know that this sounds like a Mom-and-apple-pie statement, but it's worth repeating: the tax-paying public is your ultimate customer, and they really want to know what you're doing with their money. Once they learn, they generally approve! There is a real hunger among the public to see what you're learning and to experience the thrill of astronomical discovery. If the Mars Web experience didn't convince you, nothing will! We have found through various outreach and educational programs that sometimes the public feels left out of scientific discovery. Being able to see what the scientists are seeing and read what they're thinking engages them in the discovery process. I think we have only begun to experience the potential of the information networks, but also we have the slight edge, as astronomers and physicists, of not only having visually exciting scientific data, but also of being the original pioneers on the Internet.

There is a truly desperate need for interaction and exchange between those who teach science and those who do science. While the walls are still high between the science and education departments in the colleges and universities of this country, we can still make a connection outside the somewhat insular context of teacher education by collaborating with teachers in the real world.

Classroom Web access is increasing exponentially each year as corporations and governments (including the US Federal government) sponsor classroom connectivity. The Web is a source for not only traditional education materials but also adult extension education and home school materials. In Baltimore, the Essex Community College bases an entire on-line astronomy course on Web materials served out primarily by ST ScI and Goddard. In some sense, we do not have much control about how our Web materials might be used. But we have a responsibility, as tax-supported researchers and just generally good citizens, to provide our sponsors and supporters not only with pretty pictures, but with substantive, relevant, and accurate educational content.

I found that while occasionally the technical aspects of producing a final product were exhausting and sometimes frustrating, I very much enjoyed working with Gina and Kirk and the OPO staff to make an educational product that has actually been used and tested across the country. Projects like this do take significant amounts of time and resources, but when they are leveraged into a nationally accessible Web site and thousands of posters and classroom sets, they are worth doing as well as fun.

The 1996 Amazing Space program was guided by Anne Kinney, with Trish Pengra. Flavio Mendez and Bonnie Eisenhamer are leading program evaluation and testing. Flavio and Bonnie were instrumental in compiling the statistics and scaring up handouts for this presentation, including the HDF and Life of Stars posters, sample teacher classroom products for the HDF activity, the "How to Make a Star" booklets on presentations to students and the general public, and the Planetary Trading cards. Jonathan Eisenhamer is the ST ScI Office of Public Outreach Webmaster. He also set up the computer projection system used in the presentation. Dave Paradise and Ann Feild developed the graphics and cartoons on the Web pages and the classroom handouts. Carole Rest was the Amazing Space booklet editor. Patricia Momberger supplied the statistics for poster distribution. If I have left anyone out here, I apologize. This program benefitted from the unacknowledged efforts of many individuals at the Institute who gave time, suggestions, and assistance to the project. The opinions expressed in this article are mine, and should not be attributed to anyone mentioned in the article or the acknowledgments.

The Next Generation Space Telescope: Building from the HST

By JOHN C. MATHER,[1] PIERRE Y. BELY,[2] RICHARD BURG,[3] BERNARD D. SEERY,[4] ERIC P. SMITH,[5] MASSIMO STIAVELLI,[2] AND H. S. STOCKMAN[2]

[1]NASA Goddard Space Flight Center, Code 685, Greenbelt, MD 20771

[2]Space Telescope Science Institute, 3700 San Martin Drive, Baltimore, MD 21219

[3] The Johns Hopkins University, Department of Physics and Astronomy, Baltimore, MD

[4]NASA Goddard Space Flight Center, Code 440, Greenbelt, MD 20771

[5]NASA Goddard Space Flight Center, Code 681, Greenbelt, MD 20771

The Hubble Deep Field is a startling example of the ability of a telescope in space to change the perceptions of a whole generation of astronomers. The Next Generation Space Telescope (NGST) would be the natural extension of the HST capabilities into the near infrared. Based on the requirements laid out in the *HST and Beyond* Report, it would be optimized for near IR studies, with an aperture of at least 4 m. It would be radiatively cooled to 30–70 K, enabling operation at wavelengths out to 20 or 30 μm with sensitivity limited by the zodiacal light. It could be launched by an Atlas IIAS class vehicle to the Lagrange point L2. While optical quality might not be diffraction limited at 0.5 μm, the NGST would still be a very powerful imaging system and coverage of the entire wavelength range from 0.5 to 30 μm is under consideration. Planned core instruments include a near IR camera and multiobject spectrometer covering 1–5 μm. A yardstick concept also includes a mid IR camera and spectrometer covering thermal IR wavelengths. Such a telescope would be ideal for extending the studies started with the HDF. NGST would have the sensitivity and instrumentation to detect and characterize galaxies and globular clusters in the process of formation out to redshifts of the order of 10 or more, and individual supernovae to high redshifts as well. It would also be an exceptionally powerful tool for the study of star and planet formation in our own galaxy, and the search for small objects like brown dwarfs and Kuiper belt objects.

---

1. Introduction

Better instruments and telescopes have been the basis for progress in observational astronomy since the times of Brahe and Galileo, and the Hubble Space Telescope is no exception. It is a technical marvel, that was very nearly impossible to build to the performance that was demanded of it, and without the heroic repair mission it would not be the astounding success that it is today. Some of the important surprises for cosmology include the fact that even HST does not resolve the spatial structure of the early galaxy pieces we have found, and that still most of the sky is dark. Most lines of sight do not end on the surfaces of observable galaxies even after the extremely long exposures of the Hubble Deep Field. Also, recent discoveries from ISO in the 5–15 μm band suggest that many early galaxies were very dusty, and some recent work has confirmed the contention (Fall *et al.* 1996) that significant numbers of distant quasars are obscured by intervening dust clouds. Hence, there is a need for more infrared information.

Alan Dressler and the HST and Beyond committee have written a report (Dressler *et al.* 1996) outlining the next steps for NASA in this area. They recommended that the HST be maintained as an ultraviolet and visible light observatory, that space interferometry should be developed for high angular resolution astrometry and imaging, and that a new telescope with aperture greater than 4 m should be built to extend the HST capabilities farther into the infrared. This is the Next Generation Space Telescope, or NGST. Such a telescope had been considered as early as 1989 (Bely *et al.* 1989), and proposed to NASA as the High Z mission concept (High Z, 1994). Radiative cooling was studied for the Edison proposal as well (Thronson *et al.* 1993).

Serious study of the NGST was initiated by NASA Headquarters under the leadership of Ed Weiler in late 1995. He asked NASA GSFC and the Space Telescope Science Institute to undertake a study with the following goal: Find a way to build a telescope to meet the requirements of the *HST and Beyond* report, for a construction cost of less than $500 M (1996 dollars), and that could be launched on an Atlas IIAS class vehicle. Initial study indicated that this was indeed possible, and Dan Goldin, the NASA Administrator, told the January 1996 AAS meeting that NASA would try to do this mission and that a 4 m telescope was not sufficiently ambitious. Needless to say the audience was enthusiastic, if not convinced of the feasibility.

Now, as of this writing in September 1997, much has been accomplished. Three study teams, led by Lockheed Martin, TRW, and GSFC/ST ScI, have developed concepts for the mission, and all three teams have estimated costs within the guidelines. Contracts for mission concept studies have been let to TRW and Ball Aerospace, and optical technology development contracts have been let to the University of Arizona and Composite Optics. An Ad Hoc Science Working Group has been selected, and instrument concept proposals are being solicited. The European Space Agency is considering the possibility of contributing up to $200 M of effort to NGST, and Canada is considering a contribution of the order of $50 M. A report on the three mission concept studies has been edited by Peter Stockman (Stockman *et al.* 1997) and is available in hardcopy or on the Web.

The plan is to start construction in 2003 for a launch in 2007. This is an ambitious schedule, but it is also urgent. The HST will have been in space for nearly 20 years by then, and its technology dates to the 1970's with 1990's instruments. By 2007, ground based telescopes will be using newer generations of instruments, but will still be limited by many factors. Thermal emission from the telescopes can not be avoided on the ground or on SOFIA past about 2.5 μm; atmospheric absorption blocks many portions of the near and mid IR; and adaptive optics to compensate for atmospheric turbulence can provide only small fields of view around bright stars or requires very expensive laser guide stars. By 2007, we anticipate that ground based telescopes will be close to natural limits, and further progress will still require a new generation of space telescope. At mid IR wavelengths, the ISO, SIRTF, and SOFIA will have shown us many new and wonderful phenomena, but the ISO and SIRTF are comparable in size to the 1982 IRAS mission, the SOFIA is still a warm earth-bound telescope, and the HST, although in space, will never be cold and will never be 8 m in size. An 8 m cold telescope would have speed advantages over competing ground based telescopes of the order of 10^6, and over the SIRTF of 10^4.

The Next Generation Space Telescope is the first opportunity to answer that challenge. Current concepts call for extensive use of new technology, which could lead to even further improvements in the more distant future. By 2007, this technology may no longer seem so new, so it is important to be imaginative and ambitious at this point in the program. Lightweight mirrors, only a few mm thick, coupled with adjustments to improve the imaging to diffraction limited performance, are under development. Deployment mechanisms to enable telescopes much larger than the rocket payload fairing are being developed, and have already been flown, for example on the Japanese VSOP radio telescope for VLBI, and many other radio and microwave systems. New generations of infrared detectors and spectrometers are coming on line, and there is hope that large arrays with read noises of less than a single electron may become available to astronomers. If this were to happen, we would have to consider whether it is worth while moving the observatory outside the interplanetary dust cloud, say to 3 AU from the Sun, or in an inclined orbit.

While it is advantageous to consider new technologies, it is also important to minimize the risk associated with them. The plan calls for technology development to be complete prior to the start of construction, so that the difficulties of the HST will not be repeated. Indeed, NASA management has declared that the NGST construction will simply not be started until the technology is ready.

2. Scientific goals

Clearly, astronomy will always benefit from bigger and better telescopes. A compromise must be found between technical and financial feasibility and scientific capabilities. In order to better understand the interplay of the various observatory parameters, we established a strawman observing program mostly based on the prime targets identified in the *HST and Beyond* report. This observing program was defined iteratively by testing scientific requirements against realistic capabilities and mission lifetime. This process was facilitated by formulating each type of target requirement in a computer program allowing for the testing of different instrumental parameters. The results were described by Stiavelli, Stockman, and Burg (1997).

Although the program is heavily weighted toward cosmological problems, we intend to have a balanced share between the various scientific research themes. About 70% is dedicated to a Core program directly linked to NASA's Origins Program and implementing the various aspects of the Origins of Galaxies study as outlined in the *HST and Beyond* report. This Core program is the minimum program that a mission of this type should try to achieve. The rest is devoted to science projects not related to the Origins of Galaxies studies but infeasible with other kinds of instruments. Although the selection of the scientific program elements is somewhat subjective and dependent upon current scientific interests, it is broad enough to be generically representative.

A brief description of each of the research goals and of the corresponding observational parameters is given below.

(*a*) Supernovae study (Core). The interest is twofold: one can both use the SN as standard candles to improve our knowledge of the geometry of the Universe (q_0 and Λ) and also use them to study the formation of massive stars and the heavy elements they make in the era before the birth of galaxies. These goals are achieved by identifying about 100 SNe at redshift greater than 2 (i.e., down to KAB 31), following them for two "rest frame" weeks before and two after maximum. The study includes low resolution spectroscopy. Additional SNe will be identified while monitoring the 100 main objects. 1 nJy = 31.4 magnitudes at all wavelengths on the AB scale.

(*b*) Deep Fields (Core). One deep field (down to KAB magnitude 32) and 100 less deep (KAB 30) flanking fields will be observed in several broad band filters. Several classes of objects will be identified in these fields. Galaxies at redshift greater than 2 will be studied spectroscopically at low and high resolution to derive their redshift, their spectral energy distribution, and their internal kinematics to study the early evolution of galaxies and of their dark matter content.

(*c*) Universe at redshifts $z > 2$ (Core). This includes follow-up studies of objects identified in the Deep Fields and searches for relatively bright but rare objects. In particular, primeval spheroids, birth and evolution of disks, the origin of heavy elements, birth and evolution of AGN are all included in this category.

(*d*) Cosmic Distances and Structures. Deep HST images and the NGST fields will identify many lensing systems. These can be followed up with spectroscopic and synoptic imaging programs to study the distribution of the lensing mass, time delays, and naturally magnified features of lensed galaxies. The formation of clusters and the distribution of dark matter will be studied.

(*e*) Stellar populations in the nearby universe. For both local group and Virgo Cluster galaxies color-magnitude diagrams will be obtained down to the horizontal branch luminosity both in the optical and in the near infrared. Accurate, wide field imaging of single stars in 30 galaxies will provide the star-formation fossil record for the disks and outer portions of the central bulges.

(*f*) Individual object classes. A variety of studies in both imaging and spectroscopy that can take advantage of the NGST performance. These include spectroscopy and imaging of the hidden universe (e.g., enshrouded star formation and AGN), regions of recent star formation and protoplanetary nebulae in our own galaxy. Thousands of faint objects will found by NGST deep surveys, such as very cool, low-mass white dwarfs and brown dwarfs. Collectively these projects make up 17% of the program.

(*g*) Kuiper Belt object searches. The searches will be carried out down to magnitude AB 30 in the optical and near infrared and AB 25 in the thermal infrared. This would allow for a statistically meaningful study of their properties as well as of the distribution in space beyond 40 AU. Mid-infrared signatures, such as silicate features, will be important in linking these objects, our closest proto-planetoids, to the great dust disks seen around early systems in Orion and β-Pictoris stars.

These goals will be further refined by the Ad Hoc Science Working Group, and will certainly be adjusted in response to continued progress. It is the intention of the NGST team to hold regular scientific meetings every two years to ensure that we are fully aware of progress and priorities. Additional inputs from scientists are encouraged at every time. One of the most important choices concerns wavelength range. The NGST will certainly have the optical quality to observe at wavelengths longer than 5 μm, but it will not be zodiacal background limited unless its temperature is sufficiently low, of order 600 K/$\lambda(\mu\text{m})$. Since the temperature is a fundamental engineering parameter, determined largely by the configuration and location of the observatory, it is not easily changed. On the other hand, with sufficient ingenuity the instrument package could be made to respond to changing requirements. At short wavelengths, it is unlikely that the NGST will have high UV efficiency or diffraction limited performance at visible bands.

3. System concepts

A concept resulting from the GSFC/ST ScI engineering study is shown in Figure 1. It shows a telescope at the Lagrange point L2, with an 8 m diameter primary deployed as 8 petals around a central octagon. The telescope is thermally isolated from the spacecraft bus by a deployable truss, and is radiatively cooled behind a sunshield deployed by inflatable booms. The instrument package cools to a temperature of 30 K or less. The TRW concept (Fig. 2) is similar but uses a set of seven hexagons to form the primary mirror, and the sunshield is deployed with a set of rods and pulleys. One of the Lockheed Martin concepts (Fig. 3) is quite different, using a monolithic primary and no deployments except for the solar array. To approach the sensitivity of the 8 m designs, it would require a deep space orbit (3 AU) to reduce the background radiation, or a larger rocket and payload fairing to allow a larger aperture than will fit inside the Atlas IIAS fairing. In any

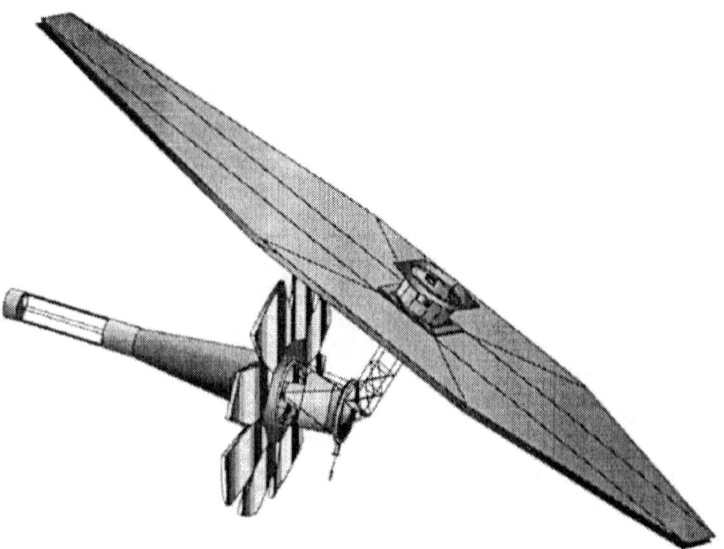

FIGURE 1. Goddard concept for an 8 m deployable telescope at L2

case it is too early to say what the design will be, but we will describe the considerations affecting the choice.

3.1. *Orbit*

There are several deep space orbit choices, all of which have the advantage of avoiding thermal and straylight problems from the low Earth orbit of the HST. An orbit around the Sun-Earth Lagrange point L2, located 1.5×10^6 km from the Earth on the Sun-Earth line, provides all these advantages. With all the heat sources on one side, a single sun shield can protect the telescope and allow strong radiative cooling. Analysis shows that a temperature of 30 K is readily achieved with four layers in the shield and a minimum of heat dissipation in and near the mirror. If an orbit as far as 3 AU from the Sun could be achieved, the zodiacal emission which provides the unavoidable photon noise floor would be reduced by about two orders of magnitude, but unless detectors are improved by a similar factor the full advantage would not be obtained. While the initial study took an Atlas IIAS launcher as a guideline, partly because of its relatively low cost (of order $100 M), it is unlikely that it will still be available in its present form in 2007. Many alternatives have been considered, ranging from smaller rockets like the Delta (which could be useful for a greatly descoped NGST), to much larger ones like the Ariane V (which could be used if it were contributed by ESA for a mission farther from the Sun, out to 3 AU, for example).

3.2. *Thermal shields and coolers*

The deep space orbits of greatest interest to NGST all share the common feature that the Earth is either in the same direction as the Sun, or is far away and negligible as a heat source. Hence, the thermal design can use "aggressive" radiative cooling very effectively. A multistage parasol between the telescope and the Sun can lower the telescope temperature to 30 K without heroic demands on materials or careful optimization. The TRW and GSFC concepts both showed this with numerical models, and the TRW design even showed an estimated secondary mirror temperature of 7 K.

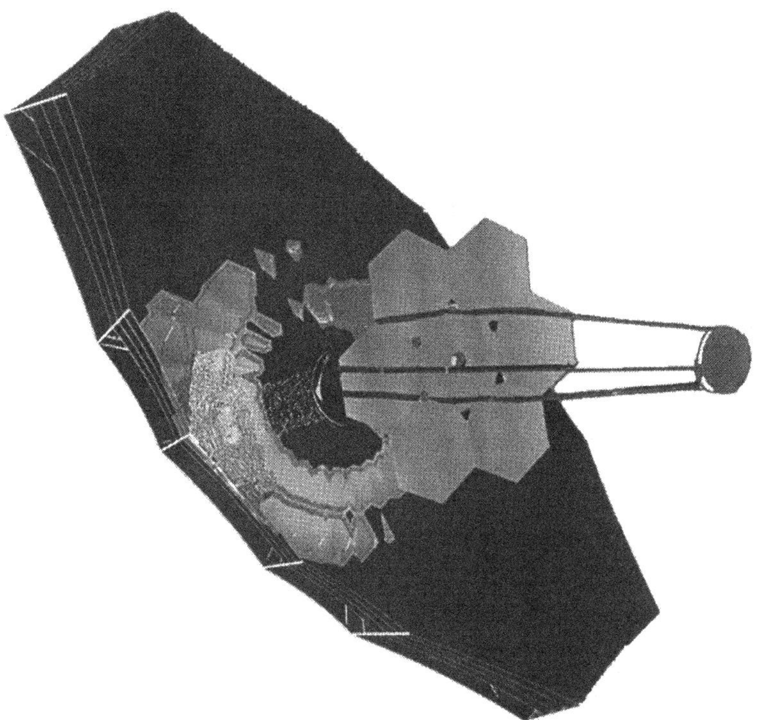

FIGURE 2. TRW concept for an 8 m deployable telescope at L2

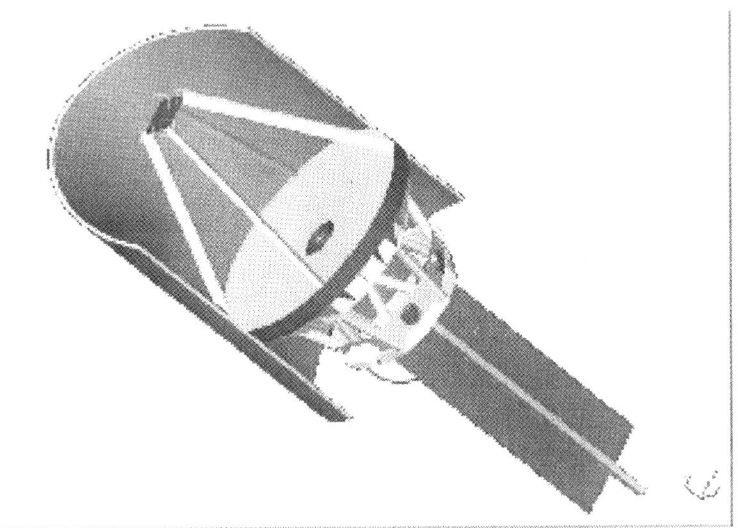

FIGURE 3. Lockheed Martin concept for a 6 m nondeployable telescope

The thermal shield could be made of many materials and deployed in many ways. The GSFC concept used inflation of tubular struts to deploy the shield, while the TRW concept used rods, cords, and pulleys. The Lockheed Martin design used a fixed shield. These are choices of engineering convenience and cost; the fundamental result is very clear: a very low temperature can be achieved by radiative cooling alone.

FIGURE 4. Test mirror, 0.5 m diameter, 2 mm thick, at U. of Arizona

As a rough rule of thumb, because of the scaling laws for dark current, band gaps, and blackbody emission, the detectors need to be at a temperature of about 200 K/λ_{max} to take full advantage of the cool optics. Clearly the InSb detectors for 5 μm can be sufficiently cooled by radiation. The available active cooler technologies include Stirling cycle coolers, reverse Turbo Brayton coolers, and sorption pumped coolers with hydrogen and helium as working fluids. These are good down to a few degrees K (McCormick *et al.* 1995, Swift 1995, Wade *et al.* 1996). The technology is still improving, but it appears possible to achieve unlimited life with small weights and volumes, eliminating the need for stored cryogens. This is an important breakthrough, with major implications for cost and configuration and operational sequences.

3.3. *Optical configuration*

Most of the concepts proposed in 1996 were fairly standard Ritchey-Chretien variants of Cassegrain telescopes, with a large, nearly circular, fast ($<$ f/1.25) nearly parabolic primary, a convex secondary, and a focal point near the vertex of the primary. In any case, whether deployed or monolithic, the primary mirror has an extremely thin face sheet ($\sim$ 2 mm thick), which could be of glass, silicon carbide, beryllium, nickel, aluminum, or a composite material. This reflecting surface is supported by an integral structure or by a separate one, which need not be the same material. In all cases adjustment is required after launch. This can be done by adjusters distributed across the mirror, or with a deformable mirror located at an image of it.

Roger Angel at the University of Arizona has already demonstrated a concept he calls MARS, for Membrane with Active Rigid Support. It is illustrated in Fig. 4. His test mirror is about 0.5 m in diameter, is made of 2 mm thick glass, and has an rms figure error of 35 nm, sufficient for diffraction limited performance at 2 μm. In this concept, the mirror shell is supported on actuators spaced about 10 cm apart. In the test device the actuators are Picomotors, piezoelectrically driven micrometer screws, available from New Focus. The actuators are supported on a graphite fiber composite lightweight backing structure. The glass shell is fabricated by first making two matching glass blocks with a spherical curvature, which are attached together with pitch. Then the convex block is ground down to have a matching concave surface, leaving a thin shell. When the thin shell is polished to the correct figure, it is removed from the other glass block by melting the pitch. The glass construction has the advantage that it is a well known, extremely homogeneous material, and standard methods produce an excellent surface in a short time.

3.4. *Optical figure control*

It is anticipated that the optical figure would be stable for long periods of time, because the space environment is likewise very stable in the deep space orbits under consideration. This is a very different environment from that of the Hubble Space Telescope, which on nearly every orbit passes from sunshine to darkness and back. Hence, a very infrequent adjustment should be required. Various wavefront sensors exist, ranging from point diffraction interferometers, to Shack Hartman tilt sensors, to other interferometers. Also, phase retrieval algorithms have been thoroughly developed and applied to the Hubble mirror problem. These demand extensive computation, but have the advantage that they can work without the need for additional focal plane instrumentation.

Figure control can be done at the primary mirror, at the secondary, or at an image of the primary mirror. Correction at the primary has the advantages that it can be done well, it is effective over the full field of view, and there is space for many actuators, but conversely the actuators may have a high weight and cost. Correction at the secondary is not effective, except for very low order errors, because it induces field dependent aberrations. Correction at an image of the primary is possible and would be enabled if a coolable deformable mirror with many actuators can be made. It has the disadvantage that there is a limit on the product of field size and error amplitude that can be corrected, so it is most valuable for small amplitude ripples.

4. Instrument concepts and systems implications

It is clearly too soon to specify the exact instrument suite for the NGST, which will not fly for about 10 years. It must be the successor for present day instruments (like the HST NICMOS) and future telescopes like the SIRTF, SOFIA, and WIRE. We also expect great progress from the Keck, Gemini, VLT, and other ground-based observatories. However, one can anticipate possible additional instruments that might be proposed. For example, a coronagraph or a rotation shearing interferometer could be used to search for planets or brown dwarf stars near brighter objects. Longer wavelength instruments would also benefit from the large cooled aperture of NGST, provided that they could be accommodated with little cost impact. To make these choices possible, the NGST study is holding a competition for instrument design concepts.

4.1. *Wavelength range and spectral resolution*

The core wavelength range of NGST is defined by the *HST and Beyond* report as 1–5 μm, with a goal of 0.5–20 μm. The shorter wavelength range would provide greatly improved sensitivity over HST, and the possibility of better angular resolution as well, depending on the mirror accuracy and stability achieved. The system engineering implications of a firm requirement for diffraction limited resolution and stability at 0.5 μm are major, and they could raise the mission cost significantly. The costs of the detectors are probably not so serious, since several excellent choices exist and CCD technology is quite mature.

The wavelength resolution of the spectrometers for NGST is currently expected to be less than $\lambda/\Delta\lambda \sim 1000$. This value is sufficient for the study of faint extragalactic objects, obtaining their redshifts and the intensities of their emission lines. Higher resolutions are obtainable from the ground with large telescopes, and the competitive advantage of a space telescope is least for high spectral resolution, because atmospheric lines can be resolved. If there are strong driving requirements that demand higher spectral resolution for NGST, it is feasible.

The requirement for longer wavelength coverage is also not firm, and could have significant cost implications. We anticipate that the experience gained from ISO, SIRTF,

WIRE, and SOFIA may greatly increase the level of interest in the 5–20 μm region. The systems engineering implications of longer wavelength coverage are primarily thermal. A better sunshade and thermal isolation system are required to keep the telescope and instrument chamber cold enough. The detector choice governs the temperature requirement, and may demand an active cooler. The requirement that the InSb be capable of excellent operation with passive cooling is the main design driver for the shields, and could result in a mirror temperature of around 50 K and an interior SI module temperature of around 25–30 K. Although the mid IR detectors would not be zodi-limited beyond around 20 μm for these conditions, their performance would still be spectacular out to their natural cutoff at 28 μm. There is no planned space mission in the NGST time frame to cover the wavelength range between NGST and FIRST, which covers wavelengths longer than 100 μm.

4.2. *Detectors*

There are many choices for implementing the cameras and spectrometers required by the scientific goals. The critical choice is the detector set. The primary wavelength range from 0.5 to 5 μm can be covered by InSb photovoltaic arrays, which operate well at 30 K. They are already very good, but larger 8192×8192 formats (achievable by mosaicing) and lower dark currents and readout noises are desired. For shorter wavelength coverage, CCDs or direct readout silicon arrays made like the InSb arrays are also of interest if they have superior performance, but currently require a higher operating temperature, of order 120 K. For wavelengths from 1 to 10 μm, and possibly longer, HgCdTe detectors are candidates. They are already in use for the NICMOS on HST for the 1–2 μm region. For wavelengths from 10 to 26 μm, Si:As photoconductive Impurity Blocked Conduction band (IBC) detector arrays are the only reasonable choice, but they require cooling to around 6–8 K, and they need improved sensitivity and larger formats (1024×1024) as well. The near infrared camera and detectors are operated at the nominal 30 K, at which temperature the Integrated Science Instrument Module can be passively cooled. At this temperature the thermal emission of the optics and the detector dark current are negligible. This is not the case in the thermal infrared and critical elements of the mid IR camera and spectrometer must be cooled actively, the filters and cold stop down to about 15 K and the detector down to at least 8 K. The temperature requirement would be eased if HgCdTe detectors with longer wavelength coverage (out to 10 μm) could be developed with low enough dark current, and if the scientific requirement for longer wavelength coverage were removed.

4.3. *Multi-object spectrometers*

Optically, the cameras and spectrometers are straightforward, except for one important innovation. Medium resolution (R = 1000) spectroscopy of a large number of faint objects is a central theme of the scientific requirements. Most of the faint objects observed by HST are already too faint to observe spectroscopically with much larger telescopes on the ground. For survey work, collecting large area images or large numbers of spectra simultaneously is as important as having a large telescope. Hence, one of the NGST instruments must be a multi-object spectrograph capable of obtaining simultaneously dozens to hundreds of individual spectra over a wide angular field. There are several concepts for such a spectrograph, including the use of microlens arrays to break a compact portion of the focal plane into hundreds of separated apertures and spectra. The Digital Micromirror Array, a remarkable innovation by Texas Instruments (Hornbeck 1996) offers a very nearly ideal possible solution. The array would be used as a switchable slit at the entrance to a grating or grism spectrometer. Properly utilized, such a device could enable simultaneous spectroscopy of hundreds of objects over a wide field of view. It is not currently available for cryogenic operation.

5. Conclusions

A Next Generation Space Telescope is scientifically necessary. No other planned or proposed equipment is capable of detecting and characterizing the first galaxies, globular clusters, and supernovae at the moment when they came into being. The NGST is also technically feasible and affordable, using new technology to reduce weight and increase aperture. It will be orders of magnitude more capable than competing facilities over a wide range of wavelengths and spectral resolutions. A substantial beginning has been made on the telescope technologies and the mission concepts, and a process has begun to continue the development with involvement from the astronomical community. International interest is high. With luck, support, and continued progress, the NGST could become a reality.

It is a pleasure to acknowledge the support of NASA Headquarters, particularly E. Weiler and H. Thronson, for the sponsorship of this work, and the HST and Beyond committee chaired by A. Dressler for their inspiring call for a new telescope. B. Seery at GSFC is the study manager, and H. S. Stockman cochairs the science working group along with the Study Scientist J. Mather. Overall scientific guidance is provided by the science oversight committee chaired by R. Kennicutt.

REFERENCES

BELY, P. Y., BURROWS, C. J., AND ILLINGWORTH, G. D., EDS. 1989 *The Next Generation Space Telescope (NGST)*, Proc. Workshop, Space Telescope Science Institute.

DRESSLER, A., *et al.* 1996 *HST and Beyond* Space Telescope Science Institute, Baltimore, MD. or http://ngst.gsfc.nasa.gov/project/bin/HST_Beyond.PDF

FALL, S. M., PEI, Y. C. & CHARLOT, S. 1996 *ApJ* **464**, L43.

1994 *High-Z: A Near-IR Space Telescope for Probing the Early Universe*, a proposal to New Mission Concepts for Astrophysics, ST ScI.

HORNBECK, L. J. 1996 *Digital Light Processing and MEMS*, SPIE/EOS Conf. on Lasers, Optics, and Vision, Besacon, France.

1996 *Lockheed Martin NGST Final Study Report*, LMMS/PO86946, http://ngst.gsfc.nasa.gov/project/text/Sept2627review.html.

MCCORMICK, J. A., *et al.* 1995, In *Cryocoolers-8* (ed. R. G. Ross, Jr.). p. 507. Plenum Press.

STIAVELLI, M., STOCKMAN, H. & BURG, R. 1997 The Next Generation Space Telescope Design Reference Mission. *ST-ECF Newsletter* **24**, 4.

STOCKMAN, H. S., ED. 1997 *The Next Generation Space Telescope: Visiting a Time When Galaxies were Young.* Space Telescope Science Institute, Baltimore. http://oposite.stsci.edu/ngst/initial-study/

SWIFT, W. L. 1995 In *Cryocoolers-8* (ed. R. G. Ross, Jr.) p. 499. Plenum Press.

H. A. THRONSON, ED. 1993 *EDISON, a M3 Proposal to ESA* Rutherford Appleton Labs.

TRW 1996 *NGST Final Study Report* http://ngst.gsfc.nasa.gov/project/text/Sept2627review.html

WADE, L. A., LEVY, A. R. AND BARD, S. 1996 In *Cryocoolers-9*. Plenum Press.

The Next Generation Space Telescope: Beyond the Hubble Deep Field

By H. S. STOCKMAN[1] AND JOHN C. MATHER[2]

[1]Space Telescope Science Institute, 3700 San Martin Drive, Baltimore, MD 21218

[2]NASA Goddard Space Flight Center, Code 685, Greenbelt MD 20771

The Hubble Deep Field is an expression of our curiosity about the stuff and laws of Nature, as well as being the realization of the best that mankind can currently accomplish. In the HDF, we see are able to observe the blue and ultraviolet light of massive stars in formation, shifted by the expansion of the cosmos into our visible reach. From this light alone, it appears that we have succeeded in seeing through the peak in the formation of stars in galaxies. In the HDF, we are beginning to peer at the universe at an age when galaxies were still growing in gas and stars. To reach greater depths, to see the earliest formation of stars and perhaps the reheating of the intergalactic medium, we must develop far greater sensitivity in the near and mid-infrared spectral bands than will be available from currently planned ground-based facilities. This paper describes these goals and the fundamental limits to ground and space-based sensitivities in the near and mid-IR. The goal of extending our vision from redshifts of $z = 4$, as seen in the most distant galaxies in the HDF, to redshifts as high $z = 20$, is a formidable and worthy one. Remarkably, it is within reach with the NGST.

1. Introduction

One of Riccardo Giacconi's tenets was "Design the next space experiment before you launch the one that you are building." Indeed, in the heyday of X-ray astronomy, he occasionally would be developing the second generation of follow-on missions before the first had flown. It was partly in this context that the ST ScI and NASA hosted the workshop "The Next Generation Space Telescope" in 1989 (Bely et al., 1989). The outcome of that workshop was an audacious call for an 8–16 m diameter, passively cooled telescope to be put in high Earth orbit (HEO) or on the lunar surface. Many astronomers provided imaginative ideas about the science that might be pursued; all were impressed by the angular resolution that such a behemoth would have—even at near infrared (NIR) wavelengths. One year later, the discovery of spherical aberration on the HST put the proceedings and the long term aspirations of optical and NIR astronomers on the shelf. The success of the first servicing mission and the remarkable sensitivity achieved in the HDF and the deep WFPC2 images by Dressler, Dickinson, Ellis, and others have been a powerful tonic. Even with his experience with aberrated images from the uncorrected HST, Dressler understood the potential of zodiacal-light limited backgrounds and diffraction-limited resolution in the study of the high redshift universe. The report of the HST and Beyond Committee which he chaired (Dressler et al. 1996) was greatly influenced by his vision and confidence.

In the companion to this paper, (Mather et al. 1997) we describe the recommendations of the HST & Beyond committee and the status of the NASA-led study of the successor to the HST, the Next Generation Space Telescope. In this paper, we describe in more detail several of the key scientific goals for this mission, goals that were outlined in the Dressler report and which have been further refined by recent theoretical work and the Design Reference Mission. It is interesting to note that many of these goals were not presented at the 1989 meeting. In retrospect, the goals for NGST in 1989 more closely resemble the goals for HST today than those in the Dressler report or the proceedings of the recent meeting held at GSFC on the topic (Smith & Koratkar, 1998). Our astronomical horizon is much closer to our current work than we imagine. We are more driven by new discoveries, such as the HDF, than by our imagination. This is an appropriate caution for the remainder of this paper.

2. Seeing the first stars and galaxies

In the HDF, we are beginning to observe galaxies in the process of formation. Large starbursts mark a dramatic increase in the stellar population of the parent galaxy; the peculiar morphologies suggest the importance of mergers in triggering these starbursts. In the local universe, we observe that merging galaxies can have several modes of star formation. Great splatters of globular-cluster sized starbursts may appear on the face of the merged galaxy, often with evidence of a wave-like or gravo-hydrodynamic origin. NGC 4038/4039 (the "antennae" galaxies, Whitmore & Schweitzer 1995) are a good example. At a later merger stage, as in Arp 220, very rapid star formation takes place in the core of the merger product, signaling its presence with extraordinarily high far infrared (FIR) luminosity. These cores may be 100–200 pc in diameter with star forming rates approaching 10^3 $M_\odot$ per year. They are shrouded by dust, optically thick even at mid infrared wavelengths (MIR, 5–30 μm). The CO emission from these cores indicates densities of 10^9 cm^{-3} and $T \sim 200$ K. These densities and temperatures are respectively 10^4 and 10 times greater than those of normal giant molecular clouds. By comparing the rates, the estimated supplies of gas, and the time scales of galaxy formation, we know that both modes of star formation can be important in a major merger. Per major merger, they convert a mass of gas to stars, $\sim 10^9$ $M_\odot$, comparable to that converted by steady star formation in a typical spiral disk (subject to the uncertainties of the initial mass function) over 10^9 years. For redshifts of z = 1–3, the task of observing these star-forming regions, the brightest lampposts of galaxies in formation, must fall to HST and large ground-based telescopes working in the optical, where the ultraviolet light is shifted, and to space submillimeter telescopes (FIRST) and millimeter arrays. At very high redshifts, $z > 10$, the problem of observing galaxies and star formation becomes more difficult as the light is shifted into the infrared but may be eased by the presumed lack of dust, at least for the earliest generation of stars.

Recently, several groups have considered the epoch and origins of the first stars (e.g., Haiman and Loeb 1997 [HL], Gnedin and Ostriker 1997 [GO], and Tegmark et al. 1997). Although their conclusions differ in detail, they agree on the overall picture. At redshifts of $z \sim 20$, the intergalactic gas has cooled by expansion to temperatures near that of the cosmic microwave background, $T \sim 57$ K at $z = 20$. At these temperatures, the principal coolants are trace amounts of molecular hydrogen. At this epoch—the details depend on the specific cosmology and the density of baryons—the fraction of molecular hydrogen is sufficient that overdensities of scale corresponding to virial temperatures of $T_{\mathrm{VIRIAL}} > 100$ K can cool, contract and become bound. Larger overdensities can also become bound but are rare, and smaller overdensities will not contract because of their own internal pressure. This sets a mass scale of $M_{\mathrm{BOUND}} \sim 10^{3-6}$ $M_\odot$ for the initially bound objects. It is expected that in the absence of strong primeval magnetic fields these bound objects will continue to collapse and will efficiently form stars. Assuming that the initial mass function (IMF) for these star forming regions is similar to a Salpeter mass function, we can calculate the flux from this star formation process using the Bruzual and Charlot (1993) models or those by Leitherer and Heckman (1995, LH). In Figure 1, we show the flux from a 10^6 $M_\odot$ starburst after 5 Myr, based upon LH model at the lowest

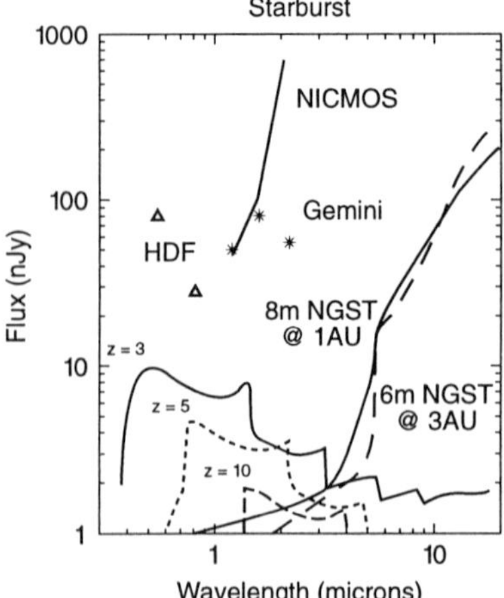

FIGURE 1. The imaging sensitivity of NGST and other facilities compared to a starburst spectrum at various redshifts. The triangles correspond to the faintest U and B dropouts in the HDF.

available metallicity (0.1 solar) and a variety of redshifts. Such a starburst may be a rare high mass event at $z \sim 20$–30, but it serves as a useful fiducial since similar masses are presumed to have been converted into stars in the formation of globular clusters. Figure 1 also shows the sensitivity of NGST designs with 6 m and 8 m diameter primary mirrors in detecting a point source at 10σ in 10^4 s and a resolution of $\lambda/\delta\lambda = 3$. The NGST is assumed to have diffraction-limited resolution at $\lambda = 2$ μm and longer wavelengths and to be limited by zodiacal light at 1 AU (deployment at L2). The NGST is therefore most sensitive at $\lambda = 3.5$ μm, in terms of λF_λ, with scattered sunlight dominating the background at shorter wavelengths and reradiated thermal radiation dominating the background at longer wavelengths. The step in sensitivity at $\lambda = 5.5$ μm is due to the change in detectors from InSb to Si BIB and an increase in the pixel size. Figure 1 also shows the performance of NICMOS/HST and a generic 8 m IR-optimized ground-based telescope with adaptive optics. The main difference in sensitivity between NICMOS/HST and the NGST at wavelengths shorter than 2 μm is due to the size of the primary mirror, 2.4 m and 8 m respectively. The difference between NGST and the generic 8 m ground-based telescope is due to atmospheric background and the temperature of the telescope optics for wavelengths $\lambda > 2.5$ μm. Finally, we show the magnitude of the faintest U and B dropout galaxies in the HDF. We make two comments about Figure 1.

- The U & B dropout galaxies in the HDF are clearly brighter than the burst that marks the formation of a single, isolated globular cluster. Indeed, it is not clear that HST or any other facility is capable of detecting such a starburst for $z > 2$–3. As a consequence, if significant star formation occurs from this mechanism, it would missed without the sensitivity of a telescope such as NGST.
- The NGST is capable of detecting such a starburst to interesting cosmological redshifts, $z \sim 20$ using integration times less than those used for the HDF.

The ultraviolet light from the first generation of massive stars will be absorbed in small H II regions at wavelengths $\lambda > 0.09$ μm but will permeate the universe and eventually photo-disassociate molecular hydrogen. Cooling of baryonic overdensities would then proceed only for those overdensities with $T_{\rm VIRIAL} \sim 10^4$ K. This corresponds to a total baryonic mass of $M_{\rm BARYON} \sim 10^8$ $M_\odot$. Since objects this size are rare at high redshift, we would naively expect a pause in the formation of stars until a redshift of $z = 10$, when the

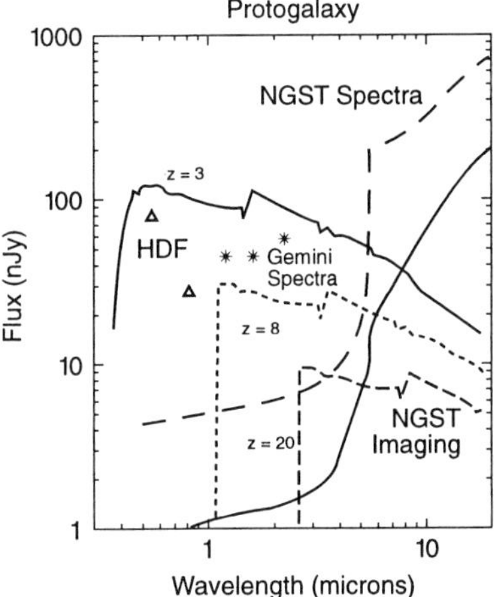

FIGURE 2. The spectrum of a protogalaxy at various redshifts. We indicate the NGST sensitivity for imaging and low-resolution spectroscopy for a point source. The protogalaxy spectrum is provided by LH for a continuous star forming rate of 1 $M_\odot$ yr^{-1} after 25 10^6 yr.

peak in the Press-Schechter function approaches this mass scale. But the transition will be smooth, with the larger scale star formation regions creating large H II regions and reheating the universe to $T \sim 10^4$ K between $z = 20$ and $z = 10$. Since the mass of bound objects increases at lower redshift and the intergalactic gas density decreases, the size and number of H II regions will rapidly increase until they overlap. These studies find that the re-ionization of the universe is essentially instantaneous, taking place in much less than a Hubble time. HL find that re-ionization occurs at $z = 10$–20 over a wide spread of model parameters, while GO find a re-ionization epoch of $z = 7$. This re-ionization epoch might be observed by the sudden drop in the absorption of bright protogalaxy light at Lyman alpha due to the Gunn-Peterson effect. In Figure 2, we indicate the flux from a protogalaxy that we simulate in terms of steady star formation at 1 $M_\odot$ yr^{-1} for 25 million years. We also show the NGST sensitivity for low resolution spectroscopy ($R \sim 100$, and an exposure of 10^5 s). In this simulation we have not included the angular size of the star forming region. In the HDF, we observe the bright cores of star forming galaxies to have angular radii of ~ 0.1 arcsec, which is comparable to the 0.06 arcsec resolution of NGST. It is clear that NGST should be capable of detecting such objects, comparable in brightness to the faintest galaxies detected by HST, to redshifts $z > 20$. NICMOS and an 8 m dia ground-based telescope such as one of the Gemini telescopes may be capable of detecting a star-forming protogalaxy of this luminosity at $z = 6$–8. Surveys to this depth should be attempted. But it is unlikely that even low resolution spectra will be possible at these flux levels, due to the rapid increase in sky background for $\lambda > 0.7$ μm. NGST must provide low resolution spectrophotometry on 10–20 nJy sources in the NIR since no other facility can.

3. Finding rare trees in a dense forest

Our initial simulations of NGST deep images indicate that each will contain tens of thousands of faint galaxies, most of which will be at moderate redshift, $z - 0$ 1. Since the most important finds will be rare, we must be able to identify and extract them with great care. An error rate of 10^{-3} would either seriously dilute the value of any deep sample or would require additional NGST observations to eliminate them. The ad hoc science working group (ASWG) will design, simulate, and analyze NGST observations in order to drive out technical requirements for the observatory. For example, we expect that high redshift galaxies will be found using the "dropout" techniques used in the HDF. For redshifts in the range $5 < z < 10$, the flux absorbed by the Lyman-alpha forest and Lyman continuum would otherwise fall in the R & I bands. Since the discovery sensitivity depends on both the on and off band sensitivities, the NGST imaging sensitivities in the R & I bands must be as good as those at J, H, & K or this technique will be limited by the short wavelength observations. We calculated the imaging sensitivity curves in Figures 1 & 2 by assuming that the image quality and effective quantum efficiency remain essentially unchanged shortward of $\lambda \sim 2$ μm. This level of performance should be retained or many NGST imaging surveys will be limited more by our inability to place an upper limit on the light that we do not expect to observe than by our ability to detect the light longward of the Lyman-alpha forest break. Of course, these capabilities would also provide significant improvements in angular resolution and sensitivity for visible imaging programs being pursued by HST.

Ironically, the plethora of background galaxies revealed in each NGST image will complicate the identification of nearby sources such as faint, dust-obscured protostars or Jupiter-mass objects in nearby star forming regions. Near the sensitivity limit, we may not be able to distinguish faint stellar objects from background high redshift galaxies solely on the basis of colors or image size. Fortunately, NGST's superb imaging resolution permits the traditional use of proper motions to establish membership in stellar associations and our Galaxy. Objects formed in the nearest dense molecular cloud complexes, such as Orion, will have apparent proper motions of 1–5 mas (milliarcseconds) yr^{-1}. If relative stellar positions per image can obtained with RMS accuracies of 3 mas—reasonable for NGST—astronomers will use time baselines of 2–8 years to confirm the association of faint sources with nearby star forming regions. Other facilities, such as SIRTF and ground-based telescopes using adaptive optics, will not have similar astrometric accuracies near their limits of sensitivity. In this regard, NGST astrometric imaging will be useful in confirming the membership of sources in nearby associations (e.g., the Hyades) discovered by these facilities.

We anticipate that the available field of view for diffraction limited imaging will be constrained by the number of pixels in the NIR imager and our ability to correct low-order wavefront errors with a deformable mirror. The field of view of a critically sampled 2.2 μm image with $10^4 \times 10^4$ pixels will be $\sim 4' \times 4'$ or ~ 0.005 square degrees. How many objects would be detected in an NGST field using single color exposures of 10^4 s, almost two orders of magnitude less than that used for the HDF? A number of authors have addressed this question, either in regard to HST, SIRTF, or NGST. We show a representative census in Table 1. Some of these estimates are reasonably sound, since they represent straightforward extrapolations of HST observations such as the HDF. In cases such as the Jupiter-sized objects in the Orion nebula, the authors provide a range of likely results. The low-mass turnoff of the IMF is unknown and can change the numbers by almost two orders of magnitude. Still other estimates are based entirely on theoretical grounds and are only accurate to an order of magnitude. What is clear, however, is that every field will hold a wealth of information, buried in a sea of background galaxies lying between redshifts $0 < z < 5$.

Astronomers will use an imaging IR spectrograph to mine these deep images. Low spectral resolution, $\lambda/\delta\lambda = 100$, will be sufficient to confirm the classification and redshift for sources approximately 10 times the detection limit. We will need moderate spectral resolutions, $\lambda/\delta\lambda \sim 1000$, to measure rotation velocities in bright protogalaxies (100 nJy), to observe changes in the Gunn-Peterson absorption coinciding with the epoch of re-

	Number in $4' \times 4'$ Field	Reference
Galaxies ($z < 5$)	$> 10^4$	Im & Stockman 1997
Protogalaxies ($z > 10$)	10^4	Haiman & Loeb 1997
AGN ($z > 10$, 10% active)	$\sim 10^2$	Loeb 1997
Pop III cold white dwarfs	0.05–6	Liebert et al. 1997
Jupiter-size Objects in Orion	2–70	Beckwith 1997
Disk M Dwarfs	> 12	Gould et al. 1997
Brown Dwarfs	~ 10	Werner et al. 1997
Type II SNe ($z > 5$)	~ 10 yr^{-1}	Miralda-Escude & Rees 1997
Protogalaxies ($>$ 100 nJy)	3	Haiman & Loeb 1997

TABLE 1. The predicted number of objects detected in a single, 10^4 s exposure with NGST using a broadband filter in the 1.0–3.5 μm band (1nJy $\sim 10\sigma$).

ionization, and for other studies such as obtaining the gravities of nearby brown dwarfs. Bright cosmological sources will be extremely rare. HL's estimate of 2–3 protogalaxies ($z > 10$) per NGST field with apparent brightness of 100 nJy in the NIR is likely optimistic. Such luminosities correspond to steady star formation rates of 20 $M_\odot$ yr^{-1} over millions of years and would probably correspond to the early formation of elliptical galaxies and spheroids. We may soon know whether this estimate is correct. As indicated in Figure 1, both NICMOS and large ground-based telescopes with adaptive optics will be capable of detecting 100 nJy sources, but their fields of view will be much smaller than the nominal NGST $4' \times 4'$ field. Hundreds of orbits with NICMOS or several weeks with a Gemini-class telescope will be required to search the equivalent of a single NGST field for such rare, but important sources.

4. Conclusions

The beauty and depth of the HDF provide us inspiration to probe the universe to greater depths and earlier times. A large, passively cooled space telescope, such as the NGST, would be capable of detecting the earliest formation of stars and galaxies. It would extend our vision into the NIR, where we can not only probe distant, cosmologically redshifted galaxies, but can also penetrate the dusty cores of nearby star forming regions and galaxies. Like astronomers using HST, we will use old techniques, such as photometry and the determination of proper motions, to sift the dozens of rare and important targets from the sand of background galaxies in every image. The promise of the NGST is similar to the promise kept by HST: new vistas and a better understanding of our universe.

We are pleased to acknowledge the support and advice of the NGST Science Oversight Committee, chaired by Rob Kennicutt, and the many volunteers who participated in the early feasibility studies and NGST designs. We also thank Ed Weiler and NASA Headquarters for their sponsorship of the NGST studies, and Alan Dressler and the HST and Beyond committee for making the scientific case beautifully clear. We particularly acknowledge and thank the speakers at the recent conference, Science with the Next Generation Space Telescope, 7–9 April 1997, at the Goddard Space Flight Center. Their ideas about the NGST science mission will influence the future of the mission for years to come.

REFERENCES

BECKWITH, S. 1997 *Science with NGST.* ASP Conf. Proceedings, (eds. E. Smith & A. Koratkar), in press.

BELY, P-Y, BURROWS, C. J., & ILLINGWORTH, G.D., EDS. 1989 *The Next Generation Space Telecope.* ST ScI.

BRUZUAL, G. B. & CHARLOT, S. 1993 *ApJ* **405**, 538.

DRESSLER, A. ET AL. 1996 *Exploration and the Search for Origins, A Vision of Ultraviolet, Optical, and Infrared Space Astronomy.* AURA.

GNEDIN, N. Y. & OSTRIKER, J. P. 1997, *ApJ* **486**, 581 (GO).

GOULD, A., BAHCALL, J. N., & FLYNN, C. 1996, astro-ph 9611157.

HAIMAN, Z. & LOEB, A. 1997 *ApJ*, in press, astro-ph 9611028 (HL).

IM, M., & STOCKMAN, H. S. 1997 In *Science with NGST.* ASP Conference Proceedings, (ed. E. Smith & A. Koratkar), in press.

LEITHERER, C. & HECKMAN, T. 1995 *ApJS* **96**, 9.

LIEBERT, J., NAJARRO, F., & KUDRTIZKI, R. 1997 In *Science with NGST*, ASP Conference Proceedings (ed. E. Smith & A. Koratkar), in press.

LOEB, A. 1997 In *Science with NGST.* ASP Conference Proceedings (ed. E. Smith & A. Koratkar), in press, astro-ph 9704290.

MATHER ET AL. 1997 In *Science with NGST.* ASP Conference Proceedings (ed. E. Smith & A. Koratkar), in press.

MIRALDA-ESCUDE, J. & REES, M. J. 1997 *ApJ* **478**, L57.

STIAVELLI, M. 1997 In *Science with NGST.* ASP Conference Proceedings (ed. E. Smith & A. Koratkar), in press.

TEGMARK, M., SILK, J., REES, M. J., BLANCHARD, A., ABEL, T., & PALLA, F. 1997 *ApJ* **1**, 474.

WERNER, M., ET AL., 1997 *SIRTF Science Requirements Document.* JPL, in preparation.

WHITMORE, B. C. & SCHWEIZER, F. 1995 *AJ* **109**, 960.

Summary

By P. JAMES E. PEEBLES

Joseph Henry Laboratories, Princeton University, Princeton, NJ, 08544, USA

1. Introduction

The evidence from observations in the Hubble Deep Field and other surveys is that the majority of the normal $L \gtrsim L_*$ spirals and elliptical galaxies had already formed at half the present lifetime of the universe as a place suitable for observers like us. Work on turning this notable advance into a theory of structure formation usually assumes the important dynamical actors at low redshift are the baryons and a class of nonbaryonic matter that may be cold (the CDM), or a family of massive neutrinos, or something even more exotic. The physics of clustering of the nonbaryonic component may prove to be simple and well adapted to numerical simulations, with initial conditions that we can deduce from general principles, as in the commonly discussed adiabatic CDM model. We won't get very far deducing the history of the baryons from pure thought, but here there is the enormous advantage of observations that yield remarkably tight and detailed constraints on what the baryons have been doing. The adiabatic CDM model has been remarkably successful in fitting together an elegant picture for the early universe—inflation—and elements of the observations, notably the Lyman-α forest and the anisotropy of the thermal cosmic background radiation. Galaxy formation may be more problematic: examples are disk formation, as noted by Efstathiou in these Proceedings, the luminosity function, and the epoch of assembly of massive halos. Observational programs in progress likely will be capable of showing whether such problems are side issues, and some variant of the CDM model is a good approximation to what happened, or something is seriously amiss. If the latter the odds are the observations will lead us to more promising ideas. And at the present frenetic rate of advance we may not have long to wait to learn which it is. Following are impressions of the state of some of the more pressing of the issues.

2. Issues

2.1. *What has become of the faint blue galaxies?*

At redshift $z \gtrsim 0.5$ a population of blue galaxies is most rapidly evolving; the evidence is discussed in these Proceedings by Ellis, Illingworth, Lilly, Windhorst, and others. Where are these galaxies now? Some have merged with each other or with normal-looking spirals and ellipticals, maybe producing the last generation of spirals, maybe adding to preexisting disks. Others faded through suppression of star formation and loss of gas by winds and collisions, maybe ending up as present-day dwarf, irregular, and low surface brightness galaxies.

Mergers certainly have consumed some of the blue population, but could this have been the common fate? If these objects were already bound and in virial equilibrium in the halos of the regular $L \sim L_*$ galaxies, and spiraling in by dynamical drag, then by analogy to the Magellanic Clouds one might expect they spiral in from galactocentric distances on the order of 100 kpc. Substantial concentrations of blue galaxies on this scale would have been noticed, but further analyses of the cross-correlation of the faint blue galaxies with normal-looking spirals and ellipticals will be interesting.

In another merging picture motivated by the adiabatic CDM model the blue galaxies are falling onto larger ones on nearly radial orbits, from maximum galactocentric distances $r_{\rm max} \sim 500$ kpc (based on the spherical collapse model). This would decrease the clustering of the faint blue galaxies, but is the factor of ten collapse from $r_{\rm max}$ to $r \sim 50$ kpc, where dynamical drag might take over, reasonable? Departures from spherical symmetry prevent it in dynamical models for the relative motions of the Local Group members that take account of gravitational interactions with galaxies outside the group.

The alternative to merging is fading to the broad range of relatively low mass galaxy types. As discussed in §2.3, our picture for the assembly of the dark halos of the $L \gtrsim L_*$ galaxies might be heavily influenced by a convincing demonstration of the fate of the blue population.

2.2. *When did the faint blue galaxies form?*

What is the epoch of assembly of the blue population observed at $z \gtrsim 0.5$? A key lead is the elegant empirical history of star formation, showing the rate is an order of magnitude larger than now at $z \sim 1$ to 2, as reviewed in these Proceedings by Madau and Fall. Illingworth shows that a relatively compact low mass component in the blue galaxy population is a substantial contributor to the star formation rate. One interpretation is that star formation is high at $z \sim 1$ to 2 because that is when many galaxies were assembled. Another possibility is that the blue galaxies that are producing stars at a high rate at $z \sim 1$ to 2 were assembled as concentrations of baryons and dark matter well before that. This is an example of the cautionary remarks by Silk, Windhorst, and others about the importance of feedback in determining star formation histories.

The issue of the origin of the faint blue galaxies illustrates the danger of confusion in the use of the phrase "the epoch of galaxy formation." One can imagine that in some types of galaxies the dominant star populations form well before assembly of the galaxies as coherent units, as in Kauffmann's analysis of elliptical galaxy formation, while in other types of galaxies the dominant bursts of star formation occur well after assembly of the mass, as maybe is the case in the blue galaxy population that contributes so strongly to the net star formation rate at $z \sim 1$ to 2. In White's phrase, "galaxy formation is a process, not an event," and as Lilly stresses the process can be quite different in different populations of galaxies.

2.3. *Assembly of massive CDM halos of galaxies*

What is the epoch of assembly of the dark massive halos of the spiral galaxies? Formation at redshift $1 \lesssim z \lesssim 3$, as predicted by the adiabatic CDM model, agrees with the history of quasar activity, though the peak in activity could instead signify the time needed to produce central engines after galaxies are assembled as mass concentrations. The high star formation rate at $z \sim 1$ to 2 also is in line with late galaxy assembly. Dickinson, Calzetti, Ellis, and others caution that one has to consider the correction for obscuration at high redshift and low—an important constraint comes from the ISO measurements described in this volume by Rowan-Robinson—but I think it is generally accepted that the star formation rate has significantly decreased since $z = 1$. The issue is whether this is because structure on the scale of galaxies has significantly evolved since $z = 1$, or because feedback has made star formation a long drawn-out process, as noted in §2.2. My preference for the latter is not entirely phenomenological; I am led to it by the following theoretical consideration.

In the standard model the mass of an $L \sim L_*$ spiral galaxy within the radius $r_{\rm g} = 10h^{-1}$ kpc is dominated by CDM, matter that is dissipationless and initially pressureless. Let us consider how this fits a scaling solution for the growth of clustering of the

CDM. Scaling requires that non-gravitational processes in the baryons are subdominant, the universe is Einstein-de Sitter or close to it, and the primeval departure from a homogeneous mass distribution is close to a scale-invariant random process with power spectrum

$$P(k) \propto k^m, \tag{2.1}$$

with $m \sim -1$ to fit the galaxy distribution. Under all these conditions characteristic comoving lengths and peculiar velocities in the mass distribution vary with time as

$$x \propto (1+z)^{-2/(3+m)}, \qquad v \propto (1+z)^{(m-1)/(2m+6)}, \tag{2.2}$$

as shown in all the better books on cosmology. At the present epoch the largest stable systems are the rich clusters of galaxies. A good approximation to the mass within the Abell radius of a rich cluster is a limiting isothermal gas sphere, $M(<r) = 2\sigma^2 r/G$, with

$$r_{\rm cl} = 1.5h^{-1}\ {\rm Mpc}, \qquad \sigma_{\rm cl} = 750\ {\rm km\ s^{-1}}, \qquad n_{\rm cl}^{-1/3} = 60h^{-1}\ {\rm Mpc}. \tag{2.3}$$

The Abell radius is $r_{\rm cl}$, the line-of-sight velocity dispersion $\sigma_{\rm cl}$ is an rms mean for Abell clusters, the mean number density of rich clusters is $n_{\rm cl}$, and Hubble's constant is $H_o = 100h$ km s^{-1} Mpc^{-1}. The parameters

$$m = -1.3, \qquad 1 + z_{\rm g} = 10, \tag{2.4}$$

in the scaling relations in equation (2.2) bring the numbers in equation (2.3) to

$$r_{\rm g} = 10h^{-1}\ {\rm kpc}, \qquad \sigma_{\rm g} = 160\ {\rm km\ s^{-1}}, \qquad n_{\rm g}(z=0)^{-1/3} = 4h^{-1}\ {\rm Mpc}. \tag{2.5}$$

The two free parameters in equation (2.4) have been chosen to satisfy two conditions, that the velocity dispersion $\sigma_{\rm g}$ is that characteristic of L_* galaxies and the physical radius $r_{\rm g} = x/(1+z)$ is characteristic of the radius within which CDM dominates the mass in a typical L_* spiral galaxy. This is not an empty exercise in parameter adjustment; there are two checks. First, we get a reasonable distance between protogalaxies. This is the quantity $n_{\rm g}^{-1/3}$ evaluated at the present epoch in equation (2.5). Second, we get a power law index, $m = -1.3$, close to that derived from the galaxy distribution on larger scales. This calculation assumes the Einstein-de Sitter cosmology; the numbers are quite similar in a low density cosmologically flat model and not greatly different in a low density open cosmology.

Baryons must dissipatively settle to form spheroids that dominate the mass near the center of a massive high surface density galaxy. This breaks scaling, of course, but the mass within $r_{\rm g}$ in an $L \sim L_*$ galaxy is thought to be dominated by dark matter. If so the gravitating mass added by baryon settling pulls only a modest additional amount of CDM through the radius $r_{\rm g}$.

Present-day clusters of galaxies show substructure and ongoing merging. In this scaling picture one similarly would expect protogalaxies at $1 + z_{\rm g} \sim 10$ are quite irregular, undergoing substantial merging events. Stellar spheroids would be assembled at $z < z_g$, as the baryons later settle, and thin disks likely would form still later, after the large baryon collapse factor needed for spin-up to rotational support.

In this scaling picture the most prominent bound and stable systems at redshift $z \sim 10$ have the properties one might look for in newly assembled protogalaxies. The initial conditions for this picture are not consistent with the adiabatic CDM model for structure formation, but an isocurvature variant shows they need not conflict with the physics of the standard hot Big Bang cosmological model, or even with inflation. The nice thing about physical science is that one can hope to put divisions of opinion on such issues to the experimental/observational test. Perhaps a test will come out of further observations in the Hubble Deep Field and other surveys in progress, perhaps in part from analyses of the following issues.

2.4. *Assembly of the elliptical galaxies*

When were elliptical galaxies and the spheroid components of the spirals assembled as the presently observed coherent star systems? I think I heard general agreement with the argument by Ellis and others that the small scatter in the color-magnitude relation for ellipticals now and back to redshift $z \sim 1$ indicates the bulk of the stars in these systems are close to coeval, the stars having formed at redshift well above unity. The similarity of ellipticals and the spheroid components of spirals argues the bulk of the spheroid stars formed early too. Ferguson finds that the HDF observations are not inconsistent with the presence of the large galaxies at $z > 1$. Zepf shows that the stars in ellipticals could not all have formed at high redshift, because that almost certainly would imply the existence of more red galaxies than is observed in the HDF and other surveys, whether the stars are already assembled in ellipticals or are in subgalaxy parts. Remnant later star formation in early-type galaxies does not seem so unlikely; it is easy to imagine the modest ongoing star formation needed to keep the population from becoming too red would be fueled by accretion and mergers. Zepf points out that the star formation history is tightly constrained by the remarkably slow evolution of the fundamental plane of the early-type galaxies at $z \lesssim 1$, as well as by the small scatter in colors. Open for discussion is which scenario more readily accommodates the constraints.

Steinmetz and Kauffmann show that within the adiabatic CDM model the stars in ellipticals and the spheroids of spirals would form at high redshift in relatively low mass subgalaxies that merge to produce the familiar Hubble sequence at $z \sim 1$. This recent assembly requires a large collapse factor: the mean mass density within the luminous parts of an $L \sim L_*$ high surface brightness galaxy is some five orders of magnitude larger than the cosmic mean density at $z \sim 1$. To repeat the argument in §2.3: if you suspect gravitational collapse factors are not large you are led to early galaxy assembly. If on the other hand you suspect the faint blue galaxies observed at $z \sim 0.5$ disappeared by falling nearly radially onto L_* galaxies, then you are entitled to argue a similar large collapse factor is capable of assembling ellipticals at $z \sim 1$. Windhorst shows a possible example of a group of subgalaxy clumps at $z = 2.4$. It will be interesting to see whether such systems are common, and whether the redshift space two-point cross-correlation function of the blue population with normal-looking spirals and ellipticals is consistent with the close to radial infall of late merging.

Giavalisco points out that the objects detected by the break in the spectrum at the Lyman limit have properties one would look for in protospheroids present at $z \sim 3$ to 4. Are the properties of these systems consistent with the picture for quite early mass assembly, $z_g \sim 10$ (eq. [2.4])? It depends on how feedback from star formation affects baryon settling and star formation, which Silk instructs us is not easy to model. The picture does predict that the gravitational velocity dispersions are characteristic of large galaxies, $\sigma \sim 150$ km s^{-1}. Under the late galaxy assembly picture from the adiabatic CDM model the Lyman-break objects would be lower mass subgalaxy clumps. The observers do not sound optimistic about the prospects for measuring gravitational velocity dispersions or rotation curves in the Lyman-break systems, but the issue certainly is critical for theories of structure formation.

2.5. *Nature of the damped Lyman-α absorbers*

Are the damped Lyman-α absorbers (the DLAs) protodisks of spiral galaxies? Wolfe, Lanzetta, and colleagues point out that they have many of the properties of puffy gaseous disks supported by rotation (and likely turbulence) in the potential wells of $L \sim L_*$ protogalaxies. If it could be shown that this is the right picture it certainly would make it easier to accept the Lyman-break systems as young spheroids. It would be particularly helpful to know the distribution of impact parameters from the quasar line of sight to the apparent dynamical center of the DLA. If the impact parameters typically were comparable to the optical sizes of present-day spiral galaxies, and we accepted that the velocity widths are gravitational, we could conclude that galaxy-size potential wells are present at $z = 3$ at the comoving number density of the large galaxies. In the adiabatic CDM model the damped Lyman-α absorbers are significantly less massive and more numerous than $L \sim L_*$ galaxies—subgalaxy fragments. Perhaps infrared imaging is the best hope for a test.

A pressing issue for the massive protodisk interpretation is the stability of a gaseous puffy disk with radius comparable to the optical radius of an L_* spiral, circular velocity $v_c \sim 200$ km s^{-1}, and HI column density $\sim 10^{21}$ cm^{-2}. The characteristic time for gas to pass from the intergalactic medium through the gaseous disk phase to stars could not be much less than the Hubble time at $z \sim 3$, because $\Omega_{\rm HI}$ in DLAs at $z = 3$ is comparable to the net mass in stars in galaxies now. I have heard strong objections to the idea that massive gaseous disks could have lasted this long, and the issue certainly deserves serious consideration.

Steinmetz and colleagues find that numerical N-body plus hydrodynamic simulations of the CDM model can account for many of the properties of the DLAs. These authors are to be congratulated on an elegant and productive treatment of a complicated issue. If confirmed by further analyses we will have a definite prediction and a critical test, the distribution of impact parameters.

2.6. *Large-scale structure*

What is the history of the clustering of matter on scales larger than that of the luminous parts of the galaxies? The power law index of the galaxy-galaxy two-point correlation function $\xi_{gg}(r, z)$ is close to constant at $\gamma \simeq 1.8$ at $z \lesssim 1$. In this range of redshift the mean mass density has dropped by an order of magnitude. What might the remarkable stability of γ mean?

The time-evolution of the amplitude of $\xi_{gg}(r, z)$ is more difficult to measure, but I understand it is not inconsistent with the assumption that galaxy clustering is stable on small scales. If the density parameter is low, $\Omega \sim 0.2$, it agrees with the idea that galaxies trace mass on scales $\gtrsim 100$ kpc, and one may be inclined to compare $\xi_{gg}(r, z)$ to the mass autocorrelation function. If the evolution of small-scale clustering is much slower than the Hubble expansion rate, and $\Omega = 1$, the small-scale power law index for the mass autocorrelation function is

$$\gamma = (9 + 3m)/(5 + m), \qquad (2.6)$$

for the power law primeval power spectrum in equation (2.1). For $m = -1.3$ (eq. [2.4]) this predicts $\gamma = 1.4$. If $\Omega \sim 0.2$ and the universe is cosmologically flat (with a term in the stress-energy tensor that acts like a cosmological constant Λ), the gravitational growth of clustering is not much suppressed from Einstein-de Sitter, and I would expect equation (2.6) still is a reasonable approximation. Numerical scaling solutions suggest the evolution of small-scale clustering increases γ on scales comparable to the clustering length, and maybe can account for the larger observed value, $\gamma \simeq 1.8$. If not, or if galaxies do not trace the mass that can cluster, what is the significance of the stability of γ? I would hate to think such a beautiful result is only a transient accident.

At low redshift intergalactic gas clouds avoid the voids defined by the high surface brightness giant galaxies. The same is true of the known classes of dwarf, irregular, and low surface brightness galaxies. If $\Omega \sim 0.2$ (ignoring the possible contribution from Λ), the mass that clusters with the galaxies is close to the total and we are entitled to speculate that the voids contain few galaxies because gravity has emptied them of galaxies and mass alike. To check that this picture is not inconsistent with the physics of the standard relativistic Friedmann-Lemaître cosmology we need to know whether there are solutions for the evolution of the mass distribution where voids are gravitationally emptied by relative peculiar velocities that now are small. This does not appear to be easy to arrange, which could mean the picture is wrong or that we have something to learn about the nature of the initial conditions. If $\Omega = 1$, on the other hand, most of the mass has to be in the voids, and the puzzle is to understand why no one has identified galaxies that survive the inhospitable but surely not inevitably destructive conditions that would have to have obtained in the voids. This is the historical argument against the biasing concept as a way to reconcile dynamical mass estimates with the high mean mass density of the Einstein-de Sitter model. Now there are other significant pieces of evidence for low Ω, but my reading of the evidence would be seriously perturbed if still deeper observations revealed a class of void galaxies.

In her contribution to these Proceedings Cohen emphasizes the strong large-scale clustering of galaxies at $z \sim 1$, and the hints of high redshift walls or filaments in the distributions of galaxies, quasars, and quasar absorption line systems. Attempts to piece together the story of how such structures evolved into what we see around us at low redshift will be informed by maps and statistical measures of the distributions of these systems as functions of time back to $z \sim 5$. Results already obtained show such measurements are feasible, an exceedingly large undertaking, and an invaluable guide to how structure formed.

2.7. *The Lyman-α forest*

Along with galaxies and galaxy clustering at $z \sim 1$ are remnants of the Lyman-α forest that fills space as a sort of froth at $z \sim 3$, a factor of two earlier in the expansion of the universe. In the adiabatic CDM model simulations, mentioned by Efstathiou in these Proceedings, the forest clouds are collapsing through first generations of structure at $z \sim 3$. The success of the simulations leads me to think they capture major elements of the truth, but the well-developed large-scale structure at $z \sim 1$ leads me to wonder if it is the whole truth. Perhaps the more straightforward reading of the evidence is that the Lyman-α forest clouds are collapsing and dissipating at $z \sim 3$ in clearings between concentrations of large galaxies that are already assembled. There do seem to be models for non-Gaussian initial conditions that can lead to this situation. In short, I have the uneasy feeling that the time we have spent searching for candidate models for structure formation is not commensurate with the rich and rapidly accumulating network of observations to which the models will be compared.

3. Concluding remarks

One hears people say that the great advances in the past few years have at last made cosmology a "real physical science." I tend to react badly because I think we have been doing real physical science in cosmology for decades. To my mind the difference is the rate of creation of knowledge.

The central story in cosmology since the 1950s has been the accumulation and sorting out of the empirical basis for our subject, to the point that we now have a remarkably clear picture of how galaxies evolved back to $z \sim 1$ and what the intergalactic medium was like back to $z \sim 5$. We have powerful lines of evidence on still earlier conditions: the thermal background probes the universe at $z \sim 1000$, light element production probes what was happening at $z \sim 10^{10}$, and baryosynthesis might yet teach us something about still higher redshifts.

The theoretical situation has changed a lot less, as I think we can appreciate by asking ourselves what one of the handful of cosmologists active the 1940s and brought to us in a time machine might make of the present state of the subject.

Our cosmologist might express surprise at the angst over the evidence that the density parameter Ω is less than unity. Bondi did mention the particular elegance of $\Omega = 1$, but neither he nor the others I have read made the value of Ω a key issue. It is a parameter, after all, and the adjustments now under discussion are a lot less fine than what Lemaître needed to get a reasonable expansion time from the early estimate of Hubble's constant. Zwicky's mass anomaly in the Coma cluster of galaxies was not heavily cried up, but it was in the literature. The big change is the very thorough exploration of the mass problem. The concept of nonbaryonic matter has been an influential theoretical advance, that I hope our cosmologist would welcome as an elegant solution to the mass problem, but we would have to admit that we have not yet identified the nonbaryonic matter. Our cosmologist might not be surprised that we have clear evidence for the evolution of the universe, from the thermal cosmic background radiation and relict deuterium and helium, as Gamow first understood, and from the evolution of cosmic structures, as Lemaître clearly recognized must happen in the relativistic cosmology. The inflation concept has been a deeply influential theoretical advance, though in my opinion we would have to admit that we don't yet have the critical evidence it happened about as pictured. I think inflation would not be entirely alien, for the concept has a lot in common with the steady state cosmology. The concerns I express over the redshift of galaxy assembly might seem strange, for advocates of the steady state cosmology found it quite easy to imagine galaxies reach their large density contrasts by collapse from the present mean density. They did not have to consider the collapse of CDM along with the baryons, however. The classical cosmological tests have been understood since the 1930s, people have been working on them ever since, and the end may be in sight: we have a remarkably good picture of how galaxies have evolved back to $z \sim 1$, at which redshift the observational effects of the cosmology are large. The demonstration of cosmic evolution contradicts the steady state cosmology of the 1940s. I think the death of this model would not seem surprising, considering the slender observational basis for cosmology at the time the model was found. To my mind the astonishing news would be that another of the cosmologies then under discussion, the relativistic Friedmann-Lemaître model, has survived a vastly enlarged network of observational constraints.

I hope my characterization of the present state of cosmology as frenetic is understood as positive; an operational definition is that the subject is advancing faster than I can follow. There is no end in sight to this happy state of affairs, for as the "easy" explorations made possible by the present generation of detectors and observatories are exhausted and people turn to "really difficult" observational programs, good management will have yielded new more powerful tools, notably the Next Generation Space Telescope. People will look back on the 1990s as a golden (though primitive) age for cosmology, and they will count the Hubble Deep Field as one of the landmark advances.

I have benefitted from advice and instruction from Daniela Calzetti, Richard Ellis, Michael Fall, Garth Illingworth, George Lake, Simon Lilly, Art Wolfe, and Steve Zepf. This work was supported in part by the US National Science Foundation.